O'Flanagan/Irle   Wörterbuch Personal- und Bildungswesen
Dictionary of Personnel and Educational Terms

Wörterbuch
Personal-
und Bildungswesen

Dictionary of
Personnel
and Educational Terms

Deutsch – Englisch
English – German

Von Rory O'Flanagan und Ruth Irle

3., wesentlich überarbeitete und
erweiterte Auflage, 2000

3rd strongly revised and
enlarged edition 2000

Publicis MCD Verlag

Die Deutsche Bibliothek – CIP-Einheitsaufnahme
Ein Titeldatensatz für diese Publikation ist bei Der Deutschen Bibliothek erhältlich

Die Deutsche Bibliothek – CIP-Cataloguing-in-Publication-Data
A catalogue record for this publication is available from Die Deutsche Bibliothek

ISBN 3-89578-154-1

3. Auflage, 3rd edition 2000

Herausgeber: Siemens Aktiengesellschaft, Berlin und München
Verlag: Publicis MCD Verlag, Erlangen und München
© 2000 by Publicis MCD Werbeagentur GmbH, Verlag, München
Das Werk einschließlich aller seiner Teile ist urheberrechtlich geschützt.
Die Verwendung außerhalb der engen Grenzen des Urheberrechtsgesetzes
ist ohne Zustimmung des Verlags unzulässig und strafbar. Das gilt
insbesondere für Vervielfältigungen, Übersetzungen, Mikroverfilmungen,
Bearbeitungen sonstiger Art sowie für die Einspeicherung und Verarbeitung
in elektronischen Systemen. Dies gilt auch für die Entnahme von einzelnen
Abbildungen und bei auszugsweiser Verwertung von Texten.

Printed in Germany

Vorwort

Die Auswirkungen der Globalisierung in der Wirtschaft, die unmittelbare Verfügbarkeit von Daten und Information sowie die zunehmende Bereitschaft einer stetig wachsenden Anzahl von Personen sich über und mit modernen Mitteln weiter zu bilden haben großen Einfluss auf die nun vorliegende 3. Auflage des Wörterbuches „Personal- und Bildungswesen" gehabt. Wenn es bei der ersten Auflage die Absicht gewesen ist, deutschen Mitarbeitern in den verschiedenen Bereichen eines Unternehmens, die mit Kollegen und Geschäftspartnern im Ausland zu tun haben, ein Hilfsmittel in die Hand zu geben, so standen bei den Änderungen für diese Auflage die Interessen und Bedürfnisse Auskunft suchender ausländischer Arbeitskräfte, die über Ausdrücke des Personal- und Schulwesens in Deutschland etwas erfahren wollen, eher im Vordergrund. Die erst kürzlich entstandene Diskussion über die „Green Card" zeigt auch, dass für eine Vielzahl von neu hinzu kommenden ausländischen Mitarbeitern in der Bundesrepublik eine solche Hilfe unerlässlich geworden ist. Wir haben uns bemüht, diesem Anliegen gerecht zu werden.

Für die aktuelle Auflage ist das Buch stark überarbeitet worden, eine Vielzahl von nicht mehr zeitgemäßen Begriffen gelöscht und durch etwa tausend neue ersetzt worden. Wir sind für die vielen Anregungen und Kritiken von Benutzern und Kollegen dankbar und haben versucht, diese so weit wie möglich zu berücksichtigen. Mein Dank gilt auch den Mitarbeitern der Terminologieabteilung des irischen Department of Education für die fruchtbaren Gespräche zum Thema „Begriffe zum Bildungswesen".

Wir bitten alle Benutzer des Wörterbuches uns mitzuteilen, in welchen Teilbereichen sie Erweiterungen begrüßen würden und uns auch weiterhin mit Ergänzungen und Vorschlägen behilflich zu sein.

München, August 2000   Rory O'Flanagan

agngring willingness of a steadily growing
number of people to make use of modern means of further education has had
strong influence on the third edition of the "Dictionary of Personnel and Educational Terms". The intention of the first edition was to give German col-
leagues in the various departments of any company which has dealings with
foreign business partners an easily usable support in terminology questions.
This aspect has changed slightly. The needs and requirements of foreign workers looking for the respective German equivalents of English expressions in the
fields of HR and education have grown. The recent discussion in Germany on
the introduction of "Green cards" illustrates that the viewpoint of foreign employees requires increased attention. We have endeavoured to provide this.

For this edition the book has been strongly revised, a number of less apt entries
being replaced by new and more appropriate ones. We wish to thank the many
users of the dictionary for their constructive criticisms and suggestions. As far
as was possible these have been taken into consideration. We would also like
to thank the staff members of the terminology department in An Gum, Department of Education, Dublin, for the exchange of ideas on educational themes.
We would like to ask all users of the dictionary to continue supporting us with
their suggestions and recommendations and also with a notification of their requirements.

Munich, August 2000   Rory O'Flanagan

Hinweise

Wörter sind zwangsweise mehrdeutig. Deshalb wurden nur Äquivalente aufgenommen, die den beiden Zielbereichen Bildungs- und Personalwesen entsprechen. Wenn der Grundwortschatz Eingang in das Wörterbuch gefunden hat, so immer mit Bezug zum Thema Personal oder Bildung.

Sollte das Wörterbuch in einer der beiden Sprachen eine Definition enthalten, ist sie der jeweiligen Sprache entnommen.

Die Abkürzungen BrE und AmE weisen auf den britischen bzw. amerikanischen Sprachgebrauch hin.

Hints

It is in the nature of words to have several meanings. Therefore, only equivalents have been entered here, which correspond to both target fields, education and personnel. If basic vocabulary has been included into the dictionary, it always has a reference to the themes personnel or education.

Should the dictionary contain a definition in either of the two languages, it is always drawn from the original language.

The abbreviations BrE and AmE refer to the British and/or the American usage of the English language.

Wörterbuch Personal- und Bildungswesen

Teil 1
Deutsch – Englisch

A

abändern – amend
Abänderung – modification
Abbruchquote – drop-out rate
Abc-Schütze – infant school child
Abendgymnasium – Abendgymnasium,
evening school leading to leaving
certificate (Das Abendgymnasium führt
Berufstätige im vierjährigen
Abendunterricht zur allgemeinen
Hochschulreife. Mit Ausnahme der
letzten drei Halbjahre ist die
Berufstätigkeit Voraussetzung für den
Besuch des Abendgymnasiums.)
(s. Graphik)
Abendkurs – evening course
Abendrealschule – Abendrealschule
(Die Abendrealschule führt
Berufstätige in drei oder vier Jahren
zum Realschulabschluss. Als
berufstätig gilt, wer seinen
Lebensunterhalt überwiegend durch
eine eigene Tätigkeit verdient.)
(s.Graphik)
Abendschule – Abendschule
(s. Graphik), evening school, night
school
**Aberkennen von ... (z.B.
Altersversorgung)** – disallowance of
... (e.g. old-age pension)
Abfahrt – departure
abfeiern – take a day in lieu of
Abfindung – dismissal compensation,
redundancy payment, severance
payment
Abfindungsplan – redundancy scheme
Abfindungsrücklagen – severance pay
reserves
Abflusslohnart – outgoing wage type
abführen – deduct
Abgabe zuständigkeitshalber –
redirected for reasons of competence
Abgabebescheid – interim reply:
referring to a further department
Abgabepflicht – liability to social
security payments

Abgang – leave; person leaving,
separation
Abgangsentschädigung – termination
payment
Abgangserklärung – exit document
Abgangsinterview – final interview, exit
interview (AmE) (Eine Befragung
durch die Personalabteilung, die
stattfindet, wenn ein Arbeitnehmer
endgültig aus dem Betrieb ausscheidet)
Abgangsrate – wastage rate
Abgangszeugnis – Leaving Certificate
Abgangszeugnis (eines(r) Angestellten)
– clearance card
abgelehnt – rejected
abgelten – compensate
Abgeltung – compensation
Abgeltungsanspruch – indemnity, right
Abgeltungsbetrag – indemnity,
redemption money
abgeschlossene Lehre – trained
apprentice
Abgleich – matching, comparison
(jobseekers/vacancies)
Abgrenzdatum – delimitation date
Abgrenzung – delimitation, demarcation
Abgrenzung, zeitliche – time-based
delimitation
Abgruppierung – downgrading
abhängig vom Lebensalter – dependent
on age
Abhängige(r) – dependent
Abhilfe – remedy, relief
Abitur – A-level, advanced level (final
secondary school examination)
Abiturfach – A-level subject
Abiturient – student who is taking or has
taken the „Abitur" examination; high
school graduate (AmE) (s. Graphik)
Abiturzeugnis – Leaving Certificate
Ablauf – expiry
Ablauf der Arbeitserlaubnis – expiry of
work permit
Ablaufmuster – schedule model
Ablauforganisation – structuring of
operations
Ableben – death
ablegen – file, store
ablehnen – reject
Ablehnung – rejection

Ablehnungsschreiben – letter of rejection
Ablenkung – distraction
abmahnen – rebuke
Abmahnung – rebuke
Abmeldung – deregistration
Abnahme – decrease
abnehmen – decrease
Abneigung – aversion
abnormal – abnormal
abordnen – assign, delegate, second
Abordnung – assignment; delegation (AmE); secondment (BrE)
Abordnungsgeld – assignment allowance (AmE), secondment payment
Abordnungskosten – assignment expenses, secondment payment
Abrechner – payroll clerk, payroll officer, expense audit clerk (AmE)
Abrechnung – statement of account
Abrechnungsanpassung – payroll customizing
Abrechnungsart – payroll type
Abrechnungsbasis – payroll basis
abrechnungsbezogen – related to payroll; related to travel expense accounting
Abrechnungscluster – payroll cluster
Abrechnungseinheit – payroll unit; accounting unit
Abrechnungsergebnis – payroll result; accounting result
Abrechnungsformular – payroll form
Abrechnungsgegenwart – current payroll; current accounting
Abrechnungskalender – payroll calendar
Abrechnungskreis – payroll subunit
Abrechnungsmonat – payroll month
Abrechnungsperiode – payroll period; accounting period
Abrechnungsprogramm – payroll program; accounting program
Abrechnungsregel – personnel calculation rule, payroll rule; accounting rule
abrechnungsrelevant – relevant to payroll; relevant to travel expense accounting
Abrechnungsrelevanz – payroll relevancy

Abrechnungsschema – payroll schema; accounting schema
Abrechnungsstatus – accounting status
Abrechnungsstunden – payroll hours
Abrechnungssystem – payroll accounting system
Abrechnungstag – payroll day
Abrechnungstreiber – payroll driver
Abrechnungstyp – payroll category
Abrechnungsvariante – payroll variant; accounting variant
Abrechnungsvergangenheit – payroll past; accounting past
Abrechnungsverwaltungssatz – payroll control record
Abrechnungszeitraum – accounting period
Abrechnungszyklus – payroll cycle
abrunden – round down
Absage – negative reply
absagen – cancel
Absatzbudget – expense budget
abschätzen – evaluate
Abscheu – aversion
Abschlag – advance pay
abschlägiger Bescheid – negative reply
Abschlagsbuchung – advance pay posting
Abschlagsverfahren – advance pay procedure
Abschlagszahlung – installment, payment on account, advance payment
Abschluss – final qualification
Abschlussdatum – closing date
Abschlussprüfung – final exam(ination)s, finals
Abschlussvollmacht – contractual power
Abschlusszeugnis – final certificate
Abschneiden – performance
abschneiden – perform
abschreiben – copy
Abschreibung – depreciation
Abschweifung – digression
Absentismus – absenteeism
Absetzbetrag – deductible amount
Absicht – intention, purpose
absolut – absolute
Absolvent – graduate
Absolvent einer weiterführenden Schule – secondary school leaver

absolvieren – graduate, complete, finish, ... from (AmE)

abstimmen – vote

Abstimmung – voting

abstraktes Denken – abstract thinking

Abteilung – department

Abteilung Arbeitsrecht – labour relations department

Abteilungsleiter – head of department; department manager

Abteilungsversammlung – department meeting

Abtretung – cession of wages

Abwanderung von Wissenschaftlern – brain drain

abwarten – await

Abweichung – variance, deviation

Abwehrmechanismus – defence mechanism

abwenden – avert

Abwerbekampagne – poaching campaign

abwerben von Mitarbeitern – entice employees, poach employees

Abwerbung – enticement; labour piracy; poaching; inducement

abwesend – absent

Abwesende(r) – absentee

Abwesenheit – absence

Abwesenheit, bezahlt – paid absence

Abwesenheit, unbezahlte – unpaid absence

Abwesenheitsart – absence type

Abwesenheitsauszählung – absence counting

Abwesenheitsauszählungsregel – absence counting rule

Abwesenheitsbewertung – absence valuation

Abwesenheitsinformation – absence data

Abwesenheitskalender – absence calendar

Abwesenheitsklasse – absence class

Abwesenheitskontingent – absence quota

Abwesenheitsmitteilung – notification of absence

Abwesenheitsmodifikator – absence modifier

Abwesenheitsrate – absenteeism rate

Abwesenheitsstunde – hour of absence

Abwesenheitstag – day of absence

Abwesenheitstyp – absence category

Abwesenheitsübersicht – overview of absences

Abwesenheitsvergütung – compensation for absence

Abwesenheitszählklasse – absence counting class

Abwesenheitszeit – absence time

Abwesenheitszeitraum – absence period

Abwicklung – administration

Abwicklungszentrum – order processing centre (Abteilung, in der Aufträge in allen Stufen durchlaufen und (maschinell) abgewickelt werden)

abziehen – subtract, deduct

Abzinsung – discounting

Abzug – deduction

Abzug wegen Bewirtung – deduction due to entertainment

Abzug, firmeninterner – company deduction

Abzug, pauschaler – lump-sum deduction

Abzug, steuerlicher – tax deduction

Abzüge, wiederkehrende – recurring deductions

Abzugsart – deduction type

Abzugsbetrag – deduction amount

Abzugsmethode – deduction method

Abzugsprinzip – deduction principle

Abzugssatz – deduction rate

Achtsamkeit – attention

a.D. – after-hours

addieren – add

Administration – administration

Adoleszenz – adolescence

adoptiertes Kind – adopted child

Adressat – addressee

Adresse – address

affektiv – affective

affektives Lernen – affective learning

Aggression – aggression

aggressiv – aggressive

Aggressivität – aggressiveness

Akademiemitglied – academician

Akademiker – university graduate

Akademiker mit erster Berufserfahrung – young professional

Akademikertrend – trend towards university graduates

akademische Ausbildung – academic background

akademische Leistung – academic achievement

akademischen Grad erwerben – take a degree

akademischer Abschluss – academic finals

akademischer Grad – academic degree, degree

akademischer Urlaub – sabbatical year

akademisches Jahr – academic year

akademisches Niveau – academic standard

Akademisierungsgrad – number of graduates

Akademisierungstrend – trend towards university studies

Akademisierungswelle – wave of graduates

Akkord – piecework, job work

Akkord(lohn)system – piecework system

Akkordarbeiter – piece-worker, job-worker

Akkordbereich – piecework area

Akkordbezahlung – piecework pay

Akkordgrundlohn – piecework wage, piecerate

Akkordlohn – piecework wage, piecerate

Akkordprämie – piecerate bonus

Akkordrichtsatz – piecework wage, piecerate

Akkordsatz – piecerate formula, contract wage payment

Akkordzettel – job ticket

akkurat – accurate, exact

akribisch – acribic

Aktennotiz – memorandum

Aktenordner – file

Aktie – share

Aktienbesitz – stock ownership

Aktienbezugsrecht – stock option

Aktienerwerb – acquisition of shares

Aktiengesellschaft – (approx. =) stock corporation, corporation, joint-stock company

Aktiengesetz – German Corporation Act

Aktiensparen – saving through equity investments

Aktionär – stockholder, shareholder (AmE)

Aktionismus – actionism

aktiv – active

aktiver Wortschatz – active vocabulary

Aktivität – activity

Alimente – alimony

Aliquotierung – monthly factoring

Aliquotierungsfaktor – monthly factor

Aliquotierungsmethode – monthly factoring method

Alkoholismus – alcoholism

Alkoholkonsum – consumption of alcohol

Alkoholverbot – order to abstain from alcohol

Alleinerziehende(r) – one-parent family (Familie mit nur einem Elternteil)

Alleininhaber (GmbH & Co.KG) – sole proprietor

Alleinverdiener – sole wage earner

Alleskönner – all-rounder

Alleswisser – know-all

allgemein – common, joint, general

allgemeinbildende Schule – general school

Allgemeinbildung – general education, liberal education

allgemeine Bemessungsgrundlage – general valuation basis

allgemeine Hochschulreife – matriculation requirements

allgemeine Lohn- und Gehaltserhöhung – across the board increase

Allgemeine Ortskrankenkasse (AOK) – (approx. =) general sickness benefit fund, local sickness benefit fund

allgemeine Personalpolitik – general personnel policy

allgemeine Schulpflicht – compulsory education, compulsory schooling

Allgemeinwissen – general knowledge

allmählich – gradual

Alltagssprache – vernacular

Alltagstrott – daily round

als freie(r) Mitarbeiter(in) tätig sein – freelancer

Alter – age

Alters- und Hinterbliebenenversorgung – old-age and survivor's insurance; pension, disability, and surviving dependents' provision

Alters- und Hinterlassenenversicherung – pension, disability, and surviving dependents' insurance

Altersaufbau – age distribution structure, age pyramid, age structure

Alterseinkaufsgeld (Arbeitgeber) – employer's share of sum to buy into retirement fund

Alterseinkaufsgeld (Arbeitnehmer) – employee's share of sum to buy into retirement fund

Altersentlastung – old-age exemption

Altersfreibetrag – age exemption, age allowance, old-age exemption amount

Altersfreizeit – age-related reduction in working hours

Altersgrenze – age limit

Altersgruppe – age group

Altersguthaben – old-age credit

Alterspyramide – age distribution structure, age pyramid

Altersrente – pension, old-age pension, retirement benefit

Altersruhegeld – pension, old-age pension, retirement benefit, retirement pension

Alterssprung – pay scale jump due to age

Altersstruktur – age structure

Altersstufe – age group

Altersteilzeit – partial retirement

Altersteilzeitarbeit – part-time employment for old-age pensioners

Altersteilzeitmodell – model for the part-time employment of old-age pensioners

Altersversorgung – pension, old-age pension, retirement benefit, retirement pension

Altersversorgung, betriebliche – company pension scheme

Alterszusatzurlaub – additional vacation for older employees

Altphilologie – classics

Alumnus – alumnus

ambulante Behandlung – out-patient treatment

ambulante Krankheitskosten – out-of-hospital expenses

Amnesie – amnesia

Amt – office

Amtsinhaber – incumbent

Amtszeit – term of office, tenure

Analphabet – illiterate person

Analphabetentum – illiteracy

Analyse – analysis

analysieren – analyse

analytisch – analytic

anbieten – offer

ändern – alter, change

Änderung der Bezüge – change in pay

Änderung der Eingruppierung – change of pay group

Änderungsbeleg – change document

Änderungskündigung – dismissal for variation of contract, notification of change in terms of employment

andeuten – indicate

aneignen – acquire, adopt, appropriate

Aneignung von Fertigkeiten – acquisition of skills

anerkannter Beruf – recognized trade, recognized profession

anerkennen – approve, acknowledge, appreciate, recognize

Anerkennung verdienen – deserve acknowledgement

Anerkennung zollen (jdm.) – pay tribute to someone

Anfängerkurs – beginners' course

Anfangseinkommen – original income

Anfangsgehalt – initial salary, starting salary

Anfangslohn – starting rate, starting wage, entrance rate

Anfangsprogrammierer – young programmer

anfechtbar – contestable

Anfeuerung – peptalk

Anfordernis – demand

Anforderung – requisition, request

Anforderungen an die Personalpolitik – requirements of personnel policies, demands made on personnel policies

Anforderungs- und Qualifikationsprofil – requirements and qualifications profile

Anforderungsart – requirement type

Anforderungsprofil – requirements profile
Anforderungstyp – requirement category
Anfrage – inquiry
Angaben zur Person – personal data
Angaben zur Tätigkeit – particulars of employment
angeben – boast
angeboren – congenital, innate
angeborene Fähigkeit – innate ability
angeborenes Verhalten – innate behaviour
Angebot – offer; supply
Angebot und Nachfrage – supply and demand
angelernter Arbeiter – semi-skilled worker
angemessen – appropriate
angesetzte Arbeitsstunde – study period (Unterrichtsstunde, die für reine Studienzwecke vorgesehen ist)
angespannt – tense, strained
Angestellte(r) – salaried employee, white collar worker, office worker
Angestelltenversicherung – social security (insurance)
Angestelltenversicherungsgesetz – salaried employees' insurance act; Social Security Act (AmE)
Angestellter – salaried employee
Angestellter, leitender – executive employee
angewandte Forschung – applied research
angleichen – adjust, adapt
Angleichung – adjustment
Angleichung der Löhne – adjustment of wages
Angst – anxiety, fear
ängstlich – apprehensive, timid
anheften – attach
Anhörungsrecht – right to be heard
Ankerfunktion – anchor function
Ankerobjekt – anchor object
Anlage – natural endowment; enclosure
Anlagen – facilities
Anlagenmechaniker – plant mechanic
Anlagentechnik – systems technology
anleiten – instruct, teach
Anleitung – schooling, instruction, guidance

Anlernberuf – semi-skilled occupation
Anlernlohnsatz – learners' wage rate, beginners' wage rate
Anlerntätigkeit – semi-skilled work
Anlernzeit – training period
Anmeldebestätigung – confirmation of (course) registration
Anmelder – applicant
Anmeldung – registration
annähern – approximate; converge
annähernd – approximate
annehmen – accept, suppose
Annuität – annuity
Annuitätenzahlung – annuity payment
annullieren – annul
anordnen – direct, dispose, order
anpassen – adjust, adapt
Anpassung – adjustment
Anpassung der Löhne – adjustment of wages
anpassungsfähig – adaptable, flexible
Anpassungsfähigkeit – adaptability, flexibility
Anpassungsmöglichkeit – customizing option
Anpassungsprüfung (bei Betriebsrente) – check necessity of an adjustment
Anpassungsschwierigkeit – adaptation problem
Anpassungszeitraum – adaptation period
Anrechnung – crediting
Anrechnungsfaktor – credit factor
anrechnungsfrei – non-creditable
Anrechnungsklausel – off-setting provision
Anrechnungsregelung – crediting regulation
Anrede – form of address
anregend – stimulating
Anregung – stimulus
Anreicherung – enrichment
Anreiz – incentive
Anreizprämie – incentive bonus
Anreizsystem – incentive system
Anruf – call
anschaulich – clear, plain, evident, obvious, eidetic
Anschauungsmaterial – illustrative material, visual aids

Anschlagtafel – notice board, bulletin board
anschneiden – mention
Ansehen – standing
Ansporn – stimulus
Ansprache – speech
Ansprechpartner – contact, contact person
Anspruch – claim, right, legal demand; entitlement
Anspruch auf Rente – right to pension
Anspruchsermittlung – calculation of entitlement
Anspruchsberechtigung – entitlement
Anspruchsberechtigung von Angehörigen (RV) – rightful claims of relatives
anspruchsvoll – demanding
anspruchsvolle Führungsaufgaben – qualified management responsibilities
Ansteigen der Löhne (allgemeine Entwicklung) – rising wages (general developments)
Anstellung – employment, state of employment, employment relationship
Anstellungsbedingungen – hiring conditions
Anstellungsverhältnis – work contract
Anstellungsvertrag – employment contract, labour contract, engagement
Anstrengung – effort, endeavo(u)r
Anteil, lohnsteuerpflichtiger – share liable to income tax
Anteilseigner – stockholder, shareholder (AmE)
Antragsfrist – deadline for application
Antragsstatus – application status
Antragsteller – applicant; claimant, plaintiff
Antragsvordruck – application form
Antrieb – incentive
Antrittsvortrag – inaugural lecture
Antwort – answer, reply, response
antworten – answer, reply, respond
Antwortheft – answer book
anvertrauen (jdm. etw.) – entrust
Anwärter – candidate, ratee (AmE)
Anwartschaft – legal right to future pension payments
Anwartschaft auf Ruhegeld – retirement pension expectancy

Anwartschaftsberechtigter – prospective beneficiary
Anwartschaftszeit – qualifying period
anweisen – authorize
Anweisung – command, directive, instruction, order
Anweisungen geben – give orders, give instructions
Anweisungen nehmen – take instructions from, take orders from
anwerben – recruit
Anwerbung – recruitment, personnel recruitment (Beschaffung neuer Arbeitskräfte (geht der Auswahl und Einstellung voraus))
Anwesenheit – attendance, presence
Anwesenheitsart – attendance type
Anwesenheitsgenehmigung – approval of attendance
Anwesenheitsgrund – attendance reason
Anwesenheitsinformation – attendance data
Anwesenheitskontingent – attendance quota
Anwesenheitskontrolliste – attendance check list
Anwesenheitsliste – roll, attendance list, list of names
Anwesenheitstyp – attendance category
Anwesenheitszeit – attendance time
Anzahlung – installment, payment on account
Anzeige – advertisement, ad
anzeigepflichtiger Arbeitsunfall – notifiable accident at work
AOK – (approx. =) general sickness benefit fund, local sickness benefit fund
Apathie – apathy
apathisch – apathetic
Aphasie – aphasia dysphasia (durch Hirnschäden verursachter Zustand)
Arbeit – work
Arbeit, bezahlte – paid work
Arbeit, für die eine Lehre nicht nötig ist – unskilled work
Arbeit, mitternachtsübergreifende – work that spans two days
Arbeit, nichtselbständige – employment (at a company)
Arbeit, wiederkehrende – recurring tasks

Arbeiter – worker, manual worker, workman, blue collar worker

Arbeiterfamilie – working class family

Arbeiterkammer – employee association

Arbeiterklasse – working class

Arbeiterrentenversicherung – labour annuity insurance

Arbeiterselbstverwaltung – worker participation in management, workers' self-management, workers' control

Arbeitgeber – employer

Arbeitgeber-Arbeitnehmer-Verhältnis – employer-employee relationship, labour relations

Arbeitgeberanteil – employer's contribution

Arbeitgeberbeitrag – employer's contribution (to benefits-related payments)

Arbeitgeberdarlehen – employer's loan

Arbeitgeberfürsorgepflichten – employer's obligation to supply benefits and medical welfare

Arbeitgeberleistung – benefits

Arbeitgeberlohnanteil – selective employment tax (BrE), payroll tax (AmE)

Arbeitgeberverband – employer's association

Arbeitgebervertreter – employer's representative

Arbeitgeberwechsel – change of employer

Arbeitgeberzuschuss – employer's allowance

Arbeitskräfteüberschuss – excess labour supply

Arbeitnehmer – employee, member of the staff

arbeitnehmerähnliches Verhältnis – quasi full-time employment situation

Arbeitnehmeranteil – employee's contribution

Arbeitnehmerbeitrag – employee's contribution (to benefits-related payments)

Arbeitnehmerberatung – employee counselling

Arbeitnehmerbeteiligung – employee participation in asset formation

Arbeitnehmerdarlehen – loan to an employee

Arbeitnehmerfreibetrag – earned-income allowance

Arbeitnehmerhandbuch – employee handbook

Arbeitnehmersparzulage – savings supplement, employee savings bonus

Arbeitnehmerüberlassung – supply of temporary workers, hiring-out of labour, manpower provision

Arbeitnehmerüberlassungsfirma – temporary employment agency

Arbeitnehmerüberlassungsgesetz – manpower provision act

Arbeitnehmervertreter – employee representative, labour representative

Arbeits- und Aufgabenanalyse – job analysis, job-task analysis, methods study

Arbeitsablauf – work flow

Arbeitsablaufplan – sequence of operation

Arbeitsamt – employment exchange (BrE), employment agency, employment office, job center

Arbeitsanalyse – job analysis, job-task analysis, methods study

Arbeitsanfang – starting time

Arbeitsanforderungen – job requirements

Arbeitsanweisung – job instruction, job-related instruction

Arbeitsaufenthalt im Ausland – work period abroad

Arbeitsaufgabe – assignment

Arbeitsausfall – working time lost (due to ...), non-productive time, loss of working hours

Arbeitsbedingungen – working conditions

Arbeitsbefreiung – release from work, time off

Arbeitsbeginn – starting time

Arbeitsbelastung – workload

Arbeitsberatung – employment counselling

Arbeitsbereich – work area

Arbeitsbereitschaft – availability for duty

Arbeitsbeschaffung – job creation

Arbeitsbeschaffungsmaßnahme –
measure targeted at job creation
Arbeitsbescheinigung – employment
certificate
Arbeitsbewertung – job evaluation
Arbeitsbewilligung – work permit
Arbeitsblatt – work sheet
Arbeitsbuch – practice book, work book
Arbeitsdirektor – Director of industrial
relations, labour director (bei einer AG
oder GmbH mit mehr als 2.000
Beschäftigten automatisch Mitglied des
Vorstands)
Arbeitsdruck – work pressure
Arbeitseinkommen – earned income
Arbeitsentgelt – remuneration, earned
income
Arbeitserfolg – work success
Arbeitsergebnis – work results
Arbeitserlaubnis – work permit
Arbeitserprobung – aptitude test
Arbeitserwartung – expectations from
work
Arbeitsessen – working dinner, working
lunch
arbeitsfähig – able to work, fit for work
arbeitsfähiges Alter – working age
Arbeitsfähigkeit – capacity to work
Arbeitsförderungsgesetz – Employment
Promotion Law
arbeitsfreier Tag – day off
Arbeitsfreude – work satisfaction, job
satisfaction
Arbeitsfrieden – industrial peace
Arbeitsgebiet – area of responsibility,
sphere of activity
Arbeitsgebietssteuer (USA) – work state
tax
Arbeitsgegenstände – work tools
Arbeitsgenehmigung – work permit
Arbeitsgericht – industrial tribunal,
labour court
Arbeitsgesellschaft – labour society
Arbeitsgesetzgebung – labour legislation
Arbeitsgestaltung – human factors
engineering, job design
Arbeitsgruppe – work group, study
group
Arbeitsinhalt – job content
arbeitsintensiv – labour intensive

arbeitsintensive Industrie – industry
with high labour costs
Arbeitskampf – industrial action,
industrial dispute, trade dispute
Arbeitsklima – working climate
Arbeitskollege – colleague, fellow
worker, co-worker
Arbeitskonflikt – industrial conflict,
labour conflict, work conflict
Arbeitskosten – labour costs
Arbeitskräfte – manpower, labour force,
workforce
Arbeitskräfteüberhang – labour surplus
Arbeitskräftebedarf – manpower
requirements
Arbeitskräftebedarfsplanung –
manpower survey
Arbeitskräfteeinsatz – manpower
assignment
Arbeitskräftemangel – labour shortage
Arbeitskräftemobilität – labour mobility
Arbeitskräftenachfrage – manpower
demand
Arbeitskräftepotential – labour potential
Arbeitskräfteüberschuss – redundant
labour, excess labour supply
Arbeitskreis – work group, study group
Arbeitsleistung – job performance,
efficiency
arbeitslos – jobless, unemployed
Arbeitslose(r) – unemployed person
Arbeitslosengeld – unemployment
benefit, earnings-related benefit
Arbeitslosenhilfe – unemployment
assistance, unemployment relief
Arbeitslosenkasse – unemployment
insurance fund
Arbeitslosenquote – level of
unemployment
Arbeitslosenrate – unemployment rate
Arbeitslosenunterstützung –
unemployment compensation,
unemployment payment
Arbeitslosenversicherung –
unemployment benefit
Arbeitslosigkeit – unemployment
Arbeitsmangel – shortage of work
Arbeitsmarkt – job market, labour
market
Arbeitsmarktaussichten – employment
prospects

Arbeitsmarktforschung – labour market research

Arbeitsmarktlage – situation on the labour market

Arbeitsmarktproblem – labo(u)r market problem

Arbeitsmedizin – occupational medicine, industrial medicine

Arbeitsmethode – method of working

Arbeitsmethodik – work methods

Arbeitsmittel – working materials

Arbeitsmöglichkeiten – job opportunities, work possibilities

Arbeitsniederlegung – work stoppage; walkout

Arbeitsort – place of work

Arbeitspause – break

Arbeitspausenplan – work break schedule

Arbeitsplan – schedule

Arbeitsplanung – work planning, task planning

Arbeitsplatz – work place, job

Arbeitsplatz- und Stellenbeschreibung – work place and job description

Arbeitsplatz- und Stellenbewertung – work place and job grading

Arbeitsplatz- und Stellenverwaltung – work place and job administration

Arbeitsplatzabbau – cutback, downsizing

Arbeitsplatzanalyse – job analysis

Arbeitsplatzanforderung – job requirement

Arbeitsplatzbefragung – job diagnostic survey, JDS

Arbeitsplatzbeschreibung – job description

Arbeitsplatzbewertung – job evaluation, work assessment, job rating (analytische Arbeitsbewertung auf der Basis der Ermittlung von unterschiedlicher Arbeitsschwierigkeit je Arbeitsplatz und Arbeitsbereich)

Arbeitsplatzbewertungsbogen – job evaluation sheet

Arbeitsplatzbewertungsstufe – job evaluation scale

Arbeitsplatzbewertungssystem – job evaluation system

arbeitsplatzbezogener Kurs – work-related course

Arbeitsplatzdaten – job data

Arbeitsplatzsicherung – job security

Arbeitsplatzsuche – search for employment

Arbeitsplatzteilung – job sharing

Arbeitsplatzvertretung – work center substitution

Arbeitsplatzwechsel – job rotation, labour turnover, fluctutation, change of job

Arbeitsplatzzulage – job bonus

Arbeitsprojektor – overhead projector, OHP

Arbeitsprozess – work process

Arbeitspsychologie – industrial psychology

Arbeitsqualität – work quality

Arbeitsrecht – labour law, industrial law

Arbeitsrechtsverfahren – labour law proceedings

Arbeitsschicht – work shift

Arbeitsschluss – stopping time

Arbeitsschutz – protection of labour, fair labor practice (AmE)

Arbeitsschutzausschuss – committee for labour protection

Arbeitsschutzgesetz – labor protection laws

Arbeitsschutztagung – labo(u)r protection meeting

Arbeitssicherheit – occupational safety

Arbeitssicherheitsexperte – on-the-job safety expert

Arbeitssicherheitsorganisation – on-the-job safety organisation

Arbeitssicherheitsprogramm – programme for improvement of on-the-job safety

Arbeitssituation – work situation

Arbeitsspitze – peak time

Arbeitsstab – staff

Arbeitsstätte – place of employment, place of work

Arbeitsstelle für die Elite – core job (wenige, hochverdichtete Arbeitsstellen für Spitzenleute)

Arbeitsstellen für die Massen – fringe jobs (massenhafte, austauschbare

Arbeitsstellen für die vielen, wenig qualifizierten)

Arbeitsstil – working style

Arbeitsstudie – job analysis, job-task analysis, methods study, time and motion study, work study

Arbeitsstudien-Ingenieur – time study engineer

Arbeitsstudienabteilung – time office

Arbeitsstufe – work level, stage

Arbeitsstunde – manhour, working hour

Arbeitssuche – job search, job hunt

Arbeitssucht – work addiction, workaholism

Arbeitstag – working day, workday

Arbeitstakt – working cycle, production cycle

arbeitsteilige Industriegesellschaft – labour division in the industrial society

Arbeitsteilung – labour division

Arbeitstempo – working speed

Arbeitstherapeut – occupational therapist

Arbeitsumfeld – work environment

Arbeitsumgebung – work environment

arbeitsunfähig – disabled, ill, incapable of work, sick, unable to work

Arbeitsunfähigkeit – inability to work, invalidity, disability

Arbeitsunfähigkeitsbescheinigung – medical certificate, certificate of disability

Arbeitsunfähigkeitsentschädigung – disability benefit

Arbeitsunfall – industrial accident, occupational accident, work accident

Arbeitsunfallentschädigung – accident benefit

Arbeitsunfallversicherung – worker's compensation

Arbeitsunterbrechung – stoppage of work, interruption of work

Arbeitsverfahren – management procedure

Arbeitsverhalten – work habits, work attitude, work behaviour

Arbeitsverhältnis – employment, state of employment, employment relationship, work relationship

Arbeitsverhältnis, ruhendes – inactive work relationship, dormant employment

Arbeitsverhältnis, unbefristetes – permanent work relationship

Arbeitsverhältnis, unterbrochenes – interrupted work relationship

Arbeitsvermittlung – employment agency, placing service

Arbeitsversuch (missglückter) – premature return to work (vorzeitige und missglückte Aufnahme der Arbeit nach einer Krankheit)

Arbeitsvertrag – employment contract, labour contract, engagement

Arbeitsverweigerung – refusal to work

Arbeitsweise – method of working, way of working

Arbeitswelt – working world, world of work

arbeitswillig – willing to work

Arbeitswissenschaft – ergonomics (Nordamerikanische Forschungsrichtung, die der deutschen Arbeitswissenschaft annähernd entspricht, wobei die Arbeitspsychologie besonders berücksichtigt wird.)

arbeitswissenschaftlich – ergonomic

arbeitswissenschaftliche Organisationsgestaltung – human factors engineering

Arbeitswoche – working week

Arbeitszeit – working hours, working time

Arbeitszeit, modulare – modular working time (Unternehmen teilen Betriebstag, -woche, -monat oder -jahr in Arbeitszeit-Bausteine ein, die die MA unter Rücksicht auf die Vorgaben des Betriebs beliebig kombinieren können.)

Arbeitszeit- und Dienstplanung – work schedule and shift planning

Arbeitszeitänderung – change of working hours

Arbeitszeitanteil – employment percentage

Arbeitszeitbeginn – start of working hours

Arbeitszeitende – end of working hours

Arbeitszeitmodell – working time model
Arbeitszeitnachweis – attendance form
Arbeitszeitplan – work schedule
Arbeitszeitplanregel – work schedule rule
Arbeitszeitregelungen – rules on working hours
Arbeitszeitübersicht – attendance survey
Arbeitszeitverkürzung – reduction in working hours
Arbeitszeitverkürzungsregelung – regulation of reduction in working hours
Arbeitszeugnis – certificate of employment; reference (from employer)
Arbeitsziele festsetzen – set objectives
Arbeitszimmer – study
Arbeitszufriedenheit – work satisfaction, job satisfaction
Arbeitszuweisung am Arbeitsplatz – task, work assignment
Archiv – archive
Ärgernis – annoyance, vexation
Argument – argument
argumentieren – argue
Arroganz – arrogance
Artikulation – articulation
Arztbesuch – doctor's appointment
ärztliche Untersuchung – medical examination
ärztliche Versorgung – medical care
ärztliches Attest – medical certificate
Assessment Center zum Erfassen sozialer Fähigkeiten – assessment center for interpersonal skills
Assessor – assessor (Jurist, mit bestandenem zweiten Staatsexamen)
Assistent(in) – assistant
Assistenzlehrer – teaching assistant
Assoziation – association; connotation
Asylbewerber – asylum seeker
AT-Angestellter – exempt employee
AT-Bereich – area where the regular pay scale does not apply
AT-Gehalt – exempt salary
Athlet – athlete
Atmosphäre – atmosphere
Attest – medical certificate
Audiovision – audio-vision
audiovisuell – audio-visual

auf den neuesten Stand bringen – update
auf eine Mindestzahl reduziertes Personal – skeleton staff
auf Firmenkosten – at the expense of the company
auf Wettbewerb eingestellt – competitive
Aufbauberuf – wider ranging profession
aufbauend – constructive
Aufbaukurs – advanced training course
Aufbauorganisation – company organization structure
Aufbauregel – layout rule
Aufbauvorschrift – layout regulation
Aufbewahrungsfrist – retention period
Aufdeckung – detection, discovery
aufeinanderfolgende Kurse – back-to-back courses
Aufenthalt – stay
Aufenthalt(sort) – residence
Aufenthaltsdauer – length of stay
Aufenthaltserlaubnis – residence permit, registration certificate
Aufenthaltsgenehmigung – residence permit
Aufenthaltsstatus – status of residence permit
auffächern – expand
Auffächerungsreport – expansion report
Auffassungsgabe – ability to grasp, comprehension, understanding
Aufforderung – request
Aufforderung zur Wiederbewerbung – request for reapplication
auffrischen – refresh, brush up
Auffrischungskurs – refresher course
Aufgabe – task, work assignment
Aufgabe übertragen – delegate tasks
Aufgabenanalyse – job analysis, task analysis
Aufgabenbereich – area of responsibility, sphere of activity
Aufgabenbeschreibung – task description
Aufgabenerfüllung – task fulfillment
Aufgabenerweiterung – job enlargement (horizontale Arbeitserweiterung auf strukturell gleichartige Aufgaben)
Aufgabengebiet – domain, purview, field of action

Aufgabengliederung – task structuring
Aufgabengruppe – task group
Aufgabenkatalog – task catalog
Aufgabenkomplex – task complex
aufgabenorientierte Führung – task-oriented leadership
Aufgabenprofil – task profile
Aufgabenstruktur – job structure, structure of the tasks
Aufgeld – extra pay
Aufgeldkennzeichen – extra pay indicator
aufgeschlossen – open-minded, receptive
aufgeweckt – intelligent, bright
Aufgliederung – breakdown, itemization
Aufgruppierung – upgrading
Aufhebungsvertrag – cancellation contract
Aufmerksamkeit – attention
Aufmerksamkeitsspanne – attention span
aufmucken – protest
Aufnahme – admission, entrance; assimilation, integration
Aufnahmebedingung – entrance qualification
Aufnahmefähigkeit – ability to take in, ability to absorb something
Aufnahmekriterien – admissions criteria
Aufnahmeprüfung – entrance exam(ination), entry test
Aufnahmeprüfung in eine weiterführende Schule – eleven plus examination (BrE)
Aufnahmeverfahren – admissions procedure
aufnehmen – absorb
aufregend – exciting
Aufregung – nervousness, stage-fright, excitement
aufrichtig – fair, just, sincere
Aufrichtigkeit – sincerity
Aufrollung – recalculation
Aufrollungsdifferenz – recalculation difference
aufrücken – put up, progress to next level
Aufruf – call
aufrunden – round up
aufs Spiel setzen – hazard
aufsässig – unruly
Aufsatz – composition, essay

aufschieben – delay, postpone
aufschiebende Bedingung – suspensive condition, condition precedent
aufschlüsseln – break down, classify, itemize
Aufschub – delay, postponement
Aufsicht – supervisor
Aufsichtsbehörde – supervisory authority
Aufsicht(sperson) – invigilator, superintendant (of exams)
Aufsichtsrat – Supervisory Board
Aufsichtsratsmitglied – Member of the Supervisory Board
Aufsichtsratsvorsitzender – Chairman of the Supervisory Board
aufstehen – get up
aufsteigen – advance
aufstellen – nominate
Aufstellen eines Lehrplans – syllabus design
Aufstellung – list, nomination
Aufstieg, beruflicher – career advancement
Aufstiegschancen – promotion prospects, promotional opportunities
Aufstiegsmöglichkeiten – career opportunities, career prospects
Aufteilung – breakdown, itemization
Auftragsvergabe – contracting out, placing an order
Auftragszeit – work order times
Auftragszentrum – order processing centre (Abteilung, in der Aufträge in allen Stufen durchlaufen und (maschinell) abgewickelt werden)
Aufwand – cost
Aufwärmübung – warming-up exercise
Aufwendungen für Berufsausbildung – cost of vocational training
Aufwendungen für Weiterbildung – cost of on-going training
Aufzeichnung – memorandum, memo, note, recording
Aufzeichnungsbogen – record sheet
aufziehen – rear, bring up
Auge – eye
Augenkontakt – eye contact
Aus- und Fortbildung – education and training

Aus- und Fortbildungsbedarf – demand for education and training

Aus- und Weiterbildung – training and continuing education

Aus- und Weiterbildung von Führungskräften – management training

Aus- und Weiterbildungsplanung – E&T Planning

ausbauen – develop

ausbeuten – exploit

Ausbilder – instructor, apprentices' teacher, trainer

Ausbildereignung – aptitude as instructor

Ausbildereignungsprüfung – instructor aptitude examination

Ausbilderzulage – instructor's bonus

Ausbildung – schooling, training

Ausbildung am Arbeitsplatz – on-the-job training

Ausbildung im Unternehmen – in-house training

Ausbildung in einer Privatschule – private education

Ausbildung innerhalb des Unternehmens – in-company training

Ausbildung vor Ort – on-site training

Ausbildung zur Führungskraft – managerial training

Ausbildung, die nicht am Arbeitsplatz erfolgt – off-the-job training

Ausbildungsabbrecher – trainee dropout

Ausbildungsablauf – training sequence

Ausbildungsabteilung – training department

Ausbildungsanspruch – entitlement to education and training

Ausbildungsaufwand – training costs

Ausbildungsbedarf – need for training

Ausbildungsbeihilfe – training allowance

Ausbildungsberatung – educational guidance, educational counselling

Ausbildungsbereich – field of training

Ausbildungsberuf – training occupation

Ausbildungsbetrieb – training firm, training company

Ausbildungsförderung – promotion of vocational training

ausbildungsfremde Arbeit – activity irrelevant to the training program

Ausbildungsgang – course

Ausbildungsgruppe – vocational category

Ausbildungsinhalt – training program

Ausbildungskonzept – training concept

Ausbildungslehrgang – course (of instruction)

Ausbildungsleiter – director of training, head of training

Ausbildungsmarkt – training possibilities on offer

Ausbildungsmaßnahme – training measure

Ausbildungsmeister – instructor, apprentices' teacher, trainer

Ausbildungsniveau – level of training

Ausbildungsordnung – training regulations, training directive

Ausbildungsplan – training schedule, training program

Ausbildungsplatz – apprenticeship, training position

Ausbildungsprofil – training profile

Ausbildungsprogramm – training programme, training scheme

Ausbildungsrahmenplan – skeleton training schedule, general training syllabus

Ausbildungsrichtung – field of training

Ausbildungsstand – level of training

Ausbildungsstätte – training centre

Ausbildungsstelle – apprenticeship, training position, place of training

Ausbildungssystem – training system

Ausbildungsurlaub – leave for further training

Ausbildungsvergütung – training allowance

Ausbildungsverhältnis – training status

Ausbildungsvertrag – training contract

ausbildungswillig – willing to be trained

Ausbildungszeit – apprenticeship, apprenticeship time, duration of education

Ausbildungszulage – training bonus

Ausbildungszweig – area of training

Ausdauer – stamina, perseverance, endurance

Ausdruck – expression, print-out (EDP)

Ausdruck der eigenen Persönlichkeit – self-expression

Ausdruck der Umgangssprache – colloquialism
Ausdrucksfähigkeit – power of expression; oral skill
Ausdrucksvermögen – power of expression
ausdrucksvoll – expressive
Ausdrucksweise – diction
Auseinandersetzung – dispute
Ausfall – loss
Ausfalllohn – idle time compensation
Ausfallprinzip – lost time
Ausfallrate – drop-out rate
Ausfallstunde – lost hour
Ausfallzeit – time off
Ausfallzeit – non-contributory period
Ausflug – outing
ausführen – execute, implement
Ausführung – execution
Ausgangsstelle – initial job
ausgebeutete Arbeiter – sweated labour
ausgebrannt sein – burnt out
ausgeglichen – balanced, equable, well-balanced, level-headed
ausgelernt – training concluded
ausgenommen – except
ausgenommen (von) – exempt
ausgetreten – left company
ausgeübter Beruf – current job
ausgewählter Wortschatz – controlled vocabulary
ausgewogen – balanced
ausgezeichnet – excellent
ausgleichen – compensate
ausgleichen, Schulden – settle (debts)
ausgleichend – compensatory
Ausgleichsabgabe – equalization levy
Ausgleichsgrenze – equalization limit
Ausgleichskasse – equalization fund
Ausgleichsstelle – equalization board
Ausgleichsvergütung – catch-up allowance
Ausgleichzulage – cost-of-living adjustment
ausgliedern – outsource (Outsourcing bezieht sich auf eine Funktion oder Dienstleistung, die nicht mehr im Unternehmen erbracht wird, sondern von externen Anbietern eingekauft wird. Es spielt dabei keine Rolle, ob dieser externe Anbieter vorher in irgendeiner Weise mit dem Unternehmen verbunden war oder nicht.)
Ausgrenzung – exclusion
ausgründen – outsource
Aushilfe – jobber, temporary personnel
Aushilfsbezüge – remuneration for temporary personnel
Aushilfskraft – casual worker, temporary worker, stopgap
Aushilfslehrer – supply teacher
Aushilfstätigkeit – temporary work
Auskunft – information
Auskunftspflicht – obligation to give information, duty to disclose information
Auslagen – cost
Ausländer – foreigner
Ausländeramt – aliens office
Ausländeranteil – number of aliens, percentage of foreigners
Ausländerfamilie – foreign family
Ausländerfeindlichkeit – hostility towards foreigners
ausländischer Arbeitnehmer – foreign worker, foreign employee
Auslandsabordnung – foreign delegation, overseas position
Auslandsaufenthalt – residence abroad, period abroad
Auslandseinsatz – foreign delegation, overseas position
Auslandsentsendung – transfer to a foreign country, assignment to a foreign country
Auslandserfahrung – foreign experience, experience abroad
Auslandsgeschäft – export trade, export business
Auslandsgesellschaft – foreign company
Auslandspraktikum – practical training abroad
Auslandsreise – trip abroad
Auslandsrente – pension paid abroad
Auslandsschule – school abroad (eine Schule, die von einem Staat außerhalb der eigenen Landesgrenze betrieben wird)
Auslandsunfallversicherung – insurance against accidents abroad
auslassen – omit

Auslese – selection, elite
Auslesedruck – selection pressure
Ausnahme – exception
ausnutzen – exploit
Ausprägung der Merkmale – validity of
  ratings
ausprobieren – test
ausreichend – sufficient
Ausrichtung – alignment
ausschalten – eliminate
ausscheiden – withdraw
ausschlachten – exploit
ausschließen – exclude
ausschließlich – exclusive
Ausschließung – exclusion
Ausschluss – exclusion
ausschreiben – contract out
Ausschreibung – job advertisement
Ausschreibungskosten – job
  advertisement costs
Ausschuss – committee
Ausschuss Arbeitsrecht – labour
  relations board
Ausschuss des Aufsichtsrats –
  committee of the Supervisory Board
Außendienst – external duty
Außenseiter(in) – misfit
Außenstände – outstanding balance
außer – except
außer Betrieb – out of order
außer Tarif – assistant manager, non-pay
  scale, exempt
außerfachliche Kompetenz – additional
  qualification
außergewöhnlich – exceptional
außerhalb der Dienstzeit – after-hours
**außerhalb des Studienplans liegende
  Tätigkeit** – extracurricular activity
äußerlich – extrinsic
außerordentliche Kündigung –
  summary dismissal (employer), leave
  without notice (employee)
außerordentlicher Professor – associate
  professor (AmE)
außertariflich – non-pay-scale
außertariflicher Mitarbeiter – member
  of the exempt staff; assistant manager
außertarifliches Personal – exempt
  personnel
Äußerung – comment
aussetzen – suspend

Aussetzen von Beschlüssen – deferment
  of decisions
Aussondern ineffizienter Mitarbeiter –
  shake-out
Aussperrung – lock-out
Aussprache – pronunciation
aussteigen – opt out
ausstempeln – clock out, clock off
„Ausstoß" der Hochschulen – graduates
  per annum
Ausstrahlung – charisma
aussuchen – choose, select
Austauschbarkeit – interchangeability,
  exchangeability
Austauschlehrer – exchange teacher
Austauschprogramm – exchange
  programme
austreten – leave
Austritte – persons leaving; termination
  of employment
Austrittsdatum – date of leaving
Ausübung – exercise
Auswahl – selection, choice, election,
  sample
Auswahl treffen – choose, select
auswählen – choose, select
Auswahlkriterium – selection criterion
Auswahlrichtlinie – selection guideline
Auswahltest – selection test
Auswahlverfahren – process of
  selection, selection procedure
ausweichen – evade
Ausweiskontrolle (Pförtner) – gate
  control
Ausweislohnart – statement wage type
Ausweisnummer – identification number
auswendig lernen – learn by heart, learn
  by rote
auswerten – evaluate
Auswertung – evaluation
Auswertung von Tests – test evaluation
Auswertungsklasse – evaluation class
Auswertungsweg – evaluation path
Auszahlung – outpayment, payout
Auszahlungsbetrag – net pay (payroll);
  amount paid (travel expenses)
Auszahlungsbetragsrest – remaining
  balance of payment
Auszahlungsperiode – payment period
Auszeichnung – award
Auszubildende – trainee

Auszubildende(r) – apprentice, trainee
Authentizität – authenticity
Autismus – autism (krankhafte Ich-Bezogenheit, das Sich-Abschließen von der Umwelt und dauernde Beschäftigung mit der eigenen Fantasie besonders bei Schizophrenie)
Autodarlehen – car loan
autogenes Training – relaxation through self-hypnosis
Automatisierung – automation
autonom – autonomous
autoritäre Methode – authoritarian method
autoritärer Führungsstil – authoritarian leadership style
autoritärer Lehrer – authoritarian teacher, disciplinarian
autoritäres Führungsverhalten – authoritarian management behaviour
autoritäres Verhalten – authoritarian behaviour
Autorität – authority, power
Autoritätshierarchie – chain of command
Azubi – apprentice, trainee

B

BAFöG (Bundesausbildungsförderungsgesetz) – Federal Law on Financial Support for Education and Training, scholarship, grant
Bakkalaureus Artium – Bachelor of Arts (B.A.)
Bakkalaureus Scientiarum – Bachelor of Science (B.Sc.)
Balkendiagramm – bar chart
Bammel haben – nervous, scared
Bande – gang
Bankfeiertag – bank holiday
Bankverbindung – bank details
Barabgeltung – payment in lieu of holidays (Eine Barabgeltung des

Urlaubs kann vom Arbeitgeber nicht verlangt werden.)
bar gezahlter Lohn – cash wages
Barentgelt – cash remuneration
Bargeld – cash, ready money
Barlohn – cash wages
Barwert – cash value, present value
Barwertfaktor – present value factor
Barzahlung – cash payment
Basaltext – basal text (komprimierte, aber vollständige Darstellung des Lehrstoffs als stoffliche Grundlage für eine Programmierte Unterweisung)
Basis – base, basis
Basisbezüge – basic pay
Basisbezugsänderung – change in basic pay
Basisbezugserhöhung – increase in basic pay
Basisbezugssplitt – basic pay split
Basisbezugszeitraum – basic pay period
Basislohnart – base wage type
Basiswissen – elementary knowledge
Bauberuf – construction job
Baubranche – construction sector
Baudarlehen – building loan
Baugewerbe – building trade
Bauingenieur – civil engineer
Bautechnik – civil engineering
Bauzeichner – draughtsman
beabsichtigen – intend
Beamtenmentalität – officialism
Beamtenstreik – civil servants strike
Beamtentum – officialdom
Beamter – civil servant
Beamter einer Schulbehörde, der Fälle von Schulschwänzen untersucht – attendance officer, truancy officer (BrE)
beanspruchen – claim benefit
beanstanden von Beiträgen – find fault with a contribution
beantragt – submitted
beantworten – answer
Beaufsichtigung – supervision
Bedarf – needs, demand, requirement
Bedarfsanalyse – assessing needs, needs assessment, needs analysis
Bedarfsart – requirements type
Bedarfsdeckung – fulfilment of demand
Bedarfsermittlung – needs analysis
Bedarfsorientierung – needs orientation

Bedeutung – importance; meaning
Bediensteter – servant, staff
Bedienungsanweisung – operating instructions
Bedingung – condition, term
bedingungslos – unconditional
Bedrohung – threat
Bedürfnisse bestimmen – determine needs
beeinträchtigen – impair
Beeinträchtigung – impairment
beendigen – finish
Beendigung – termination
befähigen – qualify someone for/as
Befähigung – ability, capability, performance ability
Befähigungsnachweis – certificate of qualification
Befähigung(szeugnis) – qualification
befangen – self-conscious, biased
Beförderung – promotion, advancement
Beförderung von Schülern in Bussen – bussing (AmE), (zum Zweck der Rassenintegration)
Beförderungschance – chance of promotion, opportunity for promotion
Beförderungsplan – promotion chart
befreien – exempt
befreiende Lebensversicherung – exempt life insurance
befreit – exempt
Befreiung – day-release, exemption
Befreiung von der Versicherungspflicht – exemption from compulsory insurance
befriedigend – satisfactory
Befriedigung – satisfaction
befristen – limit, set a time limit
befristete Professur – tenure-track-professor (eine befristete Professur für talentierte Nachwuchskräfte auch ohne Habilitation.)
befristeter Arbeitsvertrag – fixed-term employment contract, limited employment contract
befristetes Arbeitsverhältnis – limited employment, limited state of employment
Befugnis – authority
begabt – talented, gifted

Begabtenförderung – financial support for talented students, teaching the highly talented (Die Lehrer sind häufig für die Aufgabe der Begabtenförderung unzureichend ausgebildet.)
Begabtenprüfung (BRD) – entrance examination for talented students (Die Begabtenprüfung soll besonders befähigten Berufstätigen ermöglichen, die Berechtigung zu einem Hochschulstudium zu erwerben. Diese Prüfung ist für Personen gedacht, die aufgrund ihrer Begabung, ihrer Persönlichkeit und ihrer Vorbildung für ein Hochschulstudium in Frage kommen, aber wegen ihres Entwicklungsganges keinen schulischen Bildungsgang bis zur Hochschulreifeprüfung durchlaufen konnten.)
Begabung – talent
Begeisterung – enthusiasm
Begeisterung schaffen – motivate
beginnen – begin, commence, initiate, introduce
Beginntoleranz-Beginn – begin tolerance begin time
Beginntoleranz-Ende – begin tolerance end time
Beginnuhrzeit – begin time
Begleitbrief – covering letter, covering note
Begreifen durch Tun – action learning
begrenzter Wortschatz – restricted vocabulary, limited vocabulary
begriffsstutzig – slow on the uptake
Begründung – reason
Begünstigter – beneficiary
Begünstigungszeitraum – concession period; benefit period
Behaglichkeit – comfort
behandeln – treat
Beharrlichkeit – persistance
Behaviorismus – behaviourism (Eine mit dem Namen J.B.Watson in Verbindung gebrachte Richtung der Psychologie. Ausschließlich das von außen registrierbare Verhalten ist Gegenstand.)
Beherbergung – lodging
beherrscht – self-controlled

Beherrschung – mastery
behindert – disabled, handicapped
Behindertenausweis – ID for handicapped
Behindertengruppe – handicapped group
Behindertenurlaub – additional vacation for handicapped persons
Behinderter – disabled person, handicapped person
behindertes Kind – handicapped child
Behinderung – disability
Behinderungsart – handicap type
Behörden – authorities
behördliche Abmeldung – notification of change of address or residence
beide Fächer mit „sehr gut" bestanden – double first(honours)
bei einer Prüfung durchfallen – fail an exam
bei Laune halten – humour
beifügen – enclose
beigefügte Bedingungen – enclosed conditions
Beihilfe – allowance, benefit
beiläufig – incidental
beilegen – enclose
Beirat – advisory board
Beistand – assistance
beistehen – assist
Beitrag – contribution
Beitrag leisten – contribute
Beiträge nach dem wirklichen Arbeitsverdienst – contributions according to actual earnings
Beiträge zur Arbeiterrentenversicherung – contributions to the old-age pension scheme
beitragen – contribute
Beitragsabführung – remittance of social insurance contributions, deduction of contribution amount
Beitragsabrechnung – contribution accounting
beitragsbasierter Pensionsplan – defined contribution plan
Beitragsbemessungsgrenze – income threshold
Beitragsentrichtung – payment of contributions

Beitragsermittlung – calculation of contribution amount
Beitragserstattung – reimbursement of contributions
beitragsfrei – noncontributory
Beitragsfreiheit – exemption from contributions
Beitragsgruppe – contribution group
Beitragshinterziehung – defraudation of contributions
Beitragshöhe – amount of contribution
Beitragsklasse – contribution scale
Beitragskonto – contributions account
Beitragsnachberechnung – recalculation of contributions
Beitragsnachentrichtung – postpayment of contributions
Beitragsnachweis – statement of contributions paid
beitragsorientiert – contribution linked
Beitragspflicht – obligation to contribute
Beitragspflicht – liability to pay contributions
beitragspflichtiges Arbeitsentgelt – employee compensation, liable to contribution
Beitragsrückstände – contributions in arrears
Beitragssatz (allgemein) – general contribution rate
Beitragssatz (erhöht) – higher contribution rate
Beitragssatz (ermäßigt) – reduced contribution rate
Beitragssatz – contribution rate
Beitragstabelle – contribution rate table
Beitragsverrechnung – debiting contributions
Beitragszeiten – contributory period
Beitragszuschuss – contribution allowance
Beitretende(r) – entrant
bejahend – affirmative
Bekanntgabe – announcement
bekanntmachen – make known
Bekanntmachung – announcement, notification
Beklagte(r) – defendant
belastbar – resilient, able to take stress
Belastbarkeit – performance under stress
belästigen – bother, harass

Belastung – strain
Beleg – document; receipt (travel expenses)
Belegabrechnung – receipt accounting
Belegschaft – staff, employees, personnel, human resources, workforce
Belegschaftsaktien – employees' shares
Belegschaftsatmosphäre – staff relationship
Belegschaftsbericht – staff report
Belegschaftsmitglied – employee, member of the staff
Belegschaftsstruktur – workforce structure
Belegschaftsversammlung – staff meeting
Belegschaftsvertreter – employee representative
Belegung – occupancy; booking, reservation
Belegungszeit – occupancy time
Beleidigung – insult
Belehrung – instructions
Beleuchtung – lighting
Belieben – discretion
Belohnung – reward
Belohnungssystem – reward system
Belüftung – ventilation
bemerken – notice
Bemessung – assessment
Bemessungsgrenze – assessment threshold
Bemessungsgrundlage – assessment basis
Bemühen – effort, endeavour
Bemühung – effort, endeavour
benachteiligt – at a disadvantage
Benachteiligung – disadvantage, discrimination
Benehmen – behaviour
benennen – designate
benoten – mark
Benotungssystem – grading system
benutzen – use
Benutzergruppe – user group
Benutzerlohnart – user wage type
Benutzerstamm – user master data
beobachten – observe
Beobachterrolle – monitor role
Beobachtung – observation

Beobachtungsfähigkeit – observational skill
beraten – advise
beraten über – discuss
beratende(r) Direktor(in) – non-executive director
Berater – Senior Director, non-executive director, consultant
Beratervertrag – consultancy contract
Beratung – advice, counselling
Beratung durch Akademiker – academic counselling
Beratung in akademischen Fragen – academic counselling
Beratungsfunktion – advisory function
Beratungsunternehmen – consultancy
berechenbar – predictable
berechnen – calculate
Berechnung – calculation, billing, invoicing
Berechnungsgrundlage bei Pensionsrückstellungen – cash value, present value
berechtigen – authorize, empower, entitle
berechtigt – authorized, justified, entitled
Berechtigung – entitlement
Berechtigung, ablauforientierte – authorization for process organization
Berechtigung, aktionsorientierte – authorization for activities
Berechtigungsvektor – authorization vector
Bereich – Group; area
Bereicherung – enrichment
Bereichsgruppe – range group
Bereichsbetreuer – personnel assistant for a special unit
Bereichsinteresse – departmental interest
Bereichsleiter – head of department
Bereichssekretariat – department office
bereichsspezifisch – area specific, group specific
bereichsübergreifend – extending across different areas
bereichsüberschreitende Versetzung – inter-departmental transfer
Bereichsvorstand – Group Executive Management
Bereichswechsel – inter-departmental transfer
bereit – willing

Bereitschaft – readiness, willingness
Bereitschaftsart – availability type
Bereitschaftsdienst – stand-by duty, emergency service
Bereitschaftszeit – availability time
berichtigen – amend
Berichtigung – amendment
Berichtsheft – report book, log book
Berichtsweg – reporting path
Berichtswesen – reporting
berücksichtigen – take into consideration, take into account
Beruf – profession, occupation, trade, job
beruflich – occupational, professional
berufliche Aussichten – career opportunities, occupational future
berufliche Bildung – training, vocational training, qualifications
berufliche Entwicklung – career development; employment history
berufliche Kompetenz – professional competence
berufliche Umschulung – vocational retraining
berufliche Voraussetzung – professional prerequisite
berufliche Wiedereingliederung – integration into working life
berufliche Zukunft – professional future
beruflicher Erfolg – professional success
beruflicher Werdegang – career, professional career
berufliches Fortkommen – getting on in one's career
berufliches Umfeld – professional environment
Berufs... – occupational, professional
Berufsakademie – College of Advanced Vocational Studies
Berufsalltag – daily office routine
Berufsanfänger – beginner, entrant
Berufsanforderungen – occupational requirements
Berufsaufbauschule (BRD) – Berufsaufbauschule (Die Berufsaufbauschule vermittelt eine allgemeine und eine fachtheoretische Bildung, die über die Berufsschule hinausgeht) (s. Graphik)
Berufsauffassung – professional attitude

Berufsausbildung – vocational training, vocational education, occupational training
Berufsausbildungsverhältnis – employer/apprentice relationship
Berufsausbildungsvertrag – training contract
Berufsaussichten – professional prospects, job prospects
berufsbegleitender Fortbildungslehrgang – on-going vocational training
Berufsberater – career adviser, vocational counsellor
Berufsberatung – vocational guidance, career advisory service
Berufsberatung durch das Arbeitsamt – job counselling by the Labour Exchange
berufsbezogen – vocationally oriented, relevant to the job
Berufsbezug – job connection
Berufsbild – job outline
berufsbildende Schule – vocational school, technical school, trade school (s. Graphik) (Pflichtunterricht für Lehrlinge, Anlernlinge und jugendliche Ungelernte. Der Unterricht findet entweder an einem Tag pro Woche statt oder als Blockunterricht. Überwiegende Aufgabe der Berufsschule ist es, die fachtheoretischen Kenntnisse für den jeweiligen Beruf zu vermitteln.) (s. Graphik)
Berufsbildung – vocational training, vocational education, occupational training
Berufsbildungsgesetz – Vocational Training Act
Berufsbildungszentrum – vocational training centre
Berufseinstellung – attitude to work, work habits
Berufseinstieg – start in a job, job entry
Berufsentwicklung – career development
berufserfahren – professionally experienced
Berufserfahrung – professional experience

Berufserfordernisse – professional requirements

Berufsethos – professional ethics

Berufsfachschule (BRD) – full-time specialized vocational training school (s. Graphik) (Ausbildungsstätte für handwerkliche, kaufmännische, hauswirtschaftliche Berufe. Viele Berufsfachschulen vermitteln eine abgeschlossene Berufsausbildung. Der Unterricht umfasst sowohl die allgemeinbildenden und berufsbezogenen Fächer als auch die praktische Berufsausbildung.) (s. Graphik)

Berufsfeld – vocational field

Berufsgenossenschaft – employer's liability insurance association, professional association, mutual indemnity society (als öffentlich-rechtliche Körperschaft Trägerin der Unfallversicherung)

berufsgenossenschaftliche Leistungen – benefits of a professional association

berufsgenossenschaftliche Pensionskasse – pension fund of a professional association

berufsgenossenschaftlicher Beitrag – contribution to a professional association

Berufsgrundschule – full-time basic vocational school

Berufsgrundschuljahr – pre-vocational education year

Berufsgruppe – job category, occupational group

Berufsinformation – occupational information

Berufskategorie – profession group

Berufskleidung – work clothes

Berufskrankheit – occupational disease, occupational illness

Berufslaufbahn – career, professional career, work history

Berufsleben – working life, working career

berufslos – without a trade

Berufsnachweis – certificate

Berufsnörgler – professional grouser

Berufsoberschule (BRD) – Berufsoberschule (Die

Berufsoberschule umfasst zwei Jahrgangsstufen. Sie vermittelt die fachgebundene Hochschulreife. Voraussetzung für die Aufnahme ist ein mittlerer Schulabschluss und eine abgeschlossene Berufsausbildung. Der Unterricht umfasst allgemeinbildende und fachtheoretische Fächer.) (s. Graphik)

Berufsorientierung – professional orientation

Berufspause – career break

Berufsplanung – career planning

Berufspraxis – professional experience

Berufsqualifikation – skills, vocational competence, job qualification

Berufsrisiko – occupational hazard

Berufsrückkehrerin – female reentrant, woman returner, woman returning to work

Berufsschule – vocational school, technical college (Brit), trade school (Pflichtunterricht für Lehrlinge, Anlernlinge und jugendliche Ungelernte. Der Unterricht findet entweder an einem Tag pro Woche statt oder als Blockunterricht. Überwiegende Aufgabe der Berufsschule ist es, die fachtheoretischen Kenntnisse für den jeweiligen Beruf zu vermitteln.) (s. Graphik)

Berufsschule für Behinderte – technical school for the handicapped

Berufsschullehrer – vocational school teacher

Berufsschullehrgang – vocational course

Berufsschulpflicht – compulsory vocational schooling

Berufsschulzeugnis – vocational school certificate

Berufssituation – professional situation

berufsständische Körperschaft – professional body

Berufsstarter – beginner

Berufsstruktur – occupational structure

berufstätig – working, employed

Berufsunfähigkeit – occupational disability

Berufsunfähigkeitsrente – pension for
occupational invalidity, disability
pension
Berufsunkosten – professional expenses
Berufsverband – professional body (als
öffentlich-rechtliche Körperschaft
Trägerin der Unfallversicherung)
Berufsverbot – professional ban
berufsverbundenes Prestige –
occupational prestige
berufsvorbereitende Maßnahme –
pre-vocational training
Berufsvorbereitung – career preparation
Berufsvorbereitungsjahr –
pre-vocational year
Berufsvorbereitungslehrgang –
pre-vocational course
Berufsvorbereitungsprogramm –
pre-vocational training programme
Berufswahl – choice of profession
Berufsweg – occupational history
Berufswegplanung – career planning
Berufung – nomination, appointment
Berufungsverfahren – appointments
procedure
berühren – touch
beschädigen – damage, injure
Beschaffen – procurement
Beschaffung – procurement
Beschaffung von Informationen –
information procurement
Beschaffung, externe – external
recruitment
Beschaffung, interne – internal
recruitment
Beschaffungsinstrument – recruitment
instrument
beschäftigen – employ, occupy
beschäftigt – busy
Beschäftigte(r) – employee, member of
the staff, person employed
Beschäftigte, geringfügig – part-time
employees
Beschäftigtengruppe – group of
employees
Beschäftigter – employee,
Beschäftigung – employment, state of
employment, employment relationship
Beschäftigungsaussichten –
employment prospects

Beschäftigungsbedingungen –
employment conditions
Beschäftigungsbeschränkung –
employment limit
Beschäftigungsdauer – length of
employment
Beschäftigungsförderung – promotion
of employment
Beschäftigungsgarantie – job guarantee
Beschäftigungsgesetz – law governing
employment
Beschäftigungsgrad – employment level
Beschäftigungsjahr – year of
employment
Beschäftigungskatastrophe –
catastrophic employment situation
Beschäftigungsnachweis – certificate of
employment
Beschäftigungsprozentsatz –
employment percentage
Beschäftigungsschutz – employment
protection
Beschäftigungstherapeut – occupational
therapist
Beschäftigungsverbot – prohibition of
employment
Beschäftigungsverhältnis –
employment, state of employment,
employment relationship
Beschäftigungszeit – period of
employment
Bescheid – reply, notification
bescheiden – self-depreciation
bescheinigen – issue (someone with) a
certificate, certify
Bescheinigung – certification,
confirmation
beschlagnahmen – confiscate
Beschlagnahmung – attachment
beschleunigtes Lernen – accelerated
learning
Beschleunigung – acceleration
Beschluss – resolution
beschlussfähige Anzahl – quorum
beschränken – restrict
beschränkt – imbecile
beschreiben – describe
beschuldigen – accuse, blame
Beschwerde – complaint, reclamation
Beschwerdebuch – complaints book

Beschwerde(grund) – grievance; matter of complaint
Beschwerderecht – right of complaint
Beschwerdestelle – grievance committee
Beschwerdeverfahren – grievance procedure
beseitigen – eliminate
Besessenheit – obsession
besetzen von Stellen – filling a vacancy, manning
Besetzung – occupancy
Besetzungsplan – plan to fill vacancies
Besetzungsprozentsatz – staffing percentage
Besitzstand erhalten – grandfathering
Besitzstandwahrung – protection of vested rights
Besoldung – pay, wages, salary
besonders begabte Kinder – specially gifted children
Besprechung – meeting
Besserwisser – know-all, smart-aleck, smart-ass (AmE)
bestanden – passed
Bestandsmasse – point-in-time population
Bestandteil – component, part, constituent (part)
bestätigen – confirm
Bestätigung – confirmation
Bestehensquote – pass rate
bester Schüler – star pupil
Besteuerungsgrundlage – tax basis
Besteuerungsrecht – right to impose tax
bestimmen – determine
Bestimmung – regulation, guideline
Bestrafung – punishment
bestreiten – dispute
Besuch der Grundschule – primary education
besuchen – attend, participate in
Beteiligung – participation
Beteiligungen – consolidated subsidiaries and associated companies
Beteiligungsgesellschaft – associated company
Betrag – amount, sum
Betrag, hinzuzurechnender – additional amount
Betrag-Pro-Einheit – rate

Betrags-Kumulation – amount cumulation
Betragsfeld – amount field
betrauen (mit) – entrust
betreuen – have the care of, look after, maintain
Betreuer – organizer
Betreuer(in) – person who looks after someone
Betreuung (z.B. von Systemen) – in charge of, responsible for
Betreuung der Mitarbeiter – care of the employees
Betreuung neuer Mitarbeiter – buddy system
Betrieb – company, enterprise, factory, firm, works, plant
betriebliche Altersversorgung – company pension scheme, employee pension scheme
betriebliche Aus- und Weiterbildung – in-company vocational training and further education
betriebliche Ausbildung – in-company training
betriebliche Ausbildungsstätte – company training centre
betriebliche Bildungsarbeit – in-company training
betriebliche Personalpolitik – labour management
betriebliche Sozialpolitik – company social policy
Betriebsabrechnung – operational accounting
Betriebsabteilung – company department
Betriebsangehöriger – employee
Betriebsanleitung – operating instructions
Betriebsanweisungen – operating instructions
Betriebsarzt – company doctor, company medical officer
betriebsärztlicher Dienst – medical service centre
Betriebsausfall – stoppage
Betriebsausflug – staff outing
Betriebsausgaben – revenue expenditure, business expenses
Betriebsausschuss – work's committee

Betriebsbegehung – round of inspection
Betriebsbesetzung – sit-in
Betriebsbesichtigung – plant tour
Betriebsblindheit – tunnel vision
Betriebsbuchhaltung – cost accounting
Betriebsdatenerfassung – plant data collection
Betriebsdirektor – production director
betriebsfähig – operational
Betriebsferien – company holidays, plant shut-down
Betriebsfrieden – industrial peace
Betriebsgefahr – operational risk
Betriebsgeheimnis – trade secret
Betriebsgelände – plant site
Betriebshandbuch – works handbook
betriebsinterne Weiterbildung – corporate university
Betriebskindergarten – company-sponsored day care
Betriebsklima – working atmosphere
Betriebskommitee – joint committee
Betriebskosten – operating costs
Betriebskrankenkasse – company health insurance fund
Betriebsleiter – works manager
Betriebsnummer – employer number
Betriebsnummer, einheitliche – standard company number
Betriebsordnung – shop rule
Betriebsprüfung – external audit
Betriebspsychologe – industrial psychologist
Betriebspsychologie – industrial psychology
Betriebsrat – works council
Betriebsräteversammlung – meeting of works councils
Betriebsratsmitglied – member of the works council
Betriebsrente – company pension
Betriebsschließung – shutdown
Betriebsschlosser – maintenance fitter
Betriebssicherheit – industrial safety
Betriebssoziologie – industrial sociology
Betriebssport – company-facilitated sports activities
Betriebssportplatz – company sports grounds
Betriebsstätte – place of work

Betriebsstillegung – close down, plant closure
Betriebstechnik – industrial engineering
Betriebsteil – plant section, operating unit
betriebsüblich – customary
Betriebsunfall – industrial accident, occupational accident, work accident
Betriebsurlaub – plant holidays
betriebsverbundenes Praktikum – industrial secondment (praxisorientierte Arbeit in einem Unternehmen für auf Zeit freigestellte Lehrer oder Studenten)
betriebsverbundenes Studium – company-sponsored education
Betriebsvereinbarung – works agreement, works rules and regulations, single plant bargaining agreement
Betriebsverfassung – regulations governing industrial relations
Betriebsversammlung – works meeting
Betriebswirt – graduate in business administration
Betriebswirtschaft – business administration, applied economics, business economics
Betriebszeit – operating time
Betriebszugehörigkeit – period of employment, staff membership in a company or firm
Betriebszulage – company bonus
betrügerisches Einverständnis – collusion
beurteilen – judge, appraise
Beurteilung – assessment, appraisal
Beurteilung der Führungskräfte – executives' appraisal
Beurteilungsart – appraisal type
Beurteilungsbogen – personnel appraisal sheet
Beurteilungsfehler – error in judgement
Beurteilungsmerkmal – appraisal factor
Beurteilungssystem – assessment system, appraisal system, personnel review procedure
Beurteilungsverfahren – assessment system, appraisal system
Bevölkerungswachstum – population growth, increase in growth of population

Bevollmächtigter – authorized representative
Bewährung – probation
Bewährungsaufstieg – automatic progression
Beweggrund – ground, reason
Beweglichkeit – flexibility
Bewegungsfreiheit – full liberty of action
Beweis – evidence, proof
Beweisgrund – argument
Beweismaterial – evidence, proof
Beweismittel – evidence, proof
Bewerber – applicant
Bewerber auswählen – screen applicants, screen
Bewerber sieben – screen applicants, screen
Bewerberakte – applicant file
Bewerberandrang – large number of applicants
Bewerberauswahl – applicant selection
Bewerberdaten – applicant data
Bewerbererfassung – registration of applicants
Bewerbergruppe – applicant group
Bewerberkartei – applicant file
Bewerberkreis – applicant range
Bewerbernummer – applicant number
Bewerberprofil – applicant profile
Bewerbertest – applicant test
Bewerberverwaltung – applicant data administration
Bewerbervorauswahl – applicant screening
Bewerbervorgang – applicant action
Bewerbung – application, job application
Bewerbung für eine bestimmte Tätigkeit – application for a vacant position
Bewerbung für eine freie Stelle – application for a vacant position
Bewerbungsabwicklung – administration of applications, application procedure
Bewerbungsbearbeitung – processing of applications
Bewerbungsbogen – application form
Bewerbungsformular – application form
Bewerbungsfrist – application deadline
Bewerbungsgespräch – job interview, interview

Bewerbungskosten – application expenses
Bewerbungsstrategie – application strategy
Bewerbungsunterlagen – application documents, application papers
Bewerbungsverfahren – administration of applications, application procedure
Bewerbungszusatzdaten – additional applicant data
Bewerbunsunterlagen – application documents, application papers
bewerten – evaluate
Bewertung – evaluation
Bewertung durch Ebenbürtige – peer assessment
Bewertungsgrundlage – valuation basis
Bewertungslohnart – valuation wage type
Bewertungsmatrix für Führungskräfte – managerial grid
Bewertungsprinzip – valuation principle
Bewertungssatz – valuation rate
Bewertungszuordnung – valuation assignment
bewilligen – grant
bewilligte Abwesenheit – leave of absence
Bewilligung – consent, permission
Bewirtung – entertainment
Bewirtung, unentgeltliche – gratuitous meal
Bewirtungsbeleg – receipt for entertainment expenses
Bewirtungskosten – entertainment expenses
Bewirtungsspesen – entertainment allowance
Bewunderung – admiration
Bewusstseinsänderung – change of awareness
Bewusstseinsstörung – mental blackout
bezahlen – pay (for), remunerate
bezahlt – paid
bezahlte Freistellung – paid release
bezahlte gesetzliche Feiertage – paid statutory holidays
bezahlter Urlaub – vacation with pay
Bezahlung – payment, remuneration
Bezahlung, abweichende – payment, different

bezahlungsrelevant – relevant to payment
bezeichnen – designate
Bezeichnung – name
Beziehung – relation
Beziehungsebene – relational level
Bezüge – income, emoluments, payments
Bezüge, Änderung der – salary level change
Bezüge, laufende – regular pay
Bezüge, mehrjährige – payments spanning more than one year
Bezüge, sonstige – other payments
Bezüge/Abzüge – payments/deductions
Bezüge/Abzüge, einmalige – once-only payments/deductions
Bezüge/Abzüge, laufende – regular payments/deductions
Bezüge/Abzüge, variable – variable payments/deductions
Bezüge/Abzüge, wiederkehrende – recurring payments/deductions
Bezugsänderung – change of pay
Bezugsgröße – base award, basic amount
Bezugsmethode – payment method
Bezugsperson – reference person
Bezugsprinzip – payment principle
Bezugsrecht – subscription right
Bezugswert – reference value
bezuschussen – subsidize
Bibliothek – library
biegen – bend
Biennalsprung – biennial incremental increase
Bilanz – balance sheet
Bilanzieren – balancing the accounts
Bild – picture, image
bildhaftes Gedächtnis – visual memory
Bildschirm – screen
Bildung – education
Bildung und Information – training and information
Bildungsabschluss – certificate
Bildungsangebot – educational services
Bildungsanstalt – college, educational establishment
Bildungsaufwand – total costs for education and training
Bildungsbedarf – educational needs
Bildungsberater – educational consultant

Bildungsboom – educational boom
Bildungseinrichtung – educational facility, educational institution
Bildungsgang – course of education
Bildungsgesetze – education laws
Bildungsinvestition – investments in education
Bildungsmanager – training manager
Bildungsmittel – educational resources
Bildungsmöglichkeiten – educational possibilities, training possibilities
Bildungsniveau – educational standard, educational level
Bildungsnotstand – chronic shortage of educational facilities
Bildungsplanung – educational planning
Bildungspolitik – educational policy
bildungspolitischer Konsens – consensus on educational issues
bildungspolitischer Sprecher – spokesman on educational affairs
Bildungsstand – educational standard, educational level
Bildungsstruktur – educational structure
Bildungsurlaub – educational leave, study leave
Bildungsveranstaltung – educational event
Bildungssystem – training and education system
bildungsunfähig – ineducable
Bildungsweg – educational channel
Bildungswesen – educational system
Bildungszentrum – education centre
Bildungsziel – educational objective, educational goal
Bildungszweck – purpose of training
billig – cheap
Billigarbeiter – cheap labourer
billigen – approve, consent, sanction
Billigung – approval, consent, sanction
Bindung – bond
Bindungsindex – bond index (Dieser Index errechnet sich aus Mitarbeiterfluktuation, der durchschnittlichen Betriebszugehörigkeit des gegenwärtigen Personals und den Fehlzeiten.)
Biographie – biography
Biologie – biology

bisherige Karriere – track record
blamieren – disgrace
blaumachen – skip work
Blechschlosser – sheet metal worker
Blender – phoney
Bleistift – pencil
Blindbewerbung – blind application, application on spec
Blinde(r) – blind person
Blockunterricht – block release instruction
Blutalkohol – blood alcohol
Bonbonladen – tuckshop
Bonusabrechnung – bonus settlement
Bonusermittlung – bonus calculation
Bonusprozentsatz – bonus percentage
Bonuspunkt – credit point
Bonusstichtag – bonus due date
Bonusverfahren – bonus calculation procedure
borgen – borrow
Börsenkurs – stock market rate
Bote – runner
Boykott – boycott
boykottieren – boycott
Brainstorming – brainstorming (Durch gemeinsames Bemühen in einer Sitzung durch spontane Äußerung von Einfällen zur Lösung eines Problems beitragen)
Brandkategorie – fire category
Brandschutz – fire prevention
Breakeven-Analyse – breakeven analysis
breite Bildung – all-round education
Breitenbandmodell – broad band model (Mitarbeiter können im Rahmen bestimmter Bandbreiten (z.B. zwischen 15 und 40 Wochenstunden) ihre vertragliche Arbeitszeit immer wieder neu definieren. Dabei erhöht oder mindert sich das Gehalt entsprechend.)
brennbares Material – inflammable material
Brief – letter
Brückentag – bridge day
Brutalität – brutality
Brutto... – gross
Bruttoabrechnung – gross payroll accounting
Bruttoeinkommen – gross income
Bruttoeinspeisung – input into gross

Bruttoentgelt – gross pay
Bruttoermittlung – calculation of gross amount
Bruttogehalt – gross salary
Bruttokennzeichen – gross indicator
Bruttolohn – gross wages
Bruttolohn-Auswertung – gross wage evaluation
Bruttolohnfindung – gross wage calculation
Bruttolohnnachweis – statement of gross wages earned
Bruttorückrechnung – gross retroactive accounting
Bruttosammelergebnis – total gross amount
Bruttoteil – gross calculation of pay
Bruttotopf – gross bucket
Bruttoversorgungsgrad – gross level of satisfaction
Bruttozusammenfassung – cumulation of gross amount
buchen – book, reserve
Bücherbenutzung bei der Prüfung erlaubt – open book examination
Bücherweisheit – book learning
Buchführung – accounting, accountancy
Buchhalter – accountant
Buchhaltung – accounting (department), accountancy
Buchstabe – letter
buchstabieren – spell
Buchstabiertest – spelling test
Buchung – booking, reservation
Buchungsbeleg – booking note
Buchungsbestätigung – confirmation of course booking
Buchungskreis-Werk-Kombination – company code / plant combination
Buchwissen – book knowledge
Budget – budget
Budget-Planung – budget planning
büffeln – cram, swot
Bummelstreik – go slow strike
Bummeltag – go slow day
Bund – federation
Bundesamt – Federal Office
Bundesangestelltentarif – statutory salary scale
Bundesanstalt für Arbeit – Federal Labor Office

Bundesarbeitsgericht – Federal Labor Court, Supreme Labour Court
Bundesausbildungsförderungsgesetz – Federal Educational Promotion Act
Bundesdatenschutzgesetz – Data Protection Act
Bundesland – federal state
Bundessozialhilfegesetz – Federal Social Security Act
Bundesversicherungsamt – Federal Benefits Authority
Bundesversicherungsanstalt für Angestellte – Federal Social Insurance Office for Salaried Employees
Bürgerarbeit – community work (Bürgerarbeit, freiwilliges, soziales, projekt-gebundenes Engagement, das nicht entlohnt, aber durch Ehrungen oder eine Aufstockung der Rentenansprüche anerkannt wird.)
bürgerlich – middle class
bürgerliches Recht – civil law
Bürgschaft – security, surety
Burn-out – burn-out (Zustand emotionaler Erschöpfung)
Büro der Leitung – Office of Administration
Büro... – clerical, office
Büroangestellter – office worker
Büroerfahrung – office experience, office practice
Bürogehilfe – clerical assistant, office boy
Bürogehilfin – clerical assistant
Bürokauffrau – office administrator
Bürokaufmann – office administrator
bürokratische Organisation – bureaucracy, bureaucratic organization
bürokratische Organisationsstruktur – bureaucracy, bureaucratic organization
Bürokratismus – red tape
Büropersonal – clerical staff
Bürostunden von ... bis – office hours from ... to ...
Bußgeld (bei Verstößen) – fine, penalty

C

Cash flow – cash flow
Center of excellence – center of excellence
Chance – opportunity, possibility
Chancengleichheit – equal opportunity
chaotisch – chaotic
Charakter – character
Charaktereigenschaft – character trait
Charakterfestigkeit – strength of character
Charakterzug – trait
Charisma – charisma
Chef – boss
Chefberater – senior consultant
Chefetage – executive floor
Chemie – chemistry
Chemie-Ingenieur – chemical engineer
Chlorkohlenwasserstoff – chlorinic hydrocarbon
Choleriker(in) – choleric person
Clique – gang
Cluster-Struktur – cluster structure
Coach – coach
Computer – computer
Computer-Experte – computer expert
Computer-Unterricht – computer studies
Computerhersteller – computer manufacturer
computerunterstützte Unterweisung – computer-aided instruction, CAI
computerunterstütztes Konstruieren – computer-aided design, CAD
computerunterstütztes Lernen – computer-aided learning, CAL
Computerwissen – computer literacy
Controller – controller
CUL (computerunterstütztes Lernen) – CBL (computer-based learning), CAL (computer-aided learning)
cum laude – cum laude

D

Dankbarkeit – gratitude
Dankesurkunde – document of thanks
Dankschreiben – letter of thanks
Darlehen – loan
Darlehen, allgemeines – general loan
Darlehensart – loan type
Darlehensbetrag – loan sum
Darlehenskonditionen – loan terms
Darlehenssumme – amount of loan
Darlehenstilgung – loan repayment
Darlehenstilgung, allgemeine – general
  loan repayment
Darlehenstilgungsrate – loan repayment
  instalment
Darlehenstyp – loan category
darstellen – present, interpret
Darstellung – presentation
Darstellungsfähigkeit – presentation
  ability
das Wort ergreifen – address the meeting
Datei – file
Daten – data
Daten zur Person – personal data
Daten, abrechnungsrelevante – payroll
  data
Daten, beitragsrechtliche – social
  security data
Datenaufbereitung – data preparation
Datenbanksystem – data bank system
Datenbasis – data bank, database
Dateneingabeverordnung (DEVO) –
  data input regulation
Datenerfassung – data capture
Datenerfassungsverordnung – data
  entry regulation
Datenfernverarbeitung – teleprocessing
Datenschutz – data protection
Datenschutzbeauftragter – data
  protection officer
Datenschutzgeheimnis – confidential
  data
Datenschutzmaßnahmen – data privacy
  measures
Datensystem – data system
datentechnische(r) Assistent(in) – data
  processing assistant

Datenübermittlungsverordnung – data
  communications regulation
**Datenübertragungsverordnung
  (DUVO)** – data transmission regulation
Datenumsetzung – data conversion
Datenverarbeitung – data processing
Datenverarbeitungssystem – data
  processing system
Datum – date
Datum, fiktives – imaginary date
Datumsabkürzung – date abbreviation
Datumsangabe – date specification
Dauer – duration
Dauer der Bertriebszugehörigkeit – job
  tenure, (occupational seniority)
Dauerarbeitsvertrag – permanent job
  contract, permanent employment
Daueraufenthaltsgenehmigung –
  permanent residence card
Dauerauftrag – standing order
Dauerbeschäftigung – permanent
  employment
Dauercode – time-unit code
Dauerreisekosten – permanent travel
  costs
Deckung – cover, coverage
deduktiv – deductive
deduktives Denken – deductive thinking
defekt – out of order; defect
defensive Kommunikation – defensive
  communication
Defizit (persönliches) – weak point,
  deficit
degressive Prämie – reduced incentive
Dekan – dean, provost
Dekanat – deanship
delegative Führung – empowerment
delegieren – delegate
delegierte Personalverantwortung –
  delegated personnel responsibility
Delinquent – delinquent
den Arbeitsplatz verlieren – be made
  redundant
den Vorsitz führen – chair, moderate
Denken – thinking
Denkfabrik – think-tank
Denkfähigkeit – reasoning power
Denkprozess – thinking process
Denkschwäche – weak intellectual
  capacity
Denkvermögen – intellect

Denksystem – system of thought
depressiver Mensch – depressive person
Deputat – payment in kind
der Firma zur Verfügung stehen – be at the disposal of the company
Desinteresse – lack of interest
desinteressiert – lackadaisical
Detailfrage – question of detail
Detailpflege – detail maintenance
deutliche Aussprache – articulation
dezentralisierte Entscheidungsfindung – decentralized decision making
Dezentralisierung – decentralization
Dia – slide
Diagramm – chart, diagram
Dialekt – dialect
Didaktik – didactics
die Gebildeten – educated classes
die Schule schwänzen – play truant, mitch
Dienst – shift
Dienst nach Vorschrift – go slow, work to rule (Ausnutzung hemmender Dienstvorschriften)
Dienst, öffentlicher – civil service
Dienst, werksärztlicher – internal medical service
Dienstalter – occupational seniority, seniority, years of service
Dienstaltersgruppe – length of service group
Dienstalterssprung – pay scale jump due to seniority
Dienstaltertabelle – seniority roster
Dienstalterzulage – service awards
Dienstantritt – take up work, starting of work
Dienstaufnahme – take up work, starting of work
Dienstbefreiung wegen Krankheit – sick leave
Dienstenthebung – give notice, discharge, dismissal, termination of employment
Diensterfahrung – work experience
Dienstgang – off-site work
Dienstgrad – grade, service level
Dienstjahre – years of service
Dienstjahresregelung – provision for years of service
Dienstjubiläum – anniversary jubilee

Dienstleistung – services, rendition of services (AmE)
Dienstleistungsbranche – service sector
Dienstleistungsindustrie – service industry
Dienstleistungsingenieur – service engineer
Dienstleistungssektor – service sector
dienstliches Anliegen – official matter
Dienstordnung – official regulation
Dienstplan – roster, duty rota
Dienstplan, persönlicher – personal shift plan
Dienstplanung – shift planning
Dienstreise – business trip
Dienststelle – department
Dienststellenleiter – head of department unit
Diensttage – days of service
Dienstverhältnis – employment status
Dienstvertrag – service contract
Dienstvertragsrecht – employment contract law
Dienstwagen – company car
Dienstzeit – daily working hours; length of service in company
Differentialstücklohn – differential piecerate
Differenzstunden – difference hours
Differenztabelle – difference table
Diktat – dictation
Diktion – diction
Diplom – diploma
Diplom-Kaufmann – bachelor of commerce
Diplomand – student taking his diploma examination
Diplomarbeit – thesis, final year project
Diplomingenieur – engineer
Diplomkaufmann – business school graduate
direkte Arbeitskosten – direct labour costs (Lohn für die direkt am Werkstück verrichtete Arbeit, produktiver Lohn genannt, gilt als direkte Kosten, d.h. als Einzelkosten in der Gewinn- und Verlustrechnung der Firma)
direkte Methode – direct method
direktes Recruiting – referral program (Mitarbeiterwerbung durch Mitarbeiter)

Direktionsmitglied – top manager
Direktversicherung – direct insurance
Direktzusage – employer's pension commitment, pensioning warrant (Zusage mit Rechtsanspruch)
diskontinuierliche Arbeit – discontinous work
Diskrepanz – discrepancy
Diskretion – discretion
diskriminieren – discriminate
Diskriminierung – discrimination
Diskussionsgruppe – discussion group
Diskussionsleiter – moderator
Diskussionsmethode – discussion method
Diskussionsteilnehmer – panel member
diskutieren – discuss
Dispositionsmenge – amount available
Dispositionsmerkmal – availability indicator
Dissertation – dissertation
distanziert – distant
Distraktor – distractor (Vorgabe der anscheinend richtigen Lösungen bei einer Mehrfachauswahlfrage)
Disziplin – discipline
Doktorand(in) – candidate for a doctor's degree, doctoral candidate, research student
Doktorarbeit – dissertation
Doktor(titel) – doctorate, doctor's degree
Doktorvater – thesis supervisor, research supervisor
Doktorwürde – doctorate, doctor's degree
Dolmetscher – interpreter
Dominanz – dominancy
dominierend – dominant
Doppelarbeit – duplication of work
Doppelarbeitsanalyse – analysis of unnecessary duplication of work
Doppelbelastung – two-fold burden
Doppelbesteuerungsabkommen – double taxation convention
Doppelqualifikation von Facharbeitern – double qualification of skilled workers
doppelt besetzte Planstelle – doubly occupied position
doppelt gerichtete Kommunikation – two-way communication

Doppelverdienerhaushalt – double income household
Doppelverdienersatz – double income rate
Downgrading – downgrading
Dozent(in) – lecturer, assistant professor (AmE), reader (BrE)
Dr.jur. – LLD
Dr.phil. – PhD
Dreher – turner
Dreiperiodendurchschnitt – three-period average
13. Monatsgehalt – 13th monthly pay
dringender Fall – urgent case
Drittschuldnererklärung – declaration made by third party debtor
Droge – drug
Drogenabhängige(r) – drug addict
Druck unter Ebenbürtigen – peer pressure
dualer Studiengang – dual course of studies
duales System der Berufsausbildung – dual system of vocational training
dulden – tolerate
Dummkopf – dunce
Durchdringung – penetration
durchfallen – fail, flunk
Durchfallrate – failure rate
durchführbar – feasible
Durchführbarkeitsstudie – feasibility study
durchführen – execute, implement
Durchführung – execution, implementation
Durchhaltefähigkeit – stamina, perseverance, endurance
Durchschlag – copy
Durchschnitt – average, mean
durchschnittliches Bruttoarbeitsentgelt aller Versicherten – average gross pay of all insured persons
Durchschnittsberechnung – calculation of averages
Durchschnittsbewertung – valuation of averages
Durchschnittsbezahlung – average pay
Durchschnittseinkommen – average earnings
Durchschnittsgehalt – average salary

Durchschnittsgrundlage – basis for calculating average value
Durchschnittslohn – average wage
Durchschnittsprinzip – principle of averages
Durchschnittsprozentsatz – average percentage
Durchschnittstyp – average type
Durchschnittsverarbeitung – processing of averages
Durchschnittsverdiener – average earner
Durchschnittsverdienst – average earnings
Durchschnittsvergütung – average earnings
durchsetzen – enforce, force
Durchsetzungsfähigkeit – ability to assert oneself
Durchsetzungsvermögen – elbow power, power of convincing
Durchsicht – inspection
DV-Anlage – computer
DV – data processing, DP
DV-Bildungseinrichtung – DP school
DV-Erfahrung – DP experience
DV-Fachmann – DP specialist
DV-Laie – computer illiterate
DV-Organisation – DP organization
DV-Realisierung – DP-implementation
DV-Technologie – data processing technology

E

Ebenbürtige(r) – peer
Ebene – level
echt – authentic
Echtheit – authenticity
Ecklohn – basic wage
EFA-Gespräch – EFA-staff dialogue, EFA-meeting
effektiv – effective
Effektivität – effectiveness

Effektivlohn – actual wage, take-home pay
Effizienz – efficiency
egalitär – egalitarian
Egoismus – egoism
Egoist – egoist
egozentrisch – egocentric
Ehefrau – wife, spouse
Ehegatte – husband, marital partner, spouse
Ehegattenversorgungsanspruch – pension rights for marital partners
Ehegattin – wife, spouse
ehemalige(r) Schüler(in) – former pupil, alumnus/alumna
Ehepartner(in) – spouse (husband, wife)
Ehescheidung – divorce
Ehrenamt – honorary post
Ehrendoktor – honorary doctor
ehrenhalber verliehener akademischer Grad – hononary degree
Ehrenmitglied – honorary member
Ehrenmitglied des Aufsichtsrates – Honorary Member of the Supervisory Board
Ehrenpräsident – Honorary President
Ehrenvorsitzender (Verband) – Honorary Chairman of the Board of Directors
Ehrenvorsitzender des Aufsichtsrates – Honorary Chairman of the Supervisory Board
Ehrgeiz – ambition
ehrlich – honest
Ehrlichkeit – honesty
Eid – oath
eidetisch – eidetic (Fähigkeit, früher gesehenes als anschauliches Bild wieder vor sich zu sehen.)
Eifersucht – jealousy
Eigenbrötler – loner, lone wolf
Eigendünkel – egotism, sense of superiority
eigene Kündigung – employee voluntary termination notice, resignation
Eigeninitiative – self-initiative
Eigennutz – self-interest
eigensinnig – stubborn, obstinate, headstrong
eigenständig – independent, original, autonomous

Eigenständigkeit – independence, autonomy
Eigentum(srecht) – ownership
eigenverantwortlich – autonomous, personally responsible
Eigenverantwortlichkeit – personal responsibility
Eigenverantwortung – autonomous, personally responsible, self-responsibility
Eignung – aptitude, suitability
Eignungsberatung – professional qualification consultation
Eignungsgespräch – aptitude interview
Eignungsprofil – suitability profile
Eignungsprozentsatz – suitability percentage
Eignungsprüfung – aptitude test
Eignungsschwerpunkt – main aptitude
Eignungstest – aptitude test
Eignungstest für amerikanische Universitäten – Scholastic Aptitude Test (SAT) (AmE)
Eignungszeugnis – aptitude certificate
ein ins Ausland versetzter Mitarbeiter – transferee abroad, expatriate
ein Ziel erreichen – attain an objective
Ein-Personen-Unternehmen – one-man company
einarbeiten – induct, train on the job
Einarbeitung – familiarization, working-in period, orientation, practice period
Einarbeitungszeit – period of vocational adjustment, settling-in period
einbehalten (Steuern) – withhold (taxes)
Einbehaltung – retention
einberufen (eine Besprechung) – convene (meeting)
einbeziehen – include
Einbuße – loss
eindeutig – unequivocal
eine Firma verklagen – sue a company
eine lange Leitung haben – slow on the uptake
eine Lehre absolvieren – serve an apprenticeship
eine Prüfung beaufsichtigen – supervise an exam, invigilate
eine Prüfung bestehen – pass an exam

eine Prüfung machen – take an exam, sit an exam
eine Stelle besetzen – fill a vacancy
eine Tätigkeit vorübergehend einstellen – suspend
einen Antrag stellen – file an application
einen Betrieb erkunden – tour a company
einen Kurs abhalten – run a course
einen Mitarbeiter nacheinander in verschiedenen Firmenbereichen einsetzen – rotate an employee
einen Test verpfuschen – make a mess of a test
einer, der hinter den Erwartungen zurückbleibt – under-achiever
einfache Pflege – simple maintenance
einfacher Dienst – manual grade
Einfallsreichtum – wealth of imagination
einfältig – gullible
Einfluss – influence
Einfluss(nahme) – leverage
Einflussgrößen – parameters governing influence
einflüstern – insinuate
einfrieren – freeze
einfügen – insert
Einfühlungsvermögen – empathy
einführen – commence, initiate, introduce
Einführung eines neuen Mitarbeiters am Arbeitsplatz – induction
Einführungskurs – induction course
Einführungstraining – induction training
Eingangsbestätigung – confirmation of receipt
Eingangsgehalt – starting salary
Eingangsgespräch – initial talk (Ein Gespräch zwischen Professor und Student am Anfang der Studienzeit, um Interessen, Schwerpunkte usw. abzuklären.)
Eingangsklasse – first class
Eingangsprüfung – entrance examination
Eingangsvorausetzung – prerequisite
eingehend – detailed, thorough
Eingliederung – integration
Eingruppierung – classification

Eingruppierung, tarifliche – assignment
to wage group
einhalten – comply (with)
Einigung – agreement, settlement
Einigungsstelle – arbitration board,
conciliation board
Einkauf – purchasing department
Einkäufer – buyer
Einkommen – income, emoluments
Einkommen, versteuerbares – taxable
income
Einkommensanrechnung – income
imputation
Einkommensband – income scale,
income band
Einkommensbemessung – compensation
evaluation
Einkommensentwicklung –
compensation development
Einkommensfindung – regulate a just
salary
Einkommensgarantie – income
guarantee, income security
Einkommensgestaltung – income policy
Einkommensgrenze – salary limit
Einkommensliste – income list
Einkommensmitteilung – announcement
of income
Einkommenspolitik – income policy
Einkommensrahmen – income range
Einkommensstruktur – income
structure
Einkommenssystem – compensation
system
Einkommensteuer – income tax
Einkommensteuergesetz – income tax
law
Einkommensteuerrichtlinien – income
tax regulations
Einkommensteuertarif – income tax rate
Einkommenstyp – income category
Einkommensübersicht – compensation
survey
Einkommensveranlagung – means test
Einkommensvergleich – compensation
comparison
Einkommensverringerung – reduced
earnings
Einkünfte – revenue, income
**Einkünfte aus nichtselbständiger
Arbeit** – income from employment

Einladung zum Vorstellungsgespräch –
invitation for an interview
Einlernen – job familiarization
einmal gezahltes Arbeitsentgelt –
one-time employee compensation
Einmalbeitrag (z.B. zur Versicherung)
– one-time premium, single premium
Einmalbezüge – non-recurring payments
Einmalentgelt – one-time remuneration
**einmalige Auszahlung des
Rentenbetrages** – non-recurring
capital payment
Einmalprämie – one-time premium,
single premium
Einmalzahlung – once-for-all payment
Einmischung – interference
Einnahmen – revenue
Einnahmen- und Ausgabenbudget –
cash budget
einreichen – submit
einrichten – arrange, establish
Einrichter – toolsetter
Einrichtung der Erwachsenenbildung –
Adult Education Centre
Einrichtungen – facilities
einsam – lonely
Einsatz – deployment
**Einsatz und Förderung des
Mitarbeiters** – placement and
promotion of the employee
Einsatzbesprechung – briefing
einsatzfähig, beschränkt – limited work
capability
Einsatzplanung – workforce planning
Einsatzplanung, auftragsbezogene –
order-related workforce planning
einschätzen – assess
einschlägig – relevant, pertinent, relating
einschließen – include
einschränken – limit, restrict
Einschränkung – limitation, restriction
Einschreibung – enrolment, school
registration
Einschulung – school enrolment
einsetzen – install
Einsicht – insight
einsichtig – reasonable, understanding
Einsichtnahme – inspection
Einsparung von Arbeitskräften –
headcount saving

Einstell-Lohn – starting rate, entrance rate
einstellen – employ, take on, hire, engage
Einstellgehalt – starting salary
Einstellung – attitude; hiring
Einstellungsgespräch – employment interview, interview
Einstellungsinterview – employment interview, interview
Einstellungsschreiben – letter of employment
Einstellungsstop – job freeze
Einstellungstransaktion – hiring transaction
einstempeln – clock in
Einstiegschance – job opportunity
Einstiegsgehalt – starting salary
Einstiegsvoraussetzungen – initial requirements
Einstufung – classification, rating
Einstufung nach Dienstalter – service rating
Einstufung nach Leistung – rating of employees
Einstufung, tarifliche – assignment to wage level
Einstufungstest – placement test
einstweilige Verfügung – injunction
einteilen – classify
Einteilung in Leistungsgruppen – streaming
Eintritt – admission, entrance, entry
Eintritt ins Erwerbsleben – start in working life
Eintrittsalter – entry age, age-at-entry
Eintrittsdatum – date of entrance, entry date
Eintrittstag – date of hire
Eintrittstermin – date of entrance
einüben – practice
einverstanden sein – agree
Einverständnis – consent, approval
Einwanderer – immigrant
Einweg-Kommunikation – one-way communication
Einwendungen machen – object
Einzelakkord – individual piecework
Einzelakkordlohn – individual piecework rate
Einzelarbeit – single work, individual work

Einzelarbeitsvertrag – individual employment contract
Einzelaufgabe – single task
Einzelbedingungen – individual conditions
Einzelbeleg – single receipt
Einzelbuchung – single booking
Einzelerfassung – single record entry
Einzelheiten – details
Einzelkaufmann (GmbH & Co.KG) – sole proprietor
Einzelleistung – individual performance
Einzelleistungslohn – individual incentive wage
einzeln – individual
Einzelnachweis – individual statement
Einzelperson – individual
Einzelvertrag – individual contract
Einzug der Beiträge – collecting of premiums
Einzugsgebiet – catchment area
Einzugsstelle – collecting agency
Eitelkeit – vanity, conceit
eklektisch – eclectic
Elan – drive, energy
Elektriker – electrician
Elektroingenieur – electrical engineer
Elektroinstallateur – electrician
Elektromaschinenmonteur – electrical machine fitter
Elektronik – electronics
Elektroniker – electronics technician
elektronische Datenverarbeitung – electronic data processing, EDP
Elektrotechnik – electrical engineering
elektrotechnische(r) Assistent(in) – electrical engineering assistant
Elendsviertel – slums
elitär – elitist
Elite – elite
elterlicher Einfluss – parental influence
elterliche Gewalt – parental authority
elterliches Sorgerecht – parental care
elterliche Wahl – parental choice
Eltern – parents
Eltern-Lehrer-Ausschuss – parent-teacher association
Elternabend – parents' night
Elternbeirat – Parents' Council
Elternbeiratsvorsitzender – Chairman of the Parents' Council

Elternbeitrag – parental contribution
Elternberatung – parent counselling
Elternbeteiligung – parental involvement
Elternhaus – home
Elternkontakte – parental contacts
Elternteil – parent
Emotion – emotion
emotionale Intelligenz – emotional
   intelligence
emotionale Sperre – emotional block
emotionale Störung – emotional disorder
emotionale Vertriebsintelligenz –
   emotional sales intelligence
Empathie – empathy
Empfänger – recipient; payee
Empfehlungsbrief – letter of
   recommendation, testimonial
Empfehlungsschreiben – letter of
   recommendation, testimonial
Empfindsamkeit – sensitivity
empirisch – empiric(al)
empirische Lösung – trial and error
empirische Methode – empirical
   method; trial and error method
empirisches Wissen – empirical
   knowledge
Ende – conclusion, end
Endetoleranz-Beginn – end tolerance
   begin time
Endetoleranz-Ende – end tolerance end
   time
Endeuhrzeit – end time
Endeverarbeitung – final processing
Endnote – final mark
Endostruktur – endostructure (abstrakte
   Organisationselemente wie Normen
   und Überzeugungen über Systeme,
   Technologie, Umwelt usw.)
Energie – drive, energy
Energieelektroniker – power
   engineering electronics technician
Engagement – dedication
engagiert – committed
engere Auswahlliste – short list
Engineering – engineering
Engpass – bottleneck
engstirnig – narrow-minded, insular,
   parochial
Entbehrung – deprivation
Entbindung – delivery (USA);
   confinement (GB)

Entbindungsdatum – date of delivery
   (USA); date of confinement (GB)
entbürokratisieren – deburocratize
entdecken – discover
Entdeckung – detection, discovery
Entfaltung – development
Entfernungskilometer – distance in
   kilometers
Entgelt – payment, remuneration
Entgeltbeleg – employee remuneration
   information
Entgeltberechnung – calculation of
   remuneration
Entgeltbescheinigung – income
   statement
Entgeltfortzahlung – continued
   remuneration
Entgeltfortzahlungsgesetz – Continued
   Remuneration Act
Entgelthöhe – amount of remuneration
Entgeltnachweis – employee
   remuneration information
Entgeltsystem – payment scheme
Entgeltvorausbescheinigung – advance
   income statement
Entgeltzahlung – payment of
   remuneration
Entkräftung – debilitation
entlassen – dismiss, fire, give notice, sack
Entlassung – discharge, dismissal,
   termination of employment
Entlassungsentschädigung – dismissal
   compensation, redundancy payment
Entlassungsgeld – dismissal
   compensation, redundancy payment
Entlassungsgesuch – resignation (letter
   of)
Entlassungsgrund – reason for dismissal
Entlassungspapiere – letter of dismissal
Entlassungszeugnis – certificate of
   discharge
Entleiher – hirer
entlocken – elicit
entlohnen – pay (for), remunerate
Entlohnung – payment, remuneration
Entlohnung der Mitarbeiter –
   compensation of personnel
Entlohnungsart – compensation type
Entlohnungsgrundsatz – basic wage
   policy

Entlohnungsmethode – payments system
Entmündigung – interdiction, incapacitation
Entnahmebuchung – issue posting
entschädigen – compensate
Entschädigung – indemnification, compensation payment
Entschädigung für vorübergehende Entlassung – layoff benefit
Entschädigungs... – compensatory
entscheiden – decide
entscheidende Jahre – formative years
Entscheidung – decision
Entscheidungsanalyse – decision analysis
Entscheidungsfähigkeit – decision-making ability
Entscheidungsfindung – decision making
Entscheidungsoperation – decision operation
Entscheidungsschwäche – inability to make decisions
Entscheidungsträger – decision maker
Entscheidungsvorbereitung – preparation of decisions
entschlossen – determined, resolute
Entschlossenheit – determination
entschlussfreudig – decisive
entschuldigtes Fehlen – leave of absence
Entschuldigung – excuse, apology
Entsendeland – sending country
Entsendung von Arbeitnehmern – delegation of employees
Entspannung – recreation
entsprechen – comply (with)
Entstehungsdatum – date of origin
Entstellung – deformity
entwickeln – develop
Entwicklung – development
Entwicklung des Führungsverhaltens – leadership evolution
Entwicklung qualifizierter Nachwuchskräfte – development of high-potential employees
Entwicklung von Führungseigenschaften – management training
Entwicklung, berufliche – employment history

Entwicklungsalter – developmental age
Entwicklungsgrad – development level
Entwicklungsmaßnahmen – measures of development, development measures
Entwicklungsplanung – planning of promotion
Entwicklungsprozess – development process
Entwicklungsspielraum – development margin
Entwicklungsstadium – experimental stage
Entwicklungsziel – development goal, development aim
Entwurf – outline
Entziehung – deprivation
Entziehungskur – dry-out, corrective training
Entzug – deprivation
Erarbeiten von Zielen – elaboration of objectives
Erbe – heir, inheritance
Erbin – heiress
Erblasser – testator
erblich – hereditary
Ereignis – event
Erfahrung – experience
Erfahrungsaustausch – exchange of experience
Erfahrungsberichterstattung – debriefing
Erfahrungsbreite vergrößern – increase experience
Erfinder – inventor
Erfindervergütung – remuneration for employee inventions, inventor's bonus
Erfindungsgabe – ingenuity
Erfolg haben – succeed
Erfolgspotential – potential for success
erfolgreicher Hauptschulabschluss – public school examination (Jeder Schüler, der die Jahrgangsstufe 9 mit Erfolg besucht hat, erhält das Zeugnis über den „erfolgreichen Hauptschulabschluss".) (s. Graphik)
erfolgsabhängiger Einkommensanteil – profit-related income share
Erfolgsbeteiligung – profit-sharing
Erfolgskontrolle – test
Erfolgsmensch – over-achiever
Erfolgstyp – high-flier

**Erfordernisse eines
Einkommenssystems** – requirements
of a compensation system
erforschen – explore
Erfüllung – implementation, fulfil(l)ment
**Erfüllung individueller Wünsche des
Kunden** – customizing
ergänzen – amend
Ergänzungsunterricht – remedial
classes
Ergänzungsvertrag – supplementary
contract
Ergebnis – outcome (of a negotiation),
result
Ergebnisbewertung – results assessment
Ergebnislohnart – result wage type
ergiebig – productive
Ergonomie – ergonomics
(Nordamerikanische
Forschungsrichtung, die der deutschen
Arbeitswissenschaft annähernd
entspricht, wobei die
Arbeitspsychologie besonders
berücksichtigt wird.)
erhältlich – available
Erhaltung von Arbeitsplätzen –
safeguarding existing jobs
Erhebung – survey
Erhebungsbogen – inquiry sheet
erhöhen – boost, raise
Erhöhung – increase, increment, raise
Erhöhung (einer Leistung) –
improvement in performance
Erhöhungsfaktor – increase factor
Erhöhungsprozentsatz – percentage rate
of increase
Erhöhungsrelevanz – relevancy to
increase
Erhöhungstyp – type of increase
Erholung – recreation
Erholungsheim – recreation centre,
convalescence home
Erholungskur – rest cure
Erholungsplätze – recreation places
Erholungsurlaub – vacation
erinnern – remember
Erinnerung – memory
erkennen – recognize
erklären – explain
Erklärung – explanation
Erkrankung – illness, sickness

erkunden – explore
erlangen – gain, profit
Erlaubnis – permission
erlaubtes Fehlen – leave of absence
erledigen – settle, bring to a close
erledigt – completed, finished, done
Erledigung – completion, execution
erleichtern – facilitate
erlernter Beruf – initial training
erlöschen – expire
Erlöschen durch Zeitablauf – expiration
ermächtigen – authorize, empower,
entitle
Ermächtigung – authority
ermahnen – admonish
Ermessen – discretion
Ermessensentscheidung – discretionary
decision
Ermessensrente – discretionary pension
ermitteln – investigate
Ermittlung der Ertragsschwelle –
breakeven analysis
Ermittlung der Rentabilitätsschwelle –
breakeven analysis
Ermittlungsausschuss – commission of
enquiry
Ermüdung – fatigue
Ermüdungserscheinung – symptom of
fatigue
ermutigen – encourage
Ermutigung – encouragement
Ernährer – breadwinner
ernennen – appoint, nominate
Ernennung – appointment, nomination
erneuern – renew
ernst – serious
eröffnen – open
errechnen – calculate
erreichbar – attainable
erreichen – erect
Erreichung – attainment
errichten – establish
Erinnerungsstütze – mnemonic
Ersatz – replacement
Ersatzbedarf – replacement needs
Ersatzdienst – non-military service
Ersatzkasse – private health insurance
fund
Ersatzlehrer – substitute teacher
Ersatzleistung – fringe benefit
remuneration

Ersatzmitglied – deputy member
Ersatzqualifikation – alternative
  qualification
Ersatzteilnehmer – substitute attendee
Ersatzzeiten – equivalent qualifying
  periods, fictitious qualifying periods
erschöpfen – exhaust
Erschöpfung – exhaustion
Erschwernis – handicap, impediment
Erschwerniszulage – bonus for
  hazardous or unpleasant work
ersetzen – replace
ersichtlich – apparent, evident
Erstattung – reimbursement
Erstattung der Schulwegkosten –
  reimbursement of fares to/from school
Erstattungsantrag – request for refund
Erstattungsbetrag – refund amount
Erstattungsgruppe – refund group
Erstattungsklasse – refund class
Erstattungssatz – refund rate; refund
  record
Erstausbildung – initial training
Erste Hilfe – first aid
erster Arbeitsmarkt – well-paid jobs
Ersteintritt – initial entry
Ersterhebung – first survey
Ersterkrankung – first sickness
Ersuchen – requisition
Erteilung eines akademischen Grades –
  conferring of a degree, graduation
  (AmE)
Erwachsene(r) – adult
Erwachsenenbildung – adult education
Erwachsenenwelt – adult world
erwachsener Lerner – adult learner
erwähnen – mention
erwarten – expect
Erwartungen – expectations
Erweiterung – enlargement
Erwerb – acquisition
Erwerbsarbeit – gainful employment
Erwerbsfähigkeit – ability to earn a
  living, ability to work
Erwerbsleben – working life
Erwerbsminderung – reduction in
  earning capacity
Erwerbsquelle – source of income
erwerbstätig – gainfully employed
erwerbstätiger Hochschulabsolvent –
  salaried graduate

erwerbsunfähig – unable to work
Erwerbsunfähigkeit – general disability,
  invalidity, permanent incapacity to
  work
Erwerbung – acquisition
Erzieher(in) – educator, teacher
Erziehung (zu Hause) – upbringing
Erziehungs- und Ausbildungsbeihilfen
  – educational and training grant
Erziehungsberater – educational
  psychologist
Erziehungsberechtigter – parent, legal
  guardian
Erziehungslehre – pedagogy
Erziehungsrecht – right to educate
Erziehungsrente – child benefits
Erziehungsurlaub – child care-leave,
  parental leave
Erziehungswesen – education,
  educational system
Erziehungswissenschaft – pedagogy
Erzielung – attainment
erzwingen – enforce, force
es sei denn, dass – except
Essensbon – lunch coupon
Essensentschädigung – remuneration for
  meals
Essensmarke – meal voucher
Essenspause – meal break
Essenszuschuss – meal subsidy
Etat – budget
ethisch – ethical
ethnische Minderheit – ethnical
  minority
EU – European Union, EU
EU-Staaten – EU-member states
EU-weit anerkannte Lehrbefähigung –
  qualified teacher status in the EU
Europäische Kommission –
  Commission of the European
  Communities
Europäische Union – European Union,
  EU
Europäischer Binnenmarkt – Single
  European Market
Europäischer Rat – European Council
Europäisches Parlament – European
  Parliament
Europäisches Währungssystem (EWS)
  – European Monetary System (EMS)
Evaluierung – assessment, evaluation

Evaluierungsbogen – evaluation sheet
Eventualfall – contingency
Eventualitätsplanung – contingency planning
Examensarbeit beaufsichtigen – supervise an exam, invigilate
Exemplar – copy
Existenzgründung – founding one's own company, establishing one's livelihood
Existenzminimum – minimum living wage
Exit-Anweisung – exit statement
exklusiv – exclusive
exmatrikulieren – take off the university register
Experiment – experiment
experimentieren – experiment
Experimentiergerät – test equipment
Experimentierplatz – demonstration bench, demonstration place
Experte – expert
Expertengruppe – panel
Explosionsgefahr – danger of explosion
Explosionsschutz – protection in case of explosion
extensives Hören – extensive listening
extensives Lesen – extensive reading
externe Maßnahmen – external measures
externe Prüfungskommission – Secondary Examinations Council (SEC)
Externe(r) eines Internats – day pupil, non-resident participant, day boarder
Extravertierte(r) – extrovert
exzerpieren – select from, extract from

F

F-Stufe – promotion level
Fabrik – factory
Fabrikarbeiter – factory worker
Fabrikkalender – factory calendar
Fabrikleiter – works manager
Fach – subject

Fachabitur – Fachabitur
Fachabteilung – special department for....
Facharbeiter – skilled worker, craftsman
Facharbeiterausbildung – training as a skilled worker
Facharbeiterbrief – certificate of proficiency
Fachausbildung – special training
Fachausdruck – technical term
Fachberater – technical consultant
Fachberatung – consultation by specialists
Fachbereich – department specialized in ..., faculty
Fachbuch – specialized textbook
Fächerliste (Übersicht über Unterrichtsfächer) – list of subjects
Fächerwahl – choice of subjects
Fachfragen – technical questions
Fachgebiet – special area, specialist area, sector
Fachgebiet studieren – take a subject, study a subject
Fachgebietswissen – knowledge in a special field
fachgebundene Hochschulreife – matriculation requirements for the Fachhochschule (s. Graphik)
Fachgewerkschaft – craft union
Fachgymnasium – specialised secondary school
Fachhochschule (BRD) – Fachhochschule, Technical College, post-secondary Technical College, Polytechnic College, university for applied science (s. Graphik)
Fachhochschulreife – matriculation requirements for the Fachhochschule (weiterführende Ausbildungsstätte für Spezialisten) (s. Graphik)
Fachhochschulstudium – studies at a Fachhochschule
Fachidiot – person interested solely in his own subject
Fachinformatiker – systems integration engineer
Fachkenntnisse – special knowledge
Fachkompetenz – expertise
Fachkraft – specialist, skilled labour
fachkundig – competent, authoritative

Fachlaufbahn – career for specialists
Fachlehrer – specialist teacher
Fachleute – experts, specialists
fachlich – occupational, professional
fachliche Anforderungen – specific
requirements
fachliche Betreuung – specialist support
fachliche Weiterbildung – specialized
on-going training
fachliche Zuständigkeit – specialized
competence
fachlicher Rang – rank for specialists
fachliches Können – technical skill,
competence
Fachliteratur – specialist literature
Fachmann – expert
Fachoberschule (BRD) –
Fachoberschule, College of Further
Education (Die Fachoberschule umfasst
die 11. und 12. Jahrgangsstufe und
führt zu einer Aufnahmequalifizierung
für die Fachhochschule.) (s. Graphik)
Fachpersonal – skilled staff, trained staff
fachpraktische Kenntnisse – practical
experience
fachpraktischer Unterricht – practical
tuition
Fachreferent – executive, manager
Fachrichtung – subject area, special field
Fachschulausbildung – training at a
technical college
Fachschule – technical school
fachspezifisch – technical,
subject-specific
fachtheoretischer Unterricht –
theoretical tuition
fachübergreifend – interdisciplinary,
extending across disciplines
Fachverantwortung – responsibility,
accountability
Fachvorgesetzter – line manager
Fachwissen – technical knowledge,
special knowledge
fähig – capable, competent, able
Fähigkeit – ability, capability,
performance ability
Fähigkeiten – skills, competence
Fähigkeitstest – ability test
Fahrgeld – transportation expenses
Fahrgeldzuschuss – assisted passage,
transport allowance

Fahrgemeinschaft – ride-sharing group
Fahrlässigkeit – negligence
Fahrleistung – miles/kilometers travelled
Fahrtkosten – portal-to-portal pay,
transportation costs (Lohn und
Entschädigung für An- und
Abfahrtszeit von der Wohnung des
Arbeitnehmers bis zum Arbeitsplatz)
Fahrtkostenabrechnung – transportation
cost accounting
Fahrtkostenbeleg – transportation
receipt
Fahrtkostenpauschale – lump sum for
transportation
Fahrtkostenpauschalsatz – flat rate for
transportation
Fahrtkostenzuschuss – transportation
allowance
Fahrtstunde – hour of travel
Fahrzeugnutzung – use of vehicle
fair – fair, just
Faktor – factor
Fakultät – faculty
Fallstudie – case study
(Ausbildungsmethode anhand der
Besprechung von praktischen Fällen)
falsch – incorrect, false, wrong
falsche Freunde – false friends
falsche(r) Anfänger(in) – false beginner,
pseudo beginner
Familie – family
Familie und Karriere – dual-career
marriage
Familienangehörige(r) – family member
Familienarbeit – family work
Familienart – family type
Familienausgleichskasse – family
equalization fund
Familienbeihilfe – family allowance
Familienbeihilfekarte – family
allowance card
Familienhilfe – family assistance
Familienlastenausgleichsfond – family
burdens equalization fund
Familienname – family name, surname
Familienstand – marital status
Familienstandsnachweis – documents
on marital status
Familientrennung – family separation
Familienversicherung – family
insurance

Familienvorstand – head of the family
Familienwerte – family values
Familienzulage – family allowance
Familienzuschlag – family allowance
Familienzuschuss – family allowance
farbenblind – colo(u)r blind
Farblegende – color legend
faul – lazy
Faustregel – rule of thumb
FCKW – fluorine hydrocarbon
Federmäppchen – pencil case
Feedback – feedback
Feedbackgespräch – feedback session
Fehleinschätzung – erroneous
    assessment
fehlende Arbeitserlaubnis – missing
    work permit
**fehlende Zustimmung des
    Betriebsrates** – missing consent of the
    works council
Fehlentscheidung – misjudgement
Fehler – error, mistake, defect
Fehlerkultur – culture of error discussion
Fehltage – days of absence
Fehlverhalten – misconduct, incorrect
    behaviour
Fehlzeit – absence from work
Fehlzeit, sonstige bezahlte – other paid
    absences
Fehlzeiten – absent time
Fehlzeitenauswertung – evaluation of
    absent time
Fehlzeitenkalender, persönlicher –
    absence calendar, personal
Fehlzeitenquote – absenteeism rate
Feiertag – public holiday
Feiertag, beweglicher – floating holiday
Feiertag, bezahlter – paid public holiday
Feiertag, unbeweglicher – fixed holiday
Feiertage – official holidays (vgl.
    gesetzliche Feiertage, Bankfeiertag,
    Urlaub)
Feiertagsarbeit – work performed on a
    public holiday
Feiertagsart – public holiday type
Feiertagsausfallgesetz – Public Holiday
    Loss-of-Income Act
Feiertagskalender – public holiday
    calendar
Feiertagslohn – holiday pay
Feiertagszuschlag – holiday bonus

Feinmechanik – fine mechanics,
    precision engineering
Feinmechaniker – precision engineer
Feinwerktechnik – precision mechanism
    technology
Feldforschung – field study, field work
Feldtest – field test
Feldversuch – field test
Ferien – holidays, vacation
Ferien in der Mitte des Trimesters –
    half-term
Ferienadresse – vacation address
Ferienjahr eines Professors – sabbatical
    year
Fernlehrgang – correspondence course
Fernmeldeelektronik –
    telecommunications electronics
Fernmeldemonteur – communications
    equipment installer
Fernmeldetechnik – telecommunication
    engineering
Fernschulung – distant education
Fernseh-Aufzeichnungsgerät – video
    (tape) recorder
**Fernsehen für eine geschlossene
    Gruppe** – closed circuit television
    (C.C.T.V.)
Fernstudium – correspondence course,
    open University course
Fernunterricht – distance teaching
fertig – finished
Fertigkeiten – skills, competence
Fertigkeiten des Unterrichtens –
    teaching skills
Fertigung – production, manufacturing
Fertigungsaufgaben – manufacturing
    engineering
Fertigungsbetrieb – manufacturing unit
Fertigungseinheit – manufacturing unit
Fertigungsleiter – works manager
Fertigungslohn – direct labour costs
    (Lohn für die direkt am Werkstück
    verrichtete Arbeit, produktiver Lohn
    genannt, gilt als direkte Kosten, d.h. als
    Einzelkosten in der Gewinn- und
    Verlustrechnung der Firma)
Fertigungsplanung – production
    planning
Fertigungssteuerung – production
    control
Fertigungsstraße – assembly line

feste Anstellung – permanent job contract, permanent employment
feste (Renten-)Leistung – defined benefit plan
fester Lohn – stable wage
festigen – consolidate
Festlohn – fixed wage
Festnahme – detention
festsetzen – determine
Festsetzung von (gesetzlichen) Mindestlöhnen – fixing of (statutory) minimum wages
feststellen – establish
Feuerlöscheinrichtung – fire-extinguishing installation
Feuerlöscher – fire-extinguisher
feuern – dismiss, fire, sack
Fibel – reader
fiktiv – fictitious
fiktive Altersgrenze – fictitious age limit
Filzstift – felt-tipped pen
Finanzierung – financing
Finanzkonto – financial accounts
Finanzverwalter – bursar
Firmenausweis – employee I.D. card
Firmenbeitritt – accession, date of joining the company
Firmenbuchung (ohne Namensangabe) – company booking (without names)
Firmeneintritt – date of joining the company
Firmenfortführung – continued existence of a company
Firmenidentität – corporate identity (Die Substanz eines Unternehmens im Markt der Meinungen, bei Kunden, Lieferanten und Geschäftspartner.)
firmeninterner Arbeitsmarkt – internal labour market
Firmenleitung – corporate management
Firmenpension – company pension
Firmenpraktikum – practical studies in a company
Firmenrentner – company pensioner
Firmenrichtlinie – company administration guideline, management policy,
firmenseitige Kündigung – give notice
Firmenstrategie – company strategy
Firmenzeitschrift – company magazine
fixieren – settle, establish

Fixkostenstelle – cost center for fixed costs
Flächentarifvertrag – blanket agreement
flache Organisationsform – lean management
flankierende Maßnahmen – accompanying measures
flegelhaft – rude, unmannerly
Fleiß – diligence, industry
flexibel – flexible
flexible Altersgrenze – flexible pensionable age; flexible retirement; flexible retirement age
flexible Altersteilzeit – flexible partial retirement
flexible Arbeitszeit – flexible working hours
flexibles Altersruhegeld – flexible retirement benefit
flexibles Lernen – flexible learning (Universities face many challenges as they integrate ICT to support flexible learning on and off the campus.)
Fließband – assembly line
Fließbandarbeit – assembly line work; line production
fließend – fluent
Flipchart – flip chart
Flucht- und Rettungsplan – escape and rescue plan
flüchtig – cursory, superficial
Fluchtweg – escape route
Fluktuation – fluctuation, labour turnover, employee turnover
Fluktuation bekämpfen – combat fluctuation
Fluorchlorkohlenwasserstoff (FCKW) – fluorine hydrocarbon
Folge – consequence, result
Folgebescheinigung – subsequent statement
Folgekrankheit – subsequent sickness
Folgekurs – follow-up course
Folgelohnart – subsequent wage type
Folgelohnschein – subsequent time ticket
folgerichtig – consistent
Folgesatz – corollary
Folgevorgang – follow-up action
Folgezeitverfahren – measurement of times by differences
Folie – transparency, slide

Fonds – fund
Förderdatei – promotions file
Förderer – sponsor
Förderkartei – promotions file
fördern – facilitate, promote
Förderstrategie – promotion strategy
Förderstufe – promotion level
Forderung – demand, claim
Förderung – promotion
Forderung aus Vormonat – previous
month's claim
Förderung der Mitarbeiter –
promotional development of employees
Förderung der Wissenschaften –
advancement of science/arts
Förderung durch neue Aufgaben –
promotion through new assignments
**Förderung und Beratung durch
Gespräche** – coaching and counselling
Förderungsbeurteilungssystem –
promotional appraisal system
Förderungsleistung – promotion quota
(Wer am besten andere fördert, steigt
selber auf.)
Förderungsleitlinien – promotional
guidelines
Förderungsmaßnahmen – promotional
measures
Förderungsplanung – planning of
promotion
Förderungsstufe – promotion level
Förderungsversetzung – promotional
transfer
Förderungsziel – aim of a promotion
Förderunterricht – remedial teaching
formal – formal
formalistisch – formalistic
Formalstufe – formal step
Formalversicherung – formal insurance
formelle Direktive – formal policy
formelle Kommunikation – formal
communication
formelle Organisation – formal
organization
formelle Weisungsbefugnis – formal
authority
formlose Bewerbung – informal
application
Formsache – formality, matter of form
Formular – form
Formulierung – wording

forschen – do research work
Forscher – research worker
Forschergruppe – research team
Forschung – research
Forschung und Entwicklung (F&E) –
Research and Development (R&D)
Forschungsergebnisse – research results
Forschungsgebiet – field of research
Forschungskooperation – research
cooperation
Forschungsleistung – quantity of
research (oft an der Zahl der
wissenschaftlichen Veröffentlichungen
einer Person, eines Bereiches oder einer
Fakultät gemessen)
Forschungsstätte – research institute
Forschungsstipendiat(in) – research
fellow
Forschungsstipendium – fellowship
Forschungsurlaub – sabbatical year
Forschungszentrum – research centre
Fortbildung – continuing education,
on-going education, further education
(umfasst alle nachschulischen
Angebote)
Fortbildung a.D. – after-hours on-going
education
Fortbildung am Arbeitsplatz – training
on-the-job
**Fortbildung außerhalb des
Arbeitsplatzes** – training off-the-job
Fortbildung i.D. – in-hours on-going
education
Fortbildungskurs – advanced training
course
Fortbildungsprogramm – training
programme, training scheme
fortdauern – continue
fortfahren mit – continue
fortgeschritten – advanced
Fortgeschrittene(r) – advanced learner
Fortgeschrittenenkurs – advanced
course
fortlaufend – continuous
fortsetzen – continue
Fortzahlung im Krankheitsfall –
payment during illness
Fracht – shipping
Frage – question
Fragebogen – questionnaire

Fragen der (Schul)bildung –
educational issues
fraktales Unternehmen – fractal
company
Fräser – milling cutter
Frauenbeauftragte(r) – commissioner
for women's issues
Frauenförderplan – plan for the
promotion of women
Freiberufler – freelancer, self-employed
person
freiberuflich – freelance, self-employed
Freibetrag – tax allowance
Freibetrag, persönlicher – personal
exemption
freie Berufe – professions, (e.g.
architects, doctors, lawyers)
freie Entscheidung – option
freie Stelle – vacancy
freie(r) Mitarbeiter(in) – freelancer
freier Tag – day off
Freigrenze – exemption limit
Freiraum – margin, scope
Freiraum im Führungsverhalten –
management behaviour margin
Freischicht – non-working shift
Freischuss – sit a law exam before a
stipulated date (Wer beim Jurastudium
sich schon nach dem 8. Semester zum
Examen meldet hat eventuell drei statt
zwei Prüfungsversuche frei. Bei
bestandenem Examen kann er
außerdem versuchen, die im 1.
Durchlauf erzielte Note innerhalb eines
Jahres zu verbessern.)
freisetzen – displace, make redundant,
lay off, discharge
Freisetzung – day-release, exemption
Freisetzung von Arbeitskräften –
discharge of labor
freistellen – exempt
Freistellung – official absence due to
temporary reassignment, day-release,
exemption
Freistellung, betriebliche – special
company leave
Freiwerden einer Stelle – position
becoming vacant
freiwillig – voluntary
freiwillige Beendigung – voluntary
termination

freiwillige Gefolgschaft – voluntary
followship
freiwillige Mitgliedschaft – voluntary
membership
freiwillige Sozialleistungen – welfare
benefits
freiwillige Sozialleistungen der Firma –
company welfare plans
freiwillige Wiederholung – voluntary
repetition
Freizeit – free time, leisure time, spare
time
Freizeitausgleich – time in lieu
Freizeitbeschäftigung – leisure activity
Fremdfirma – outside company
Fremdsprache – foreign language
Fremdsprachenkenntnisse – foreign
language skills
freundlich – friendly
Friedenspflicht – obligation to keep
peace
Frist – time limit, deadline
Fristenberechnung – calculation of
cut-off dates
fristlos – without time limit
fristlose Entlassung – summary
dismissal, fired on the spot
fristlose Kündigung – summary
dismissal (employer), immediate
resignation (employee)
Frontalunterricht – teacher-centred
lesson
frühe Kindheit – infancy
Frühjahrstrimester – Hilary term (BrE)
Frühlingstrimester – spring term
frühreif – precocious
Frührente – early retirement
Frühschicht – morning shift, early shift
Frühstückspause – breakfast break
Frühverrentung – early retirement
Frustration – frustration
frustrieren – frustrate
führen – conduct, lead
Führung – leadership
Führung durch Beispiel – leadership by
example
Führung in vier Richtungen –
four-directional management (Aufgabe
einer Führungskraft, für effektive
Zusammenarbeit zwischen sich, seinen
Vorgesetzten, seinen gleichrangigen

Kollegen, und seinen Untergebenen zu sorgen)

Führungsalltag – everyday managerial life

Führungsaufgabe – management task, executive duty

Führungsbefugnis – managerial authority

Führungsebene – management level, management group

Führungseigenschaft – ability to lead, leadership ability, management capability

Führungsentscheidungen – management decisions

Führungserfahrung – management experience

Führungsetage – executive floor

Führungsfähigkeit – management capability, leadership ability

Führungsfunktion – management function

Führungsgabe – ability to lead, leadership ability

Führungsgrundsatz – principle of management

Führungshilfen – managerial resources, managerial aids

Führungsinstrument – management tool

Führungsinventar – management inventory (Stellenbesetzungsplan, Qualifikationsnachweis und andere Einzelangaben für Führungs- und Führungsnachwuchskräfte)

Führungskompetenz – leadership competency

Führungskonzept – management concept

Führungskraft – executive, manager

Führungskräftebedarf – demand for executives

Führungskräftenachfolge – management succession

Führungskräfteschulung – management training

Führungskreis – management level, management group

Führungskultur – management culture, personnel management, HR-culture

Führungsmodell – leadership model

Führungsnachwuchs – management trainee

Führungsnachwuchskraft – management trainee

Führungsposition – leadership position, leadership responsibility, management position

Führungspotential – management potential

Führungspraxis – managerial practice

Führungsprinzip – principles of management

Führungsqualitäten – management skills

Führungsspanne – span of management

Führungsstelle – supervisory job

Führungsstil – managerial style

Führungstalent – leadership talent

Führungsteam – management team

Führungstechnik – management technique

Führungsverhalten – managerial style

Führungswerkzeug – management tool

Führungswissen – management know-how

Führungszeugnis – letter of recommendation, testimonial

Füllfederhalter – fountain pen

Fundbüro – lost property office

fundiert – well-founded

Funktion – function

funktionale Weisungsbefugnis – functional authority

Funktionsbereich – functional area

Funktionsbewertung – function assessment

Funktionsbezeichnung – job title, position title

funktionsbezogen – function-based

funktionsbezogene Weiterbildung – function-related training

Funktionscharakter – function character

Funktionseinheit – functional unit

funktionsmäßig – functional

Funktionsordnung – functional structure

Funktionsstruktur – functional structure

Funktionsstufe – function level

Funktionswert – function value

für nichtig erklären – annul, cancel

furchtsam – timid

Fürsorgepflicht – welfare responsibility

Fürsorgestelle – welfare office
Fürsorgeverantwortung des Arbeitgebers – welfare responsibility of the employer
Fusion – amalgamation
fusionieren – amalgamate
Futterneid – jealousy

G

Gabe – gift, ability
ganzheitlich – holistic
ganztägige Gleitzeitentnahme – all-day flexi leave
Ganztagsarbeit – full-time job
Ganztagsbeschäftigung – full-time employment
Ganztagshort – full-day nursery, crèche
Ganztagsschule – all-day school, whole day school
Ganztagsunterricht – whole day classes
Garantielohnart – guaranteed wage type
garantierter Mindestlohn – guaranteed minimum wage
garantiertes Minimum (bei Akkordentlohnung) – guaranteed minimum rate
Garantiewert – guaranteed value
Gastarbeiter – immigrant worker, migrant labour
Gastdozent – visiting lecturer, guest lecturer
Gasthörer – guest
Gastland – host country
Gastlehrer – visiting teacher
Gastprofessor – visiting professor
Gastsprecher – guest speaker
GBR-Ausschuss – committee of the Central Works Council
ge(zer)brochen – broken
Gebärdensprache – deaf-mute sign language
Gebiet – region
Gebietskrankenkasse – regional health insurance fund

Gebietsverkaufsleiter – area sales manager
gebildet – educated
Gebrauch – use
Geburt – birth
Geburtenprämie – maternity grant
Geburtenzulage – maternity grant
Geburtsart – type of birth
Geburtsdatum – date of birth
Geburtshelfer – incubator
Geburtsland – country of birth
Geburtsname – name at birth
Geburtsort – place of birth
Geburtstagsliste – birthday list
Geburtsurkunde – birth certificate
Geburtszulage – bonus paid after birth of child
Gedächtnis – memory
Gedächtniskunst – mnemonics
Gedächtnisstütze – mnemonic
Gedanke – thought
Gedankenaustausch – exchange of ideas
Gedankengang – idea, thought
Gedingelohn – piecework pay
gedrucktes Material – printed material
geduldet – tolerated
geeignet – suitable
Gefahr – danger
Gefährdungsanalyse – danger analysis
Gefahrenklasse – class of risk
Gefahrenstoffe – dangerous materials
Gefahrenzulage – hazardous duty pay, danger money
gefährlich – dangerous, hazardous
gefährliche Arbeitsstoffe – dangerous materials
gefragter Beruf – hot skills (Leute, die sich zu Vertretern der *„Hot Skills"* zählen, werden schnell abgeworben.)
gefragter Experte – hot skills
Gefühl – feeling, sentiment
gegenseitig – mutual, reciprocal
gegenseitig informieren – inform on a mutual basis
gegenseitige Abhängigkeit – interdependency
gegenseitiges Vertrauen – mutual confidence
Gegenseitigkeitskennzeichen – reciprocity indicator
Gegenstand – object

gegenständliches Denken – concrete thinking
gegenwärtig – current
Gegenwartswert – cash value, present value
Gegenwert – equivalent
Gehalt – salary
Gehalt nach Familienstand und Kinderzahl – salary according to marital status and number of children
Gehaltsabrechnung – pay slip; salary statement, payroll accounting for salaried employees
Gehaltsabtretung – cession of wages.
Gehaltsänderung – change in salary
Gehaltsanpassung – salary adjustment
Gehaltsart – salary type
Gehaltsbestandteil – pay element
Gehaltsempfänger – salaried employee
Gehaltserhöhung – salary increase
Gehaltserhöhungsprogramm – pay increase programme
Gehaltserwartung – salary expectations
Gehaltsfindung – salary finding
Gehaltsforderung – salary demands
Gehaltsfortzahlung – continued payment of salary
Gehaltsgefälle – salary differential
Gehaltsgruppe – salary bracket, salary group
Gehaltsgruppenbezeichnung – pay scale grouping
Gehaltskonto – current account (BrE), checking account (AmE)
Gehaltskorrektur – salary adjustment
Gehaltskürzung – cut in salary
Gehaltslesungsliste – salary review list
Gehaltsliste – payroll, payroll sheet
Gehaltsmatrix – salary matrix
Gehaltsmodifikator – wage type modifier
Gehaltsnachzahlung – back-payment
Gehaltsnebenleistungen – voluntary social contributions
Gehaltsniveau – salary level
Gehaltspolitik – salary policy
Gehaltsrahmen – salary frame
Gehaltsspanne – salary range
Gehaltssprung – salary jump
Gehaltssteigerung – pay increase
Gehaltsstruktur – salary structure

Gehaltsstufe – salary level
Gehaltsüberprüfung – salary revision
Gehaltsüberweisung – pay remittance
Gehaltsvorschuss – salary advance
Gehaltsvorstellung – desired salary
Gehaltswünsche – salary demands
Gehaltszahlung – salary payment, payment of salaries
Gehaltszettel – salary slip, pay slip
Gehaltszusatz – fringe benefit
geheime Abstimmung – secret ballot
geheime Wahl – voting by secret ballots
Geheimhaltung – secrecy; concealment
Geheimnisträger – person with security clearance
Gehenbuchung – clock-out entry
Gehilfe – assistant
Gehirn – brain
Gehirn... – cerebral
Gehördefekt – hearing deficit
gehörlos – deaf
gehobener Dienst – executive grade
Gehorsam – obedience
Gehorsamsverweigerung – disobedience, refusal to obey
Geist – mind
Geistesblitz – brainwave
geistesgestört – mentally defective
Geisteskrankheit – mental illness
Geistesschwache(r) – moron
Geisteswissenschaften – liberal arts, humanities
geistig behindert – mentally handicapped
geistige Fähigkeit – mental ability
geistige Flexibilität – flexibility in thinking, mental flexibility
geistiges Eigentum – intellectual property
geizig – mean
gelangweilt – bored
gelassen – cool, calm, collected
Geldakkord – piecerate work
Geldentschädigung – compensation in money
Geldlohn – cash wages
Geldmittelzuteilung – resource allocation
Geldstrafe – forfeit, fine
geldwerter Vorteil – imputed income
Gelegenheit – opportunity, possibility

Gelegenheitsarbeit – casual work, jobbing
Gelegenheitsarbeiter – casual worker, temporary worker, stopgap
gelegentlich – incidental
Gelehrtenwelt – academic community
Gelehrter – scholar, man of learning
Geltendmachung – enforcement, insistence
Geltungsbereich eines (Tarif)vertrages – scope of agreement
Gemeinde – municipality
Gemeinkosten – overheads
gemeinnützig – non-profit making, charitable
gemeinnütziges Bildungswerk – non-profit educational establishment
gemeinsam – common, joint
gemeinsam getragene Vision – shared vision
gemeinsame Verhandlungen (der Tarifpartner) über Löhne und Gehälter – joint bargaining
Gemeinschaft – community
Gemeinschaft der Fellows in einem College oder Universität – fellowship
gemeinschaftlich – common, joint
Gemeinschaftserziehung – co-education
Gemeinschaftssinn – esprit de corps
gemischte Schule – mixed school
Gemütsbewegung – emotion
genau – accurate, exact
Genauigkeit – accuracy
genehmigen – authorize, empower, entitle
genehmigt – approved
Genehmigung – permission, approval
genehmigungspflichtig – subject to approval
genehmigungspflichtige Anlage – plant subject to approval
Genehmigungsprozentsatz – approval percent
Genehmigungszeitraum – approval period
Generalist – all-rounder
Generationskonflikt – generation gap
Generierungsregel – generating rule
Genesungsurlaub – convalescent leave
Genie – genius

Genossenschaftskasse auf Gegenseitigkeit – cooperative mutual pension fund
Genugtuung – satisfaction
geöffnet von... bis... – open from... to...
Geometrie – geometry
geplant – planned
Gerätepark – equipment pool
gerecht – fair, just
Gerechtigkeit – justice
geregelte Kompetenzordnung – formal organization
gerichtliche Anordnung – injunction
Gerichtskosten – court costs
Gerichtsverfahren – lawsuit
geringe geistige Fähigkeit – low mental ability
geringfügig Beschäftigte – parttime employees
geringfügige Beschäftigung – limited part-time work
Geringverdiener – low-income earner
Geruchssinn – sense of smell
Gerücht – gossip, rumour
Gerüchteküche – grapevine
Gesamt... – total
Gesamtarbeitszeit – total working hours
Gesamtbelegschaft – total work force
Gesamtbetriebsrat (GBR) – central works council
Gesamtbetriebsratsausschuss – committee of the central works council
Gesamtbeurteilung – record of achievement
Gesamtbrutto – total gross amount
Gesamtdauer der Abordnung – duration of delegation
Gesamteinkommen – total income
Gesamtentwicklung eines Kindes – social growth of a child
Gesamtgehalt – total salary
Gesamthochschule – comprehensive university; integrated university
Gesamtleistung – total output, overall achievement
Gesamtlohn – total wages
Gesamtschule – comprehensive school
Gesamtsozialversicherungsbeitrag – total social security contribution
Gesamtstellenbesetzungsplan – overall staffing schedule

Gesamtunternehmen – concern (as a whole)

Gesamtverantwortung – overall responsibility

Gesamtziel – key objective

Geschäft(e) – business

geschäftlich – on business

Geschäfts-Neugründung – Start-up

Geschäftsbericht – annual report

Geschäftsbetrieb – business

Geschäftsentwicklung – business development

geschäftsfähig – legally competent

Geschäftsfeldplanung – business development and strategy planning

Geschäftsführer – managing director

Geschäftsführung – conduct of business, management

Geschäftsgebiet – Division

Geschäftsgebietsleiter – head of operational sector

Geschäftsgeheimnis – trade secret

Geschäftsinhaber – owner, manager, proprietor

Geschäftsjahr – business year, fiscal year

Geschäftsleitung – management

Geschäftsmann – businessman

Geschäftsordnung – procedural rules; standing orders

Geschäftsplanung – business planning

Geschäftspolitik – strategic planning

Geschäftsprozess – business process

Geschäftsreise – business trip

Geschäftsschluss – closing time

Geschäftsstrategie – business strategy

Geschäftsteilhaber – co-owner, partner

geschäftsunfähig – legally incompetent

Geschäftswertbeitrag – economic value added (EVA)

Geschäftszeichen – company reference number

Geschäftszweck – organizational purpose

Geschäftszweig – branch of business, line of business

geschasst – chucked out, booted out

gescheit – intelligent, bright

Geschichte – history

Geschick – skill

Geschicklichkeit – dexterity

geschieden – divorced

Geschiedenen-Witwenrente – widow's pension for divorced partner

Geschlecht – sex

Geschwindigkeit – pace, speed

Geselle – journeyman, little master (BrE)

Gesellschaft – association, company, corporation, firm

Gesellschaft mit beschränkter Haftung – limited liability company (AmE)

Gesellschafter – stockholder, shareholder

Gesellschaftspolitik – social policy

gesellschaftspolitisch – socio-political

Gesetz – law

Gesetz gegen Diskriminierung wegen Alters – Age Discrimination Act

Gesetz über Copyright – Copyright law

Gesetzespflicht – mandatory; bound by law

Gesetzgeber – legislator; law-maker

Gesetzgebung – legislation

gesetzliche Haftpflicht – legal liability

gesetzliche Krankenversicherung – statutory health insurance

gesetzliche Regelung – regulation, provision, regularization

gesetzliche Rentenversicherung – statutory pension insurance fund

gesetzlicher Feiertag – legal holiday, public holiday, official holiday (allgemeine gesetzl. Feiertage in der BRD [Stand 1996] sind: Neujahr, Karfreitag, Ostermontag, 1. Mai, Christi Himmelfahrt, Pfingstmontag, 3. Oktober, 25. Dezember. Weitere regionale Feiertage sind z.B.: Heilige Drei Könige, Fronleichnam, Mariä Himmelfahrt)

gesetzlicher Vormund – legal guardian

gesetzmäßig – legal

gesetzwidriger Streik – illegal strike

gesondert – separate(d)

Gespräch – conversation, discussion

Gespür – feel, feeling

gestaffelte Arbeitszeit – staggered hours

gestatten – allow, permit

gesteuertes Lesen – controlled reading

gestiegene Anforderungen bewältigen – master increased requirements

gesunder Menschenverstand – common sense

Gesundheit – health

gesundheitliche Eignung – adequate health

gesundheitsgefährdend – injurious to health

Gesundheitsgefährdung – health hazard

Gesundheitsrisiko – health hazard

Gesundheitsschutz – health protection

Gesundheitsvorsorge – preventive medical programme(s)

Gesundschrumpfen – downsizing

getrennte Wahl – voting by separate ballots

gewagt – hazardous, daring

gewählter Durchschnittszeitwert – average selected time

gewähren – award, grant

Gewährungszeitraum – allowance period

Gewalt – compulsion, coercion

Gewaltbereitschaft – propensity to violence

Gewandtheit – dexterity

Gewerbe – craft, trade

Gewerbeaufsichtsamt – trade inspection board

Gewerbeaufsichtsbeamter – trade supervisory officer

Gewerbeaufsichtsbehörde – trade supervisory authority

Gewerbeordnung – Trade Regulation Act

Gewerbeschule – trade school

Gewerbesteuer – trade tax

Gewerbetechnik – industrial engineering

gewerblich – industrial

gewerblich Tätiger – wage-earner, non-salaried employee

Gewerbliche – industrial worker

gewerbliche Ausbildung – industrial training

gewerbliche Wirtschaft – industry

gewerblicher Arbeitnehmer – industrial employee

Gewerkschaft – trade union

Gewerkschaftsbeiträge – union dues

Gewerkschaftsbewegung – trade union movement

gewerkschaftspflichtiger Betrieb – closed shop (Betrieb, der ausschließlich Gewerkschaftsmitglieder einstellt)

Gewerkschaftspolitik – union policy

Gewerkschaftsvertreter – union representative

Gewerkschaftsvertreter im Unternehmen – shop steward

Gewinn – gain, profit

Gewinnbeteiligung – profit-sharing

gewinnen – gain, profit; win

Gewinner – winner

Gewinnung von Führungskräften aus den eigenen Reihen – management recruitment from within the company

Gewissen – conscience

gewissenhaft – conscientious

Gewohnheit – habit

Gewohnheitsrecht – customary law, common law

Gewöhnung – habituation

gezielte Einstellungspolitik – targeted employment policy

Gilde – guild

Gitterorganisation – lattice organisation

glauben – believe

Glaubwürdigkeit – credibility

gleich – equal

Gleichaltrige(r) – the same age

Gleichbehandlungsgrundsatz – principle of equal treatment

gleichberechtigt – equal

Gleichberechtigung – equal rights

Gleichberechtigung am Arbeitsplatz – equal employment opportunity (EEO)

Gleiche(r) – peer, equal

gleicher Wert – equivalent

gleichgeschlechtlich – homosexual

Gleichgewicht – balance

gleichgültig – indifferent

Gleichheit – parity

gleichmäßig – consistent

Gleichstellung – equalization (mit Beamten)

gleichwertig – equal

Gleichwertigkeit – equivalence

gleitende Altersruhe – flexible pension-age approach (Mitarbeiter können über einen längeren Zeitraum Arbeitsstunden auf einem Langzeitkonto „einzahlen". Diese können sie im Alter langsam abbauen. Der Vorteil gegenüber Altersteilzeit: Die Rente wird nicht verkürzt.)

gleitende Arbeitszeit – flexible working hours

gleitende Lohnskala – sliding wage scale

Gleitzeit – flexible working hours, flextime, flexitime

Gleitzeitausgleich – compensation for accumulated flexitime hours

Gleitzeitguthaben – positive flexitime balance

Gleitzeitmodell – flexitime model

Gleitzeitregelung – flexitime regulation

Gleitzeitsaldo – flexitime balance

Gleitzeitsaldoüberschuss – positive flexitime balance

Gleitzeitschulden – negative flexitime balance

Gleitzeitspanne – flexi timespan

Gleitzeittagesprogramm – day program for flexitime

Gleitzeitüberschuss – flexitime excess

Gleitzeitunterschreitung – flexitime deficit

Gliedertaxe (Unfallversicherung) – dismemberment grading

global – comprehensive, across the board, global

GmbH – limited liability company (AmE)

GmbH & Co.KG – limited partnership with limited liability company as general partner

Goldkragen-Mitarbeiter – gold-collar worker (MA werden als Goldkragen-MA, als wertvolle Ressourcen, besser noch, als Investitionen angesehen, in deren Aus- und Weiterbildung man langfristig investieren und die man langfristig an sich binden muss.)

Goodwill – goodwill

Grad – degree

Gradlinigkeit – straight

graduierter Ingenieur – engineer

Grammatik – grammar

graphische Darstellung – chart, diagram

Gratifikation – gratuity

Gratifikation, feste – fixed gratuity

Gratifikationsbasis, kumulierte – cumulated gratuity basis

gratulieren – congratulate

Grenzarbeitnehmer – border worker, frontier worker

Grenzbelastung – marginal tax burden

Grenzfall – borderline case

Grenzgänger – international commuter, border worker

Grenznutzen – border value (das, was einer besser kann als die anderen)

Grenzregion – region near international border

Grenzübertritt – border crossing

Grenzübertritt, Hinreise – border crossing, trip out

Grenzübertritt, Rückreise – border crossing, trip home

Grenzwert – critical value

GRID-Muster – managerial grid

grober Schnitzer – howler

große Pause – play time

Großindustrie – big industries

Großunternehmen – large-scale company, corporation

großziehen – rear, bring up

großzügig – generous

Grund zur Klage – grievance

Grundablauf – basic procedure

Grundausbildung – basic training

Grundausbildungslehrgang – basic training course

Grundbetrag – base award, basic amount

Grunddaten zur Person – basic personal data

Grundeinkommen – basic income

gründen – establish, found

Grundfach – foundation subject

Grundfertigkeiten – basic skills, basic competence

Grundgedanke – principle

Grundgehalt – basic salary

Grundinformation – basic information

Grundkenntnisse – basic knowledge

Grundkurs – foundation course

Grundlage – base, basis

Grundlagenforschung – basic research, pure research

Grundlagenwissen – basic knowledge

gründlich – thorough

Gründlichkeit – thoroughness

Grundlohn – basic wage

Grundordnung des Hauses Siemens (SAG) – basic corporate policies of Siemens AG
Grundsätze des Arbeitsschutzes – basic labour protection policies
Grundsatzfragen – basic policies
Grundschulalter – primary school age
Grundschulbildung – elementary education
Grundschule – primary school (BrE), elementary school, grade school (AmE) (s. Graphik)
Grundschullehrer – primary school teacher
Grundschulunterricht – primary education
Grundschulwesen – primary education
Grundstudium – basic studies
Grundstunde – basic hourly pay
Grundwissen – basic knowledge
Grundwortschatz – core vocabulary
Grundzeit – basic time
Grüne Karte – green card
Gruppe – group
Gruppen-Leistungsprämie – group incentive
Gruppen-Prämiensystem – group bonus system
Gruppenakkord – group piecework
Gruppenakkordlohn – group piecework rate
Gruppenaktivität – group activity
Gruppenarbeit – group work, teamwork
Gruppendiskussion – group discussion
Gruppendynamik – group dynamics
Gruppenentscheidung – group decision
Gruppenführer – group leader
Gruppengeist – team spirit
Gruppenleistungslohn – group incentive wage
Gruppenleiter – group leader
Gruppenlohn – group payment
Gruppennorm – group norm
Gruppenprozess – group process
Gruppenübung – group exercise
gültig – valid
gültiger Lohnsatz – prevailing wage
Gültigkeit – validity
Gültigkeitsdauer – term, validity period
Gültigkeitszeitraum – term, validity period

günstig – convenient
Guru – guru
gut abschneiden – do well
Gutachten eines Sachverständigen – expertise, expert opinion
guter Einfall – brainwave
gutgläubig – credulous, trusting
gutheißen – approve
gymnasiale Oberstufe – senior grammar school level
Gymnasialwesen – secondary education
Gymnasium – grammar school, high school (AmE) (s. Graphik)

H

Habilitand – person writing post-doctoral thesis to qualify as a university lecturer
Habilitation – post-doctoral lecturing qualification
Habilitationsschrift – post-doctoral thesis required for qualification as a university lecturer
hadern – be at odds with, quarrel
Haft – detention
haftbar – liable
Haftpflicht – liability
Haftpflicht des Arbeitgebers – employer's liability
Haftpflichtversicherung – third-party liability insurance, civil liability insurance
Haftung – liability
Haftungsausschluss – nonliability
Haftungsbefreiung – exemption from liability
Halbbelegung (RV) – half-cover
halbdynamische Versorgungszusage – half-dynamic employer's pension commitment
halbtags – half-time
Halbtagsarbeit – part-time work, part-time job, part-time employment

Halbtagskraft – half-time worker, half timer
Halbwaisenrente – half-orphan benefit
Halbwertszeit – decay meter
Haltung – attitude
Handarbeit – manual work, work done by hand
Handarbeiter – manual worker
Handbuch – manual, guide, handbook
Handel – commerce
Handelsbilanz – published financial statements
Handelsgesetzbuch – code of commercial law, german commercial code
Handelskammer – chamber of commerce
Handelsrecht – commercial law
Handelsunternehmen – business
Handelsvertreter – commercial agent, commercial representative
Handgepäck – carry-on luggage
Handgreiflichkeit – fisticuffs
handhaben – administer, handle
Handicap – health exclusion, handicap
Handlungsbedarf – necessity to take action
Handlungsspielraum – room for maneuver, range of competence, latitude
Handlungsträger – person who has commercial power of attorney
Handlungsvollmacht – contractual power, commercial power of attorney
Handschrift – handwriting
Handwerk – craft trades
Handwerker – worker, workman, labourer, tradesman
handwerklicher Beruf – craft, trade
handwerkliches Können – craftsmanship, manual dexterity
Handwerksbetrieb – workshop
Handwerksordnung – crafts code
hänseln – tease
Härtefallregelung – settlement of hardship cases
Härteregelung – settlement of hardship cases
Hätte-Prinzip – "as if" principle
häufig – frequent
häufiger Stellenwechsel – job hopping

häufiges unentschuldigtes Fehlen – absenteeism
Häufung – concentration
Hauptabteilung – central department
hauptamtlich – full-time
Hauptarbeitsgebiet – main work area
Hauptarbeitsschutz-Kommission – general committee for labour protection
Hauptausschuss eines Verbands – Executive Committee
hauptberuflich – full-time
Hauptfach – main subject, core subject, major (subject) (AmE)
Hauptfürsorgestelle – main welfare office
Hauptgeschäftsführer (GmbH & Co.KG) – General Executive Manager
Hauptlaufbahn – primary career
Hauptpersonalabteilung – main personnel department, personnel division (AmE)
Hauptpersonalbüro – main personnel office
Hauptredner – keynote speaker
Hauptschema – main schema
Hauptschule – secondary modern school (BrE), junior high (school) (AmE) (Die Hauptschule umfasst die Jahrgangsstufen 5 mit 9. Der Lehrplan und die Fächerauswahl kommen den Begabungen und Neigungen des Schülers einer weiterführenden Schule entgegen. Sie orientieren sich überwiegend am praktischen Umgang mit den Dingen und geben so dieser Schulart ihr eigenes, unverwechselbares Profil.) (s. Graphik)
Hauptstadtzuschlag – capital city allowance
Hauptstelle für Bewerbungen – main office for applications
Hauptstudium – main course of studies
Haupttätigkeit – principal occupation
Hauptveranstaltung – main (business) event
Hauptverwaltung – headquarters
Hauptziel – principle aim, top goal, major target, primary objective
Hausarbeitstag – household day
Hausaufgabe – homework
Hausbote – interoffice messenger

Hausfrau – housewife
Haushälterin – housekeeper
Haushaltsgehilfe – domestic help
Haushaltszulage – household allowance
häusliches Milieu – family environment
Hausmeister – caretaker, janitor (AmE)
Haustarif – in-company wage agreement
Hauswirtschaftlehre – domestic science
Hauswirtschaft(slehre) – home economics
Hauszeitung – house journal, company magazine
Hautkrankheit – skin disease
Headhunter – head-hunter
Hebelwirkung – leverage
Heilbehandlung – therapeutic treatment
Heim – home
Heimarbeit – outwork
Heimarbeiter – outworker
Heimatanschrift – home address in country of birth
Heimatkunde – local history
Heimatland – home country
Heiratsurkunde – marriage certificate
helfen – assist
hemmend – restrictive
Hemmschuh – stumbling block
hemmungslos – uninhibited
Herabsetzung – cut, cutback
Herabsetzung der Löhne – cutting of wages
Herabstufung – downgrading (Verdrängung gelernter oder angelernter Arbeitskräfte aus ihren bisherigen Tätigkeiten in weniger qualifizierte, außerhalb ihres bisherigen Berufes liegende und schlechter bezahlte Tätigkeiten)
herangehen – approach
herausfordernd – challenging
herausfordern – challenge
Herausforderung – challenge
Herbstferien – autumn half-term holidays
Herbsttrimester – Michaelmas term
Herkunftskennzeichen – origin indicator
Herkunftsland – country of origin
hervorbringen – create
hervorragen – surpass, exceed, excel
hervorragende Eigenschaft – strong point

heuristisch – heuristic
Hierarchie – hierarchy, ranking, rank order
Hierarchie der Bedürfnisse – hierarchy of needs
Hierarchie der Weisungsbefugnis – chain of command
Hierarchiedenken – hierarchism
Hilfsarbeit – auxiliary work
Hilfsarbeiter – unskilled worker
hilfsbedürftig – in need of help
hilfsbereit – helpful
Hilfsmittel – resources
Hilfsquellen – resources
Hilfstabelle – utility table
hinauswerfen – dismiss, fire, sack
Hindernis – hindrance, stumbling block
Hinderungsgrund – impediment
Hinfahrt – outward journey
Hingabe – dedication; commitment
Hinreise – outward journey
Hinterbliebenenrente – survivor's pension
Hinterbliebenenversorgung – surviving dependents' insurance
Hinterbliebener – surviving dependent
Hintergrund – background
Hintergrundwissen – background knowledge, insider knowledge
Hinweis – indication
Hinzurechnungsbetrag – additional amount
Hinzuverdienst – additional income
Hirnschaden – brain damage
Hoch- und Tiefbau – civil engineering
Hochbau – structural engineering
Hochbegabte(r) – highly gifted
Hochbegabtenförderung – furtherance of gifted students
Hochmut – pride
hochqualifizierte Kraft – highly qualified employee
Hochrechnung – projection
Hochschul(aus)bildung – university education, academic training, higher education (Hochschulausbildung umfasst die Hochschulbildung bis zum 1. bzw. 2. Staatsexamen, die von Universitäten, Polytechniken und colleges of higher education angeboten wird.) (s. Graphik)

Hochschulabschluss – university degree
Hochschulabsolvent – university graduate
Hochschuldiplom – higher-education diploma
Hochschule – university, college
Hochschulkurs – post-secondary course, university course
Hochschulkurse für nicht eingeschriebene Hörer – extramural courses
Hochschullehrer – university lecturer
Hochschulrat – university board of trustees (Der Hochschulrat setzt sich aus Interessenvertretern aus Wirtschaft und Wissenschaft zusammen.)
Hochschulreife – certificate of aptitude for higher education
Hochschulrekrutierung – college recruitment
Hochschulverband im Nordosten der U.S.A. – Ivy League (Ein Verband bestehend aus Brown, Columbia, Cornell, Dartmouth, Harvard, The University of Pennsylvania und Yale.)
hochsprachliche Aussprache – received pronunciation
Höchstalter – maximum age
Höchstarbeitsstunden – maximum working hours
Höchstbeitrag – maximum contribution
Höchstbetrag – maximum amount
höchste Rente – maximum pension
Höchstleistung – peak performance
Höchstlohn – top wages, top wage rate, maximum wage rate, wage ceiling
Höchstsatz – maximum rate
Höchstsatz, ermäßigter – maximum rate, reduced
Hoffnungslosigkeit – hopelessness
höflich – polite, courteous
Höflichkeit – politeness, courtesy
höher einstufen – upgrade
höhere Berufsstände – professional classes
höhere Privatschule mit Internat – Public School (s. Graphik)
höhere staatliche Schule – secondary grammar school, high school (AmE) (s. Graphik)
höherer Dienst – administrative grade

höheres Lehramt – secondary school teaching
Höherstufung – upgrading
Höherversicherung – increased insurance
hohes Ziel – aspiration
homogen – homogeneous
Honorar (freie Berufe) – fee
Horde – gang
hören – listen
Hörer – student
horizontale Dezentralisierung – horizontal decentralization
horizontale Information – horizontal information
Hörsaal – auditorium, lecture room
Hörsinn – sense of hearing
Hort – nursery, créche
Hörtest – aural test
Hörvermögen – power of hearing
Hörverständnis – listening comprehension
Hörwahrnehmung – auditory perception
Hotelbeleg – hotel receipt
humanistisch – humanistic
humanistische Bildung – classical education
humanistisches Gymnasium – grammar school emphasizing Latin and Greek
Humankapital – human capital
Humor – humo(u)r
Hygienefaktor – hygiene factor (Mit diesem Begriff bezeichen Motivationsforscher eine Voraussetzung, die von Mitarbeitern als selbstverständlich angesehen wird und kaum jemand zum Wechseln oder Bleiben bewegt.)
Hypothek – mortgage
hypothetisch – hypothetical
hysterischer Mensch – hysterical person

I

i.D. – in-hours
Idee – idea
Ideenklau – plagiarism
Ideenpotential – reservoir of ideas
Identifikation – identification
Identifikator – identifier
Identität – identity
illegale Beschäftigung – illegal employment
im Auftrag von – on behalf of (BrE), in behalf of (AmE)
im Ausland – abroad
im Außendienst arbeiten – work as a sales representative
im Einklang mit – compatible
im In- und Ausland – at home and abroad
im Interesse von – on behalf of (BrE), in behalf of (AmE)
im Wettbewerb stehen – compete
Immatrikulation – enrolment, school registration
immatrikulieren – enrol, matriculate
Immigrant(in) – immigrant
Immissionen – noxious air
Immunität – immunity
implizit – implicit
Imponiergehabe – exhibitionism
improvisieren – improvise
Impuls – impulse
impulsiv – impulsive
in anderen Umständen – pregnant
in den Ruhestand treten – retire
in den Ruhestand versetzt werden – be retired
in der Forschung tätig sein – carry out research work
in der Regel – as a rule
in gutem Glauben – bona fide, in good faith
in Managementaufgaben ausgebildete Führungskraft – professional manager
in Verbindung bringen – associate
in Zweifel ziehen – dispute
inadäquat – inadequate
inaktiv – inactive

Incentive-System – incentive system
Index – index
Index der Familienfreundlichkeit – family friendly index
Indexlohn – pegged wages
indirekt – indirect
indirekte Arbeitskosten – indirect labour costs
Individualisierung – individualization
Individuallohn – individual wage
Individualphase – independent learning phase
individuelle Bedürfnisse – individual needs
individuelle Könnerschaft – individual skill, individual ability
Individuum – individual
Indolenz – indolence
induktiv – inductive
Industrie – industry
Industrie- und Handelskammer – Chamber of Industry and Commerce
Industrie... – industrial
Industriearbeiter – factory worker, industrial worker
Industrieelektroniker – industrial electronics technician
Industrieerfahrung – industrial experience
Industriefernsehen – closed circuit television (C.C.T.V)
Industriegewerkschaft – industrial union
Industriekauffrau (mit abgeschlossener Lehre) – industrial clerk
Industriekaufmann (mit abgeschlossener Lehre) – industrial clerk
industriell – industrial
Industrielle Revolution – industrial revolution
Industrieller – industrialist
Industriemechaniker – industrial mechanic
Industrietechnologe – industrial technologist
Industriezweig – branch of industry
ineffektiv – ineffective, inefficient
ineffizient – inefficient
Ineffizienz – inefficiency
infantil – infantile

Infantilität – puerility
Infektionskrankheit – infectious disease
Informatik – computer science
Informatik-Unterricht – computer science instruction
Informatiker – computer scientist
Information aus erster Hand – first-hand information, straight from the horse's mouth
Information nach oben – information to superiors
Information nach unten – information to staff
Informations- und Kommunikationstechnologie – information and communication technology (ICT)
Informationsfluss – flow of information, information flow
Informationsgesellschaft – information society
Informationsmangel – information deficit
Informationspflicht – obligation to inform employees
Informationsquelle – source of information
Informationsrecht – right to information
Informationstechnik – information technology
informatorische Rolle – informational role
informelle Direktive – informal policy
informelle Kommunikation – informal communication
informelle Organisation – informal organization
informelle Weisungsbefugnis – informal authority
Informieren (das) – information
Infrastruktur – infrastructure
Ingenieur – engineer
Ingenieurassistent(in) – engineering assistant
Ingenieurkreis – Siemens engineering student's program
Inhaber – owner, proprietor, holder
Initiative – initiative, enterprise
in Kraft treten – take effect, come into force
inländisch – domestic

Inlandsaufgaben – domestic tasks, domestic matters
Inlandseinkommen – domestic income
Inlandsreise – domestic trip
Innendienst – office duty
innerbetrieblich – intercompany
innerbetriebliche Zusammenarbeit – joint consultation
innere Kündigung – mental withdrawal, restricted identification
innerhalb der Dienstzeit – in-hours
Innovationsdynamik – dynamics of innovation
Innovationsprozess – innovation process
Innung – guild
Innungskrankenkasse – guild health insurance fund
ins Leben rufen – commence, initiate, introduce
Inserat – advertisement, ad
Insolvenz – insolvency
Insolvenzversicherung – insolvency insurance
Inspiration – inspiration
installieren – install
Instandhaltungspersonal – maintenance employees, maintenance people
Instinkt – instinct
instinktiv – instinctive
Institut – institute
Institut der Stadt und der Zünfte von London – City and Guilds of London Institute
Instrumente zur Selbsteinschätzung des Mitarbeiters – career anchor assessment
Integrität – integrity
Intellekt – intellect
intellektuell – intellectual
Intellektueller – egg-head
intelligent – intelligent, bright
Intelligenz – intelligence; intelligentsia
Intelligenzquotient – intelligence quotient
Intelligenztest – intelligence test
Intensität – intensity
intensiv – intensive
Intensivkurs – crash course, intensive course, total exposure course
interdisziplinär – inter-disciplinary
Interesse – interest

Interessenausgleich – reconciliation of interests
Interessenlage – interests
Interessent – interested party
Interessentest – interest test
Interessenvertretung – lobby, interest group
interkulturelle Erfahrung – cross-cultural experience
interkulturelle Studie – cross-cultural study
interkulturelles Training – cross-cultural training
Internat – boarding school
internationaler Berufswettbewerb – international vocational training competition (IVTC) (Es werden überwiegend handwerkliche Tätigkeiten nach berufstypischen Unterlagen geprüft.)
interne Maßnahmen – internal measures
Interne(r) – boarder
interner Personalwechsel – mobility within the firm
interpretieren – interpret, paraphrase
interpretieren – interpret
Interview – employment interview, interview
Intoleranz – intolerance
Introvertierte(r) – introvert
Intuition – intuition
Invalide – disabled person
Invalidenkasse – disability fund
Invalidenrente – disability pension
Invalidenzusatzrente – supplementary disability pension
Invalidität – disability
Invaliditätsrente – disability pension
Invaliditätsversicherung – disability insurance
Inventar – inventory
Investitionsbudget – capital expenditure budget
ionisierende Strahlen – ionizing rays
IRC – intercultural relations and communications (IRC)
irreführen – mislead
Irrglauben – heresy
Irrtum – error, mistake
Isolation – isolation
Isolierung – isolation, insulation

Ist-Gehalt – actual salary
Ist-Kopfzahl – actual number of employees
Ist-Arbeitszeit – actual hours worked
Ist-Plan – actual plan
Ist-Zeit – actual time
Ist-Zeiterfassung – actual time recording

J

Jahrbuch – annual, year book
Jahres... – annual
Jahresabschlussvergütung – year-end bonus, year-end premium
Jahresabschlusszahlung – year-end bonus, year-end premium
Jahresarbeitsentgeltgrenze – annual employee compensation limit
Jahresarbeitslohn – annual earned income
Jahresarbeitsverdienst – annual earnings
Jahresarbeitsverdienstgrenze – annual earnings threshold
Jahresarbeitszeit – annual working hours (Die pro Woche vorgeschriebene Arbeitszeit wird auf das Jahr hochgerechnet. Je nach Arbeitsanfall und persönlichen Wünschen sind die Mitarbeiter im Betrieb anwesend. Während die Arbeitszeit im Jahresverlauf variiert, bleibt das Gehalt jeden Monat gleich.)
Jahresausgleich – annual adjustment
Jahresbeitrag – annual contribution
Jahresbrutto – gross annual amount
Jahreseinkommen – annual income
Jahresentgeltbescheinigung – annual remuneration slip
Jahresgehalt – annual salary
Jahreshauptversammlung – annual general meeting, AGM
Jahreskumulation – annual cumulation
Jahreslohnkonto – annual payroll account

Jahreslohnsteuertabelle – annual income tax table
Jahresmeldung – annual report
Jahresmeldung – annual income statement to social insurance agent
Jahresumlage – annual contribution amount
Jahresurlaub – annual holiday, annual leave
Jahreszahlung – annual payment, annuity
Jahreszeugnis – school report, report, credit (AmE), grade (AmE)
Jahrgangsstufe – year, form, grade
jährlich – annual
jährliche Zahlung – annuity
Jargon – jargon
jemanden ausschimpfen (für etwas) – tell (someone) off (for something)
jemanden in den Ruhestand versetzen – retire
jemanden um Rat fragen – consult
Job Enlargement – job enlargement (horizontale Arbeitserweiterung auf strukturell gleichartige Aufgaben)
Job Enrichment – job enrichment
Job-sharing – job sharing
Job rotation – job rotation
Jobanbieter – employer
Jobholder – jobholder (Der Jobholder von heute ist ein flexibel einsetzbarer Experte, anpassungsfähig und anpassungsbereit.)
Journal – journal
Jubilar – jubilarian
Jubilarfeier – anniversary, jubilee
Jubiläum – anniversary, jubilee
Jubiläumsgeschenk – jubilee present
Jubiläumsgratifikation – anniversary bonus
Jubiläumszahlung – jubilee payment
Jugendarbeitslosigkeit – youth unemployment
Jugendarbeitsschutzgesetz – Young Persons Employment Act
Jugendausbildungsprogramm – Youth Training Scheme (YTS)
Jugendklub – youth club
jugendliche Aggressivität – juvenile aggressiveness

Jugendliche(r) – young person, youngster, juvenile, adolescent
jugendlicher Straftäter – juvenile delinquent
jugendliches Alter – adolescence
Jugendstrafanstalt – detention centre (BrE), detention home (AmE)
Jugendversammlung – juvenile employees' meeting
Jugendvertreter – youth representative
Jugendvertretung – representation of juvenile employees
Jugendzentrum – youth centre
Junior College – junior college (AmE) (College, an dem man die ersten zwei Jahre eines 4-jährigen Studiums absolviert.)
Justitiar – head of the legal department, corporation lawyer, general counsel

K

Kaffeeküche – coffee corner
Kalendertag – calendar day
kalendertagübergreifend – lasting longer than one calendar day
Kalkulation – calculation
kalkulierte Mindestleistung – calculated minimum (Nur das tun, was unbedingt sein muss.)
kalter Verstand – cold reasoning
Kanal – channel
Kandidat – candidate, examinee, ratee (AmE)
Kantine – canteen, lunchroom (AmE)
Kantinenessen – canteen food
Kanzler (Univ.) – chancellor
Kapazität – expert
Kapazität, maximale – capacity, maximum
Kapazität, minimale – capacity, minimum
Kapazität, optimale – capacity, optimum
Kapazitätskopfzahl (in Vollzeit umgerechnete Kopfzahl) – number of

employees in terms of full-time employment

Kapazitätsplanung – planning of capacity

Kapitalerhöhung – capital increase

Kapitalertragssteuer – capital gains tax

Kapitalgesellschaft – joint-stock company, limited (liability) company

kapitalstark – financially strong

Kapitalzahlung – capital payment

Kapitalzahlung (einmalig) – once-only commutation

Kapitel – unit, chapter

Kappung – capping

Kappungssatz – cut-off rate

Karenzentschädigung – compensation for competitive restriction

Karenztag – unpaid day of sick leave

Karenzurlaub – sick leave grace period

Karriere – career, professional career

Karriere- und Nachfolgeplanung – Career and Succession Planning

Karriereerwartung – career expectations

Karrierekandidat – career candidate

Karriereknick – career interruption

Karriereleiter – career ladder

Karrieremöglichkeiten – career opportunities, career prospects

Karriereplanung – career planning

Karrieresprung – jump forward in a career

Karriereverzicht – sacrifice a career

Karriereziel – career goal

Karte – business card, calling card, visiting card, card

Kasino – company restaurant

Kasse – cash office

Kategorie – category

Kaufkraft – purchasing power

Kaufkraftausgleich – purchasing-power allowance

Kaufkraftparität – purchasing-power parity

kaufmännisch – commercial, business

kaufmännische Ausbildung – commercial training

kaufmännische Grundausbildung – basic commercial training

kaufmännische Vertriebsaufgaben – sales and marketing functions

kaufmännischer Angestellter – office worker

kaufmännischer Beruf – commercial occupation

Kaufmännischer Bildungsausschuss (KBA) (SAG) – committee on commercial training

Kausalität – causality

Kenntnis(se) – knowledge

Kenntnisnahme – notice, cognizance, acknowledgement

kenntnisreich – knowledgeable

Kennzahl – classification figure

Kennzahlen – management ratios

Kernarbeitszeit – core working hours

Kernbelegschaft – core team of employees

Kernkompetenz – core competence

Kernnachtarbeit – core night work

Kernzeit – core working hours, core time

Kettenarbeitsvertrag – chain employment contract

Kfz-Pauschale – tax-deductible amount for car

Kfz-Verrechnung – breakdown of compensation for mileage

Kilometeraufteilung – miles/kilometer distribution

Kilometergeld – mileage money

Kilometergeldregelung – miles/km allowance rule

Kilometerleistung – miles/kms travelled

Kilometerlimit – miles/kilometer limit

Kilometerpauschale – mileage rebate

Kilometersatz – miles/km rate

Kind – child

Kind, das an Liebesentzug leidet – deprived child

Kind im Vorschulalter – pre-school child

Kinder von Einwanderern – immigrant children

Kindererziehung – child education

Kinderfreibetrag – child allowance

Kindergarten – kindergarten, nursery school, play school

Kindergeld – child allowance, child benefit, family allowance, childrens' allowance

Kindergeldberechtigter – person entitled to child allowance

Kindergeldkasse – child benefit fund; family allowance
Kindergeldnummer – child benefit number
Kinderheim – children's home
Kinderhort – full-day nursery, crèche
Kinderkrippe – day nursery, créche
Kinderstube – upbringing
Kindertagesstätte – day nursery, créche
Kinderzulage – child allowance
Kindheit – childhood
kindisch – childish, infantile
Kirchensteuer – church tax
Kirchensteuergebiet – church tax area
Kirchensteuerhebesatz – church tax rate
Kirchensteuerkappung – church tax reduction
Kläger – claimant, plaintiff
klar – clear, evident, plain, obvious
klare Ziele setzen – establish clear objectives
Klassenarbeit – class test, test paper
Klassenbester – top of the class
Klassenbuch – class-book, class diary
Klassengröße – class size
Klassenlehrer – class teacher, form master
Klassensprecher – class prefect, class spokesman
Klassenstärke – class numbers
Klassenzimmer – classroom
Klassifikationskriterium – classification criterion
klassifizieren – classify
klassische Organisationstheorie – classical organisation theory
Klausel – clause
Klausur – exam, paper
Kleingruppenarbeit – work in small teams
Klub – club
Klubhaus – club house
Klugheit – acuity
Klüngel – clique
Klüngelwirtschaft – nepotism
knapp – barely enough, scarce
knappschaftliche Rentenversicherung – miner's pension scheme
knausrig – mean
Know-how – know-how
Koedukation – co-education

kognitiv – cognitive
kognitive Fähigkeit – cognitive ability
kognitive Störung – cognitive disorder
kognitives Lernen – cognitive learning
Kolleg – secondary school for adults (Das Kolleg führt Erwachsene in dreijährigem Unterricht (ganztägig) zur allgemeinen Hochschulreife.) (s. Graphik)
Kollege – colleague, fellow worker, co-worker
kollegiale Leitung – cooperative management
Kollegialität – cooperativeness
kollegiales Führungsverhalten – collegial management behaviour
Kollegstufe – sixth form (s. Graphik)
Kollegstufenschüler – sixth form student (s. Graphik)
Kollektivvertrag – collective agreement, collective contract
Kollisionsprüfung – collision check
Kollusion – collusion
Kombilohnart – multi-purpose wage type
Kombinationsgabe – reasoning power
Komfortzone – comfort zone, comfortable zone, agreeable working group
Kommanditgesellschaft (KG) – limited commercial partnership
Kommanditist – limited partner
Kommenbuchung – clock-in entry
Kommentar – comment
Kommilitone – fellow student
Kommt-Buchung – key in
Kommt-Geht-Zeiten – clock-in/clock-out times
Kommunikation – communication
Kommunikation innerhalb einer Führungsebene – lateral communication
Kommunikation innerhalb einer Gruppe – group communication
Kommunikation zwischen verschiedenen Führungsebenen – diagonal communication
Kommunikationsart – type of communication
Kommunikationselektroniker – communications electronics technician

Kommunikationsfähigkeit – ability to communicate
kommunikative Fähigkeiten – communicative skills
kommunizieren – communicate
Kompaktformular – summarized form
kompatibel – compatible
Kompendium – compendium
kompetent – competent, authoritative
Kompetenz – competence
Kompetenzbereich – competence area
Kompetenzen überschreiten – overstep the boundaries
Kompetenzentwicklung – competency development
Komplementär (KG) – general partner (limited partnership)
Komplott – conspiracy
kompromisslos – uncompromising
Konferenz – conference
konferieren – confer
Konfession – religious denomination
konfessionelle Oberschule – denominational high school
konfessionelle Schule – denominational school
Konflikt – clash, conflict
Konfliktbewältigung – managing conflict
Konflikte bereinigen – clear differences
Konflikte lösen – clear differences
Konfliktfähigkeit – ability to cope with conflict
Kongress – congress, convention
Konjunktur – market prospects, state of the economy
Konkurrent – competitor
Konkurrenzdruck – competition
konkurrenzfähig – competitive
Konkurrenzklausel – competition clause, restraint clause
Konkurrenzverbot – restraint of trade
konkurrieren – compete
Konkurs (Insolvenz) – bankruptcy
Können – skills, competence
Konnotation – connotation
Konrektor – deputy headmaster
Konsens – consensus
konservativ – conservative
Konsequenz – consequence, result
konsolidieren – consolidate

Konsonant – consonant
konstant – constancy
Konstrukteur – design engineer
Konstruktionslehre (Lehrfach) – design
konstruktiv – constructive
konstruktive Kritik – constructive criticism
konstruktive Rückmeldung – constructive feedback
Kontakt – contact, connection
Kontaktaufnahme – initial contact (eine derartige Recherche könne dann auf eine gezielte *Kontaktaufnahme* mit dem zukünftigen Chef hinauslaufen.)
kontaktfreudig – sociable
Kontaktstudium – further education courses for graduates
Kontenklärung – documentation of contributory and non-contributory periods
Kontierung, symbolische – symbolic account assignment
Kontierungszuordnung – account assignment
Kontingentabtragung – quota deduction
Kontingenttyp – quota type
Kontingentverarbeitung – quota processing
Kontingenzplanung – contingency planning
Kontingenztheorie – contingency theory
kontinuierliche Zeitmessung – continuous reading method
kontinuierliche Zeitnahme – continuous reading method
kontinuierlicher Verbesserungsprozess (KVP) – Kaizen
Konto – account
Konto, symbolisches – symbolic account
Kontrolle – check, control; review
Kontrolle (an der Pforte) – gate control, security control
Kontrolle durch Aufsicht – control by inspection
Kontrolle durch Planabweichung – control by exception
Kontrollgruppe – control group
kontrollieren – check, control; review
kontrolliert – controlled
Kontrollliste – checklist

Kontrollspanne – span of control, span of supervision
konventionell – conventional
Konzentration – concentration
konzentrationsarm – poor concentration
Konzentrationsfähigkeit – power of concentration
Konzentrationsspanne – span of attention
Konzept – concept
konzeptionell – conceptional
Konzern – group of affiliated companies (BrE), trust (AmE)
Konzernabschluss – consolidated financial statement
Konzernbetriebsrat – combine works council; group works council
Kooperationsbereitschaft – willingness to cooperate
kooperative Konkurrenz – coopetition (Verschmelzung von Kooperation und Konkurrenz, Kooperation mit dem Wettbewerber.)
kooperativer Führungsstil – people-centred leadership
kooperatives Führungsverhalten – cooperative management behaviour
Koordination – coordination
koordinieren – coordinate
Kopfhörer – earphones
Kopfzahletat – headcount budget (AmE)
Kopie – copy
kopieren – copy
körperbehindert – physically handicapped
körperliche Arbeit – manual labour
körperliche Fähigkeit – physical ability
körperliche und geistige Entwicklung – physical and mental development
körperlicher Schaden – physical defect
Körperschaft des öffentlichen Rechts – body corporate, public body
Körperschutz – protection against body damage
Körpersprache – body language
korrekt – accurate
Korrektur – marking
Korrekturbeleg – correction sheet, correction document
Korrekturlauf – correction run
Korrekturperiode – correction period

Korrespondenz – correspondence
korrigieren – correct
Kosten – costs
Kosten tragen – to bear the cost
Kostenabbau – cost reduction
Kostenbetrag – cost
kostenbewusst – cost-conscious
Kostenbewusstsein – cost consciousness
Kostendenken – cost consciousness
Kostenerstattung – refund of costs, reimbursement of costs
Kostenplan – costs projection
Kostenplanung – cost planning
Kostenrechnung – cost accounting
Kostenstelle – cost center
Kostenträger – sponsor
Kostenübernahme – cost transfer
kraft Gesetzes (von Amts wegen) – ex officio
Kraftfahrzeugmechaniker – motor mechanic
Kramschachtel – lost property box, lost and found
krank – ill, sick
krank sein – be ill
Krankengeld – sickness benefit, sick pay
Krankengeldzuschuss – sick pay supplement
Krankengeldzuschussfrist – sick pay supplement period
Krankenhaus – hospital
Krankenhausbeihilfe – in-hospital benefit
Krankenhaustagegeld – hospital per diem allowance
Krankenkasse – health insurance fund
Krankenschein – sickness certificate
Krankenscheinausdruck – sickness certificate printout
Krankenschwester – nurse
Krankenstand – ill, sick
Krankentagegeld – per diem sickness indemnity
Krankentagegeldversicherung – per-diem allowance insurance
Krankenurlaub – sick leave
Krankenversicherung – health insurance
Krankenversicherung der Behinderten – health insurance for the disabled
Krankenversicherung der Rentner – health insurance for old age pensioners

Krankenversicherung der Studenten – health insurance for students
Krankenversicherungsbeitrag – health insurance premium
Krankenversicherungsgesellschaft – health insurance company
krankfeiern – malinger, swing the lead
krankgeschrieben – certified unfit for work due to illness
Krankheit – illness
krankheitsanfällig – prone to illness
Krankheitsbeschreibung – description of sickness
Krankheitsfall – case of sickness
Krankheitsquote – sickness rate
Krankheitstag – sick day
Krankheitsurlaub – sick leave
Krankheitsverfolgung – sickness tracking
Krankheitszeiten – absences due to illness
Krankmeldung – notification of illness, notification of sickness
krankschreiben – put on the sick list
kreativ – creative
Kreativität – creativity
Kreativitätstechnik – creativity technique
krebserzeugender Arbeitsstoff – carcinogenic agent
Kreide – chalk
Kreislauftrainingskur – blood circulation training
kriecherisch – servile, grovelling, bootlicking
Kriegsbeschädigtenrente – war disablement pension
Kriegsbeschädigung – war disablement
Krise – crisis
Kriterienbank – criteria base
Kriterium – criterion
Kritikfähigkeit – ability to accept criticism
kritisch – critical
Kugelschreiber – ball-point pen, biro
kühn – bold
Kultur – culture
kulturell Benachteiligte – culturally deprived
kulturelle Unterschiede – cultural differences

kulturelle Veränderung – cultural change
kulturelles Erbe – cultural heritage
Kulturvolk – civilized nation
Kultusminister – Minister of Education
Kultusministerium – Department of Education and Science
Kultusministerkonferenz – Permanent Committee of the Ministers of Education
Kumpel – pal, chum, mate, buddy
Kumulation – cumulation
Kumulationskennzeichen – cumulation indicator
Kumulationslohnart – cumulation wage type
Kumulationsregel – cumulation rule
kündbar – subject to notice
Kündbarkeit – terminableness
Kunde – customer
Kundenkartei – customer file
Kundenloyalität – customer loyalty
Kundenorientierung – customer orientation
kundenspezifisch – customer specific
Kundenzufriedenheit – customer satisfaction
kündigen – give notice
Kündigung – termination, notice of dismissal, notice of termination
Kündigung (Arbeitsverhältnis) – dismissal
Kündigung (Vertrag) – give notice
Kündigung durch Arbeitgeber – dismissal
Kündigung durch den Arbeitnehmer – employee voluntary termination notice, resignation
Kündigung durch die Firma – give notice, discharge, dismissal, termination of employment
Kündigung, außerordentliche – extraordinary notice of dismissal
Kündigung, fristlose – instant dismissal
Kündigung, ordentliche – ordinary notice of dismissal
Kündigungsdatum – date of notice of dismissal
Kündigungsfrist – period of notice
Kündigungsgrund – reason for giving notice

Kündigungsklausel – cancellation clause
Kündigungsrecht – right to give notice
Kündigungsrücknahme – withdrawal of termination
Kündigungsschreiben – written notice
Kündigungsschutz – protection against unwarranted notice, protection against unlawful dismissal
Kündigungsschutzklage – action against wrongful dismissal
Kunsterziehung – art education
Kunsthandwerk – handicraft
Kunsthochschule – College of Art
Künstler – artist
Kur – cure
Kuratorium – committee, curatorship
Kurheim – cure home
Kurs – training course
Kursablauf – course schedule
Kursabsage – course cancellation
Kursangebot – range of courses, course offer
Kursauswertung – course evaluation
Kursbelegung – course registration
Kursbesucher – participant
Kursbewertung – course evaluation
Kursbroschüre – course brochure
Kursbuchung – course booking
Kursgebühr – course fee
Kursgruppe – course group
Kursmitteilung – course notification
Kursort – course location
Kursteilnehmer – participant in a course
Kursteilnehmer aus ethnischen Minderheiten – ethnic minority pupils
Kurstyp – course type
Kursverfahren – course procedure
Kursvoraussetzung – course prerequisite
Kursziel – objective of course
Kurszusage – guarantee of a slot in a course
kurz wiederholen – recapitulate
Kurzarbeit – short-time work, reduced working hours
kurzarbeiten – work short-time
Kurzarbeiter(in) – short-time worker
Kurzarbeitergeld – short-time allowance
Kurzarbeitsabrechnung – payroll accounting for reduced working hours
kurzatmig – short-winded, feeble
Kurzfassung – abstract, summary

kurzfristig – short-term
Kurzprüfung – test
Kurzschrift – shorthand, stenography
Kürzung – cut, cutback, reduction
Kürzungsmethode – reduction method
Kürzungsregel – reduction rule
Kürzungsvorschrift – reduction regulation
Kürzungszähler – reduction numerator
Kurzurlaub – short leave
kurzweilig – entertaining
Kurzzeitbeleg – short-term document
Kurzzeiterkrankung – skip work
Kurzzeitgedächtnis – short-term memory
kurzzeitiges Beschäftigungsverhältnis – short-term employment
Kurzzeiturlaub – short-time vacation
Kybernetik – cybernetics

L

Labor(atorium) – lab(oratory)
Laborant – laboratory worker
Lager – storehouse, warehouse
Lagerarbeiter – storeman
Lagerung – storage
Lagerverwalter – store-room clerk
Lagerwesen – stores
Lampenfieber – nervousness, stage-fright
Land – state
Länder-Modifikator – country modifier
länderabhängig – country-dependent
Ländergruppe – country group
Ländergruppenschlüssel – country group key
Ländergruppierung für Personal – personnel country grouping
Länderkennzeichen – country indicator
Länderreport – country report
länderübergreifend – cross-national
länderunabhängig – international
Länderversion – country version

Landesarbeitsamt – regional employment office
Landesarbeitsgericht – regional labour court of appeal
Landessprache – vernacular
Landesversicherungsanstalt – regional social insurance office
Landkarte – map
Langeweile – boredom
langfristige Einkommensentwicklung – long-term compensation development
langfristige Personalplanung – long-term personnel planning
Langsamkeit – tardiness
langweilig – boring
Langzeit-Arbeitslosigkeit – long-term unemployment
Langzeitarbeitslose(r) – long-term unemployed person
Langzeitbeleg – long-term document
Langzeitgedächtnis – long-term memory
Langzeitstudent – long-term student
Langzeiturlaub – sabbatical (Wenn ein Mitarbeiter z.B. auf ein zwölftel des Jahresgehaltes verzichtet aber weiterhin voll arbeitet, hat er Anspruch auf zusätzliche vier Wochen Urlaub. Wird dies über drei Jahre angespart, hat er Anspruch auf einen 3-monatigen bezahlten Urlaub.)
Lärm-Beurteilungspegel – noise rating level
Lärmschutz – noise prevention
Laserstrahlenschutz – laser beam protection
Laufbahn – career, career development
Laufbahnmodell – career model
laufend – current, continuous
laufende Geschäftsangelegenheiten – day-to-day business matters
laufende Leistungen – current benefits
laufende Verwaltungsgeschäfte – current administration
laufendes Arbeitsentgelt – current employee compensation
Laufzeit (z.B. einer Versicherung, Rente) – term
Laune – humour
Laune, bei guter ~ halten – humour
launenhaft – moody
launisch – moody

laut vorlesen – reading aloud
Leben – life
Lebensalter – age
Lebensarbeitszeit – working life
Lebenserfahrung – experience, experience of life
Lebenshaltungsindex – consumer price index
Lebenshaltungskosten – cost of living
Lebenshaltungskosten-Ausgleich – cost-of-living allowance (COLA)
lebenslanges Lernen – life long learning
Lebenslauf – curriculum vitae (cv), personal history, biodata, resume
Lebensstandard – standard of living
Lebensumfeld – personal environment
Lebensunterhalt – livelihood
Lebensversicherung – life insurance
Lebensversicherungsvertrag – life insurance policy
Lebenswandel – line of conduct, way of living
Lebenszeit – life
ledig – single, unmarried
Leerlaufzeit – down time, idle time
Lehramtsbewerber – teacher candidate
Lehranstalt – college
Lehranstalt einer Universität – college
Lehrauftrag – lecturing post
Lehrbefähigung – teacher's diploma, qualification to teach
Lehrberuf – teaching job, teaching profession
Lehrbetrieb – firm where apprentices are trained
Lehrbrief – certificate of apprenticeship
Lehrbuch – textbook
Lehre – apprenticeship, apprenticeship time
lehren – instruct, teach
Lehrer – teacher
Lehrer mit anerkanntem Studienabschluss – qualified teacher
Lehrer, der an mehreren Schulen unterrichtet – peripatetic teacher
Lehrer, der für ein Gruppenhaus zuständig ist – house master (BrE)
Lehrer-Schüler-Verhältnis – staff-student ratio
Lehrerarbeitsmarkt – employment opportunities for teachers

Lehrerhandbuch – teacher's book
Lehrerhandreichungen – teacher's book
Lehrerkollegium – teaching staff
Lehrerverband – teachers' organization
Lehrervertretung – replacement teacher
Lehrerzimmer – common room, staffroom
Lehrgang – course (of instruction)
Lehrgang außerhalb der Dienstzeit (a.D.) – after-hours course
Lehrgang innerhalb der Dienstzeit (i.D.) – in-hours course
Lehrgangsleiter – training supervisor
Lehrgegenstand – subject
Lehrinhalt – course content
Lehrjahr – year of apprenticeship
Lehrjahre – apprenticeship, apprenticeship time
Lehrkraft – teacher
Lehrling – apprentice, trainee
Lehrlingsgehalt – apprentice payment
Lehrlingsvergütung – apprentice payment
Lehrmaterial – teachware
Lehrmedien – teaching media
Lehrmeister – instructor, apprentices' teacher, trainer
Lehrmethode – teaching method
Lehrmittel – teaching aids
Lehrplan – curriculum, syllabus
Lehrplan für die Hauptfächer – core skills curriculum
lehrreich – educational, instructive
Lehrstätte – seat of learning
Lehrstelle – apprenticeship place, trainee place
Lehrstellenbewerber – applicant for an apprenticeship
Lehrstellenmangel – lack of apprenticeships, lack of training positions
Lehrstellenvermittlung – apprenticeship agency
Lehrstoff – course content
Lehrstuhl – chair
Lehrstuhlinhaber – incumbent, chair
Lehrveranstaltung – course
Lehrvertrag – apprenticeship contract, indenture
Lehrwerkstatt – training workshop, vestibule school (AmE)

Lehrzeit – apprenticeship, apprenticeship time
Leibesübungen – physical education
leicht – facile
leichtgläubig – gullible
Leichtigkeit – facility, ease
Leidenschaft – passion
leidenschaftlich – passionately
leidenschaftsarm – dispassionate
Leiharbeit – job leasing, temporary employment
Leiharbeitsfirma – temporary employment agency
Leiharbeitskraft – hired help, temporary help (AmE), worker on temporary loan
Leiharbeitsverhältnis – temporary employment on loan basis
Leihbücherei – lending library
leihen – borrow
Leihgabe – object on loan
leisten – perform
Leistung – performance, achievement, output, merit
Leistung steigern – increase efficiency, increase performance
Leistungen – benefits
Leistungen an Unterhaltsberechtigte – dependents' benefits
Leistungen anrechnen – offset benefits
Leistungen beanspruchen – claim benefits
Leistungen beziehen – draw benefits
Leistungsabfall – decline in performance
leistungsabhängig – performance-related
Leistungsabrechnung – performance expense report
Leistungsanreiz – incentive
Leistungsanspruch – right of benefits
Leistungsausgleich – compensation for rendered services
leistungsbedingt – according to merit
leistungsbedingte Gehaltserhöhung – performance related salary promotion
Leistungsbereich – benefit area
Leistungsbereitschaft – performance readiness
Leistungsbeurteilung – rating of employees, perfomance evaluation
Leistungsbeurteilungssystem – performance appraisal system

Leistungsbewertung – rating of employees
leistungsbezogene Einkommensgestaltung – performance related compensation
leistungsbezogene Gehaltserhöhung – merit increase
leistungsbezogene Lohnerhöhung – merit increase
Leistungsempfänger – recipient of benefits
Leistungsentgelt – efficiency payment
Leistungsentlohnungssystem – incentive system
Leistungsergebnis – performance results
leistungsfähig – efficient, powerful (e.g. computer), competitive
Leistungsfähigkeit – ability, capability, performance ability
Leistungsgarantie – performance guarantee
leistungsgerechtes Einkommen – performance-related income
Leistungsgrad – standard of attainment, standard of performance
Leistungsgrenze – threshold performance
Leistungskategorie – benefit category
Leistungskostenkomponenten – benefit cost criteria
Leistungskriterium – standard of performance, performance criterion
Leistungskurs – advanced level course
Leistungslohn – incentive payment, incentive wage
Leistungslöhner – incentive wage earner
Leistungslohnschein – production time ticket
Leistungsmaßstäbe – performance measures
Leistungsminderung – reduced output
Leistungsmotivation – achievement motivation
Leistungsniveau – standard of attainment, standard of performance
Leistungsnorm – standard of attainment, standard of performance
Leistungsnote – mark
leistungsorientiert – performance linked
leistungsorientierter Mensch – over-achiever

Leistungsparameter – achievement parameter
leistungspflichtig – liable for payment
Leistungsrückstand – performance deficit
Leistungsplan – benefit plan
Leistungsprinzip – performance principle
leistungsschwach – underperforming
Leistungsstandard – standard
Leistungssteigerung – increase in performance, speedup
Leistungssystem – performance system
Leistungstest – performance test
Leistungsträger – performer
Leistungsübertragung – vesting
Leistungsverbesserung – improvement in performance
Leistungsvergleich – performance comparison
Leistungsverlust – decrease in efficiency
Leistungswettbewerb – efficiency contest
Leistungswille – motivation
Leistungsziel – attainment target
Leistungszulage – performance linked bonus, merit bonus
leiten – conduct, lead
leitend – managerial
leitender Angestellter – executive employee
Leiter – head of department
Leiter der Personalabteilung – head of the personnel department
Leiter der Rechtsabteilung – head of the legal department, corporation lawyer, general counsel
Leiter des Personalwesens – head of personnel
Leiter einer Hauptabteilung – Vice, Head of ...
Leiter eines Zentralbereiches – Executive Vice President
Leitfaden – manual, guide, handbook
Leitgedanke – guiding idea
Leitsatz – guiding principle
Leitsätze für Führungskräfte – guidelines for management, guiding principles for managerial staff
Leitungskreis – management level, management group

Leitungsstelle – management job
Leitungsstellenbesetzungsplan – staffing schedule for management
Lernatmosphäre – learning atmosphere
lernbehindert – educationally handicapped
lernbereit – willing to learn
Lernbereitschaft – willingness to learn
Lernbetreuung – coaching, tutoring
lernen – learn
Lernen am Arbeitsplatz – on-the-job training
lernen am Computer – electronic learning
Lernen bei Bedarf – just-in-time learning
Lernen durch entdecken – discovery method
Lernen durch Erfahrung – experimental learning
Lernen durch Hören – aural learning
Lernen durch Tun – learning by doing; action learning
Lernen mit einem interaktiven Lernprogramm – computer-based training, CBT
lernende Organisation – learning organisation
lernendes Unternehmen – learning company (ein Unternehmen, das eine Wissenskultur durch systematisches Wissensmanagement etabliert und diese Kultur permanent optimiert)
Lernerfahrung – learning experience
Lernerfolg – learning achievement
lernfähig – capable of learning
Lernfortschritt – learning progress
Lernfreude – enjoyment of learning, eagerness to learn
Lerngeschwindigkeit – learning pace
Lerngesellschaft – learning society
Lernkurve – learning curve
Lernmaterialien – learning materials
Lernmuster – pattern of learning
Lernmittel – learning aids
Lernmittelfreiheit – free provision of schoolbooks and equipment (Die Lernmittelfreiheit gilt z.B. für die öffentlichen Schulen in Bayern. Danach haben die Schüler einen Rechtsanspruch darauf, alle Schulbücher, die unter die Lernmittelfreiheit fallen, kostenlos von der Schule zu erhalten. Den Trägern privater Schulen ist es freigestellt, die Lernmittelfreiheit einzuführen.)
Lernphase – learning phase
Lernprozess – learning process
Lernressourcen – learning resources
lernschwach – poor learner, weak learner
Lernschwierigkeiten – learning difficulties
Lernstil – style of learning
Lernstrategie – learning strategy
Lernsystem – learning system (The potential impact of the new mobile data services on learning systems will be assessed.)
Lerntechniken – learning techniques
Lerntempo – speed of learning
Lerntheorie – learning theory
Lernzeit – learning time
Lernziel – learning objective
Lesbarkeit – legibility
Lesebuch – reader
Lese- und Schreibfähigkeit – literacy
Lesefähigkeit – reading ability
Lesefertigkeiten – reading skills
Lesegeschwindigkeit – reading speed
Lesegewohnheit – reading habit
Leselernkarte – flash card
Leseliste – reading list
lesen – read
Lesen, Schreiben, Rechnen – three R's (reading, (w)riting, (a)rithmetic)
Leseschwierigkeit – reading difficulty
Lesestoff – reading material
Leseverständnis – reading understanding
Lesewortschatz – reading vocabulary
letztes Schuljahr – final school year
letztes Studienjahr – senior year
Leumund – reputation
Lexik – lexis
liebenswürdig – charming, kind
lineare Programmierung – linear programming
Linien-Führungskräfte – line management
Linienmanager – line manager
Linienmann – line man
Linkshänder – left-handed person
Liquidation – liquidation

liquide Mittel – available funds, liquid funds
lispeln – lisp
Literat – man of letters
Literatur – literature
Lizenzgebühren – royalties
loben – praise
Lockerungsübung – relaxation exercise
Lockmittel – lure
logische Schlussfolgerung – logical inference
logisches Denken – logical thinking
Logopädie – speech therapy
Lohn – wage, compensation, pay
Lohn beziehen – draw wages
Lohn- und Gehaltsabrechnung – payroll accounting
Lohn- und Gehaltsabrechnung (fremde) – third-party payroll accounting
Lohn- und Gehaltsentwicklung – wages and salary history
Lohn- und Gehaltserhebung – survey of earnings
Lohn- und Gehaltsliste – payroll, payroll sheet
Lohn- und Gehaltsvergleich – pay survey
Lohn- und Salärstatistik – remuneration statistics
Lohn-Preis-Spirale – wage-price spiral
Lohn-Preis-Verhältnis – wage-price relationship
Lohnabbau – cutting of wages
lohnabhängig – wage-dependant
Lohnabkommen – wage agreement
Lohnabrechnung – payroll accounting for the hourly paid; wages slip
Lohnabrechnungszeitraum – payroll period
Lohnabstufung – wage scale
Lohnabtretung – wage assignment
Lohnabzüge – payroll deductions
Lohnabzugsverfahren – checkoff system (AmE)
Lohnangleichung – wage adjustment
Lohnanteil – share in wages, portion of wages
Lohnarbeit – paid labour, wage labor
Lohnart – wage type
Lohnart, abgeleitete – derived wage type

Lohnart, summierte – cumulated wage type
Lohnart, technische – technical wage type
Lohnarten-Jahresentwicklung – annual wage type development
Lohnartenbewertung – wage type valuation
Lohnartengenerierung – wage type generation
Lohnartenmuster – wage type model
Lohnartennachweis – wage type statement
Lohnartenschlüssel – wage type key
Lohnartenschlüsselung – wage type coding
Lohnartenstruktur – wage type structure
Lohnartentabelle – wage type table
Lohnartentext – wage type text
Lohnartentyp – wage type category
Lohnauftrieb – wage drift
Lohnausfall – loss of wages
Lohnausgleich – wage adjustment, compensation for wage deficiencies
Lohnausweis – wage statement
Lohnauszahlung – payment of wages
Lohnberechnung – calculation of wages
Lohnbescheinigung – pay statement
Lohnbestandteil – wage element
Lohnbewegungen – wage fluctuation
Lohnbezüge – wage payments
Lohnbüro – payroll department, payroll office
Lohndrift – wage drift (Lohnbewegung auf Grund Arbeitsmarktschwierigkeiten und Anpassung an veränderte wirtschaftliche Bedingungen.)
Lohndruck – wage pressure
Lohnempfänger – wage-earner, non-salaried employee
Lohnentwicklung – wage trend
Lohnerhöhung – wage increase
Lohnersatzleistung – earnings replacement benefit
Lohnfestsetzung – wage finding
Lohnfindung – wage calculation
Lohnforderungen – wage demands, wage requirements
Lohnform – wages system
Lohnfortzahlung – continued pay

Lohnfortzahlungsanspruch – entitlement to continued pay
Lohnfortzahlungsfrist – period of continued pay
Lohngefälle – wage differential
Lohngefüge – wage structure
Lohngestaltung – wages policy
Lohngleichheit (für Mann und Frau) – equality of wages (for men and women)
Lohngleitklausel – escalator clause
Lohngruppe – wage group
Lohngruppe, tarifliche – standard wage group
Lohnindex – wage index
lohnintensive Industrie – industry with high labour costs
Lohnjournal – payroll journal
Lohnkampf – free collective bargaining
Lohnkarte – time card
Lohnkonflikt – wage dispute
Lohnkonto – payroll, payroll account, payroll sheet
Lohnkontoformular – payroll account form
Lohnkosten – wage costs
Lohnkurve – wage curve, wage graph
Lohnkürzung – wage reduction, wage cut
Lohnnachgenuss – payments from retroactive accounting
Lohnnebenkosten – non-wage labour costs, payroll fringe costs, additional wage costs
Lohnniveau – wage level
Lohnpause – pay pause
Lohnpfändung – attachment of wages, garnishment of wages
Lohnpfändungsbeschluss – garnishment ruling (USA); court order (GB)
Lohnpolitik – wage policy
Lohnrahmentarif – overall wage scale
Lohnsatz – wage rate, pay rate
Lohnschein – time ticket
Lohnschere – earnings gap
Lohnschwankungen – wage fluctuation
Lohnsenkung – wage reduction
Lohnskala – wage scale
Lohnspanne – wage spread
Lohnsteigerung – wage increase
Lohnsteuer – income tax

Lohnsteuerabzug – income tax deduction
Lohnsteueranmeldung – income tax notification
Lohnsteueraufkleber – label for income tax card
Lohnsteuerausweis – income tax form
Lohnsteuerbescheinigung – income tax certificate, withholding statement
lohnsteuerfrei – tax-free income
Lohnsteuerfreibetrag – employee's withholding exemption, income tax allowance
Lohnsteuerjahresausgleich – annual income tax return
Lohnsteuerkarte – wage tax card
Lohnsteuerklasse – income tax class
Lohnsteuerstatistik – income tax statistics
Lohnsteuervoranmeldung – wage tax notification
Lohnstopp – wage freeze
Lohnstreifen – wage slip
Lohnstruktur – wage structure
Lohnstückkosten – unit labour costs, unit wage costs
Lohnstufe – wage scale
Lohnstufe, tarifliche – standard wage level
Lohnstufenstatistik – wage scale statistics
Lohnstunden – wage hours
Lohntabelle – wage table
Lohntarif – wage rate
Lohntüte – wage packet
Lohnübersicht – payroll summary (AmE)
Löhnung – payment of wages
Lohnunterlage(n) – wage-related document, payroll records (AmE)
Lohnvergleich – wage comparison
Lohnverhandlungen – wage negotiations
Lohnverwaltung – payroll department
Lohnverzicht – renunciation of pay
Lohnvorschuss – wage advance
lohnwirksam – pay effective
Lohnzahlung – payment of wages
Lohnzettel – pay slip
Lohnzettelgrenze – pay slip limit
Lohnzulage – extra pay

Lohnzuschlag – extra pay (Teil des vertraglich vereinbarten Arbeitsentgelts für schwere oder schmutzige Arbeiten oder aus Rücksicht auf die sozialen Verhältnisse des Arbeitnehmers)
Lösung – solution
Luftfracht – air freight
lustig – amusing
lustiger Versprecher – Spoonerism
lustlos – lackadaisical
Lyzeum – girls' grammar school, girls' high school

M

Macho – macho
Macht – authority, power
Macht durch Anerkennung – referent power
Macht durch Beziehungen – connection power
Macht durch Einschüchterung – coercive power
Macht durch funktionelle Weisungsbefugnis – legitimate power
Macht durch Informationsvorsprung – information power
Macht durch Sachkompetenz – expert power
Machtanspruch – claim to power
Machtausübung – exercise of power
Machtbefugnis – power, authority
Machtverlust – loss of power
Mädchen – girl
Mädchengymnasium – girls' secondary school
Magister – master's degree
Magister Artium – Master of Arts (M.A.)
Magister Scientiarum – Master of Science (M.Sc.)
magna cum laude – magna cum laude
Magnettafel – magnetic board
Mahndatum – reminder date
mahnen – warn, admonish

Mail-Kennzeichen – message indicator
Management menschlicher Ressourcen – workforce management
Management-Buyout – management buyout (Form der Ausgliederung, bei der das Eigentum an der neu gegründeten Gesellschaft auf das bisherige Management oder die Mitarbeiter übergeht.)
Managementfunktion – management function
Managementgrundsatz – principle of management
Managementinformationssystem – management information system
Managementkonzept – management concept
Managementnachwuchs – young managers, young executives
Managementprinzip – principles of management
Managementprogramm (Schulung und Förderung) – management programme
Managementprozess – management process
Managementsystem – management system
Managementtechnik – management technique
Managementverhalten – management style
Mangel – deficiency, lack, shortage
Mängelanalyse – deficiency analysis
Mangelberuf – understaffed profession
mangelhaft – inadequate
mangelnde Anpassung – maladjustment, inability to adjust adequately
mangelnde gesundheitliche Eignung – inadequate health
Manipulation – manipulation
Manuskript – manuscript
Manteltarifvertrag – collective agreement, umbrella agreement
Marketing – marketing
Markt für Teilzeitkräfte – part-time labour market
Markt-Nachfrage – market demand
Marktaufgabe – strategic mission
Marktentwicklung – market development
Marktforschung – market research

Marktkenntnisse – insider knowledge, knowledge of the market
Marktprognose – market forecast
Maschinenbau – mechanical engineering
Maschinenbauingenieur – mechanical engineer
Maschinenschlosser – machine fitter
Massenarbeitslosigkeit – mass unemployment
Massenentlassung – mass redundancy
maßgebend – decisive
maßgeschneidert – made-to-measure
mäßig – moderate
mäßigen – moderate
Maßnahme – measure, action, step
Maßnahmen zur Fort- und Weiterbildung – training and development measures
Maßregelung – disciplinary punishment, crackdown (AmE)
Materialwirtschaft – materials management
Mathematik – mathematics
Mathematiker – mathematician
Mathetik – mathetics (Lehr- und Lernmethode nach T.F.Gilbert. Man geht vom Ganzen aus, um zu den einzelnen Lernschritten zu kommen.)
Matrix-Organisation – matrix organization
Matrix-Struktur – matrix structure
Mcjob – unqualifed work
mechanistische Vorgehensweise – mechanistic approach
Medien – media
Mediendidaktik – media didactics
Medieneinsatz – use of media
Medienerziehung – media education
Medienmarketing – marketing via the media
Medienpädagogik – media pedagogy
Medium – medium
Mehr... – excess
Mehrarbeit – overtime
Mehrarbeitsausgleich – overtime compensated by time in lieu of payment
Mehrarbeitsberechnung – calculation of overtime
Mehrarbeitsbetrachtung – overtime perspective

Mehrarbeitsgenehmigung – overtime approval
Mehrarbeitspause – overtime break
Mehrarbeitsstunden – hours of overtime
Mehrarbeitsverrechnung – overtime compensation
Mehrarbeitszuschlag – overtime bonus
Mehraufwand – additional expenses
Mehrbelastung – extra load, additional burden
Mehrfachabrechnung – multiple payroll
Mehrfachauswahl – multiple choice (m.c.)
Mehrfachauswahlfrage – multiple choice question
Mehrfachauswahlübung – multiple choice exercise
mehrfachbehindertes Kind – multiply handicapped child
Mehrfachbeschäftigung – multiple employment
Mehrfachbewerber – multiple applicant
Mehrfachbewerbung – multiple application
Mehrfachbezug – multiple benefits
Mehrfacherkrankung – repeated sickness
Mehrfachqualifikation – multiple qualification
Mehrheitswahl – majority vote
Mehrkosten – excess costs
Mehrlingsgeburt – multiple birth
mehrsprachig – multilingual, polyglot
Mehrzeitausgleich – overtime compensation
Meinungsbefragung – opinion poll
Meister – master (of a trade), master craftsman
Meister-Dienstvertrag – employment contract for foremen
Meisterbrief – master craftsman's certificate
Meisterfachschule – master craftsmen's college
Meisterschule – training school for master craftsmen
Melancholiker(in) – melancholic
Meldepflicht – obligation to register
Meldeverfahren – reporting procedure

Meldeverfahren in der Sozialversicherung – reporting procedure for social insurance
Mengenleistung – quantitative output
Mensa – refectory, dining-centre, canteen
Menschenkenntnis – knowledge of human nature
Mentor – mentor
Merkblatt – leaflet, prospectus
Merkmal – feature, attribute, characteristic
Merkmalsgruppe – category, ability category (AmE)
Messe – trade fair
Messung – measurement
Metallurgie – metallurgy
Methode – method, procedure, technique
Methode der rangmäßigen Einordnung – ranking method
Methode der selbständigen Einarbeitung – "pick-up" method (ohne Anleitung durch einen Lehrmeister; der Arbeiter muss sich die nötigen Kenntisse durch Beobachtung erfahrener Kollegen aneignen)
Methode des kritischen Pfads – critical path method
Methoden und Zeitstudien – methods time measurement (MTM) (ein Verfahren, das, aus den Erkenntnissen von gefilmten Bewegungsstudien entwickelt, sich zur Analyse von Arbeiten und zur Bestimmung von Arbeitszeiten auf 19 Grundbewegungen stützt)
Methodik – methodology (Wissenschaft vom planmäßigen Vorgehen z.B. beim Unterrichten)
methodisch – methodical
methodische Arbeitsweise – methodical procedure, methodical manner
methodisches Verfahren – methodical procedure, methodical manner
methodisches Vorgehen – methodical procedure, methodical manner
Microteaching – microteaching (spezifische Verhaltensfertigkeiten des Unterrichtens werden isoliert und einzeln geübt)
Miete – rent

Mietzuschuss – housing allowance, rent allowance
Mikrosystemelektroniker – microsystems electronics technician
Milde – lenience
mildern – moderate
Minderheit – minority
Minderheitengruppe – minority group
Minderjährige(r) – minor
Minderung der Erwerbsfähigkeit – diminution of the earning capacity
Minderwertigkeit – inferiority
Minderwertigkeitskomplex – inferiority complex
Minderzeit – shortfall
Mindestalter – minimum age
Mindestanwesenheitszeit – minimum attendance time
Mindestarbeitsstunden – minimum working hours
Mindestbeitrag – minimum contribution
Mindestlohn – minimum wage
Mindestlohn für Aufforderung und Erscheinen am Arbeitsplatz (selbst wenn nicht gearbeitet wird) – call-in pay
Mindestnote – minimum mark
Mindeststunden – minimum hours
Mindestzeit – minimum time
Ministammsatz – mini-master record
Ministerialbeauftragter – ministry representative
Minorität – minority
MIS – management information system
Missachtung – disregard; flouting; disrespect
Missbildung – deformity
Mission – strategic mission
Misslingen – failure
Misstrauen – distrust
mit großem Lob – magna cum laude
mit höchstem Lob – summa cum laude
mit Lob – cum laude
mit Verwaltungsaufgaben befasste Führungskraft – administrator
Mitarbeit – cooperation, assistance, collaboration; coursework (Ähnlich wie bei der Kollegstufe wird die aktive Beteiligung im Unterricht während der zwei Vorbereitungsjahre bei der

Benotung in der Abschlussprüfung in
Betracht gezogen.)
mitarbeiten – collaborate, work together,
cooperate
Mitarbeiter – colleague, fellow worker,
co-worker, employee
Mitarbeiter anleiten – instruct
subordinates
**Mitarbeiter mit einem
Sonderleistungsvertrag** – non-salaried
employee with a merit-based contract
**Mitarbeiter mit einem
Sonderlohnvertrag** – non-salaried
employee with a special wage contract
Mitarbeiter-Stammdatei – employee
master file
Mitarbeiterabwerbung – enticement,
labour piracy, poaching, inducement
Mitarbeiterbefragung – employee
attitude survey
Mitarbeiterbeteiligung – staff
participation
Mitarbeiterbeurteilung – assessment,
appraisal, perfomance appraisal
Mitarbeiterbindung – staff bond
Mitarbeitereinsatz – deployment of
employees
Mitarbeitergespräch – staff dialogue
Mitarbeitergruppe – employee group
Mitarbeiterkontokorrentkonto –
employee's customer/vendor account
Mitarbeiterkreis – staff
mitarbeiterorientierte Führung –
people-centred leadership
Mitarbeitersollarbeitszeit – employee's
planned working time
Mitarbeiterstab – staff, employees,
personnel, human resources
Mitarbeiterverzeichnis – list of
employees
Mitarbeiterzahl – employee count
Mitbestimmung – co-determination
Mitbestimmungsgesetz – Law on
Co-determination (regelt Fragen
bezüglich der Besetzung und
Verantwortung im Aufsichtsrat)
Mitbestimmungsrecht –
co-determination right
Mitbewerber – competitor, rival
Mitfahrer – passenger
Mitglied – member

Mitglied des Aufsichtsrates – Member
of the Supervisory Board
Mitglied des Bereichsvorstands – Vice
President, Member of the Group
Executive Management
Mitglied des Vorstands – Member of the
Managing Board
**Mitglied einer wissenschaftlichen
Vereinigung** – fellow
Mitglied eines Colleges (BrE) – fellow
Mitgliederabwerbung – raiding
Mitgliedschaft – membership
Mitgliedsnummer – membership
number
Mitgliedsverzeichnis – list of members
Mitschüler – classmate
Mittagspause – midday break, lunch
break
Mitteilung – notification
Mitteilungskürzel – notification code
mittelfristig – medium-term
Mittelmaß – mediocre
Mittelschule – grammar school
mittelständisch – middle-class
Mittelstufe – junior high school (AmE),
junior cycle, intermediate school
(s. Graphik)
mitternachtsübergreifende Arbeit –
work that spans two days
mittlere Reife – O-Level, ordinary level
mittlerer Dienst – clerical grade
mittlerer Führungskreis – middle
management
Mittlerer Schulabschluss – General
Certificate of Secondary Education
(GCSE) (s. Graphik)
mittleres Management – middle line
Mitwirkung – participation
**Mitwirkung der Arbeitnehmer in der
Geschäftsführung** – worker
participation
Mnemotechnik – mnemonics
Möbel – furniture
Mobilität – mobility
Mobilität der Arbeitskräfte – labour
mobility
Modellversuch – model
Moderation – moderation
moderieren – moderate
modernisieren – modernize, rationalize,
streamline

Modifikator – modifier
modulares Ausbildungsprogramm – modular training program
mogeln – cheat
Möglichkeit – opportunity, possibility
monatliche Rate – monthly installment
Monatsarbeitszeit – monthly working hours
Monatsarbeitszeitplan – monthly work schedule
Monatsgehalt – monthly salary
Monatslohn – monthly wages
Monatsmeldung – monthly report
Monatssaldo – monthly balance
Monatsultimo – last day of month
Monatsverdienst – monthly earnings
monoton – monotonous
Montan-Mitbestimmungsgesetz – Co-determination in the Mining Industry and in the Iron and Steel Production Industry
Mores – manners
Morgenandacht – morning assembly
Motivation – motivation
Motivationstechnik – motivation technique
Motivationstheorie – motivation theory
motivieren – motivate
Multimomentaufnahme – period time sheet
multiple Regression – multiple regression
multiplizieren – multiply
mündig – of age
mündliche Prüfung – oral examination
mündlicher Bericht – oral report
musikalische Intelligenz – musical intelligence
Muss-Buchung – essential course booking
Muster – sample; pattern
Musterabkommen – pattern-setting model
Musterbeispiel – paradigm
Musterkatalog – model catalog
Musterlohnart – model wage type
Mustermappe – standard file, sample file
Musterung – physical examination
Mustervereinbarung – pattern-setting model

mutlos – discouraged, disheartened, despondent
Mutter, werdende – expectant mother
Mutterschaft – maternity
Mutterschaftsfall – maternity case
Mutterschaftsgeld – maternity benefit, maternity pay
Mutterschaftsurlaub – maternity leave
Mutterschutz – maternity protection
Mutterschutzfrist – maternity protection period
Muttersprache – mother tongue
muttersprachlicher Unterricht für ausländische Kinder – mother-tongue classes for foreign children (Ausländische Kinder, die dem Unterricht in deutscher Sprache folgen können, werden in eine deutsche Regelklasse aufgenommen, sofern die Eltern nicht eine zweisprachige Klasse bevorzugen. In Regelklassen wird der Unterricht nur in deutscher Sprache erteilt. Ausländische Kinder, die dem Unterricht in deutscher Sprache nicht folgen können oder deren Eltern verstärkten Unterricht in der Muttersprache wünschen, werden in zweisprachige Klassen aufgenommen. Sie wurden bisher nur für die Sprachen Griechisch, Italienisch, Portugiesisch, Serbokroatisch, Spanisch und Türkisch eingerichtet.)

N

N.N.-Buchung – no-name course booking
nach allen Abzügen – net
nach Büroschluss – after-hours
nachahmend – eclectic
Nachahmung – imitation
Nacharbeit – make-up-work
nacharbeiten – redo
Nachbarschaft – neighbourhood, vicinity
Nachbearbeitung – error handling

nachbereiten (eines Kurses) – perform
  follow-up course work
Nachbesserung – subsequent
  improvement
nachdenklich – thoughtful
nacheifern – emulate
Nachfolgekandidat – succession
  candidate
Nachfolgeplanung – succession planning
Nachfolger – successor
Nachhilfeunterricht – private tuition
**Nachhilfeunterricht während der
  Ferien** – private lessons during the
  holidays, grinds
Nachhilfeunterrichtsgruppe – private
  tuition group
Nachholbedarf – backlog demand
nachlassende Aufmerksamkeit –
  flagging attention
nachlässig – lackadaisical
Nachprüfung – verification
Nachrückkandidat – move-up candidate
**nachschulische Angebote in der
  Erwachsenenbildung** – continuing
  education, ongoing education, further
  education (umfasst alle
  nachschulischen Angebote)
Nachsicht – lenience
nächstgelegene Schule – nearest school
nächsthöherer Vorgesetzter – next
  higher ranking manager
Nachtarbeit – night-work
Nachtarbeitszuschlag – night-work
  bonus
Nachteil – disadvantage
nachteilig – adverse
Nachteilsausgleich – deficiency
  compensation, shift change
  compensation
Nachtschicht – night shift, late shift
Nachtschichtgesetz – night shift
  regulations
Nachtschichtschwerarbeit – night shift
  heavy work
Nachtschichtzulage – night shift bonus
Nachuntersuchung – follow-up
  examination
Nachvergütung – further compensation
Nachversicherung – payment of
  retrospective contributions
Nachweis – evidence, proof

Nachweisliste – documentary list
Nachwuchsförderung – promotion of
  young people
Nachwuchsplanung – succession
  planning
Nachwuchspotential – pool of junior
  executives, capacity of management
  trainees
Nachwuchssorgen – recruitment
  problems
Nachzahlung – subsequent payment,
  supplementary compensation,
  back-payment
Nähe – neighbourhood, vicinity
näher angeben – specify
Namensliste – roll, attendance list, list of
  names
Namensschild – name tag
Namenszusatz – name affix
narzistischer Mensch – narcissistic
  person
Nationalität – nationality, citizenship
Nationalitätskennzeichen – country ID
Naturalleistung(en) – payment in kind
Naturalzulage – non-monetary bonus
natürliche Begabung – natural
  endowment
natürliche Führungspersönlichkeit –
  natural leader
natürliche Neugier – natural curiosity
natürlicher Abgang von Arbeitskräften
  – natural wastage
Naturwissenschaften – natural sciences,
  physical science
Naturwissenschaftler – natural scientist
Nebenabrede – subsidiary agreement
Nebenberuf – sideline, minor
  occupation, second job
nebenberufliche Tätigkeit – sideline,
  minor occupation, second job
Nebenbeschäftigung – second job
Nebenbetrieb – auxiliary plant
Nebeneinkommen – supplementary
  income
Nebeneinnahmen – supplementary
  income
Nebenfach – additional subject,
  subsidiary subject minor (subject)
  (AmE)
Nebenkosten – incidental costs

Nebenlaufbahn – secondary career model
Nebenleistungen – goodies
Nebenprodukt – spin-off
Nebensächlichkeit – irrelevant detail
Nebentätigkeit – second job, side job
Nebenverdienst – additional income
necken – tease
Negativerfassung – negative time recording
Neigung – inclination
Nenner – denominator
Nervenzusammenbruch – nervous breakdown
nervliche Belastung – stress
nervös – nervous
nervöse Störung – nervous disorder
netto – net
Netto, gesetzliches – statutory net amount
Nettoabrechnung – net report payroll
Nettobetrag – net amount
Nettobezüge – net salary, net earnings
Nettoeinkommen – net income
Nettoentgelt – net income, net remuneration
Nettoermittlung – net calculation
Nettolohn – net wage, take-home pay
Nettolohnabrechnung – net payroll
Nettoprämie – net premium
Nettorückrechnung – net retroactive accounting
Nettoteil – net part
Nettozusage – guaranteed net amount
Nettozusammenfassung – cumulation of net amount
Netzplantechnik – network planning technique
neu ordnen – readjust
Neubewertung – reclassification
Neubildung – new formation
neue Aufgaben übernehmen – assume new tasks, take over new tasks
Neueinschreibung – reenrolment
Neueinstellung – new employee
Neueinstufung – reclassification
neuere Universität – red-brick university (eine später als Oxford/Cambridge gegründete englische Universität)
neues Mitglied – entrant, new member
Neufassung – revised edition

Neugier – curiosity
Neuling – newcomer
Neuordnung – reorganization
Neuphilologie – modern languages
neurotisch – neurotic
Neusprachliches Gymnasium (NG) – grammar school emphasizing modern languages
nicht anerkennen – repudiate
nicht benotet – no marks given, pass/no pass
nicht eigenständig – eclectic
nicht engagiert – uncommitted
nicht fortgeschriebenes Budget – zero-base budget
nicht gewerkschaftspflichtiger Betrieb – open shop (Betrieb, der sowohl Gewerkschaftsmitglieder als auch Nicht-Mitglieder beschäftigt; Betrieb, der nur Nicht-Gewerkschaftsmitglieder beschäftigt)
nicht verfallbar – non-forfeitable
Nicht-Arbeit – non-working period
Nicht-Feiertag – normal working day
Nicht-Warteliste-Buchung – non-waiting list booking
Nicht-Warteliste-Priorität – non-waiting list priority
Nichtakademiker – non-academic
Nichtanerkennung – repudiation
nichtselbständige Arbeit – employment (at a company)
nichtige Vereinbarung – void contract
Nichtvorliegen – non-existence
Niedriglohn-Beschäftigung – low-paid employment
Niveau – standard
nominieren – nominate
Nominallohn – nominal wages
Nominalwert – nominal value
Nominierung – nomination
non-verbale Antwort – non-verbal response
non-verbale Intelligenz – non-verbal intelligence
non-verbale Kommunikation – non-verbal communication
non-verbales Lernen – non-verbal learning
Nonsenswort – nonsense word
Nörgelei – nagging, grousing, grumbling

Nörgler – querulous person, troublemaker, crab, grumbler, fault-finder
Norm – standard, norm
normal – regular
Normalarbeit – normal work
Normalarbeitszeit – normal working time
normierter Test – standardized test
Normierung – standardization
Notadresse – address to contact in case of emergency
Notausgang – emergency exit
Notbeleuchtung – emergency lighting
Notbudget – emergency budget
Note – mark, grade
Notendurchschnitt – average marks (BrE), grade point averages (AmE)
Notiz – memorandum, memo, note
Notlösung – temporary solution
Notstandsbeihilfe – emergency relief
notwendig – necessary
Numerus Clausus – restricted entry, student limitation
Nutzen – benefit
Nutznießer – beneficiary

O

obere Betriebsführung – senior management
obere Führungsebene – senior management
obere Führungsschicht – senior management
oberer Führungskreis – senior management
oberflächlich – cursory, superficial
Obergrenze – upper limit
Oberingenieur (SAG) – Senior Engineer
oberster Führungskreis – corporate management
Oberstudiendirektor(in) – headmaster (BrE), principal (AmE) (Der Oberstudiendirektor leitet ein

Gymnasium und ist für die pädagogische und verwaltungs-technische Abwicklung zuständig.)
Oberstudienrat(rätin) – senior teacher
Oberstufe – sixth form college, senior high school (AmE), (s. Graphik)
Oberstufenschüler – pupil at a senior high school, pupil in senior cycle (s. Graphik)
Objektbezeichnung – object name
objektiv – business-like, objective
Objektivität – objectivity
Objektkürzel – object abbreviation
objektorientiert – object-oriented
Objekttyp – object type
obligatorisch – obligatory, compulsory
obsolet – obsolete
Ochsentour – through the ranks
offenbar – apparent, evident
offene Stelle – vacancy
offene Universität – open university
offene Veranstaltung – open access
offener Betrag – outstanding balance
Offenheit – openness
öffentliche Beziehungen – public relations, PR (Maßnahmen zur Herstellung eines positiven Verhältnisses des Unternehmens zur Öffentlichkeit, zur Werbung um öffentliches Vertrauen.)
öffentliche Mittel – public funds
öffentliche Schule – state school, public school (AmE) (s. Graphik)
öffentliche Versorgungsleistung – public service
öffentliche Verwaltung – public administration
öffentlicher Bildungsbereich – public education sector
öffentlicher Dienst – public service
Öffentlichkeitsarbeit – public relations
Öffnungszeit – opening time
Ohr – ear
Operation – operation
operational – operational (Eine Definition ist dann operational, wenn sie grundsätzlich für jeden beliebigen Beobachter und zu jeder beliebigen Zeit überprüfbar ist.)
operationales Lernziel – operational learning objective

operative Planung – operational planning
operatives Ziel – specific objective
optimale Stellenbesetzung – optimal filling of vacancies
Optionsrecht – option right, stock option
ordentliche Hauptversammlung – ordinary shareholders' meeting
ordentliche Kündigung – dismissmal with notice, termination with notice
Ordentliches Vorstandsmitglied – Executive Vice President
Ordinarius – professor (Ordentlicher Professor an einer Hochschule)
Ordner – file
Ordnung – order
Ordnungsaufgaben – administrative services
ordnungsgemäß – regular
ordnungsgemäße Kündigung – due notice
Ordnungswidrigkeit – non-criminal offence, irregularity
Ordnungszahl – reference number
Organigramm – organigram
Organisation – organization
Organisation und Planung – Organization and Planning
Organisationsdesign – organization design
Organisationsebene – organizational level
Organisationseinheit – organizational unit
Organisationsentwicklung (OE) – organization development (OD)
Organisationsentwicklungsplanung – organizational development planning
Organisationsfähigkeit – organizational ability
Organisationsform – organizational form
Organisationsgabe – organizational ability
Organisationsgestaltung – organization design
Organisationsgrundstruktur – simple organization structure
Organisationsmanagement – organizational management

Organisationsplan – organizational plan, organization chart
Organisationsplanung – organizational planning
Organisationsschema – organization chart
Organisationsschlüssel – organizational key
Organisationsstruktur – organizational structure
Organisationstalent – organizational talent
Organisationstechnik – organization methods
Organisationszuordnung – organizational assignment
organisatorisch selbständige Einheit – internal profit center, business unit (Unternehmenseinheiten, die im Unternehmen bleiben, aber eigenständig wirtschaften, eine eigene Rechnungslegung haben und für ihr Ergebnis selbst verantwortlich sind.)
organisatorische Bedürfnisse – organizational needs
Orientierung – orientation
Orientierung und Integration neuer Mitarbeiter – orientation and integration of new employees
Orientierungshilfe – orientation aid
Orientierungsprogramm – orientation program
Orientierungsstufe – orientation level
Original – original
Originalabrechnung – original payroll run
Ort – location
Ortswechsel – change of locality
Ortszuschlag – cost-of-living allowance
Outplacement – outplacement
Oxbridge – Oxbridge (Bezeichnung für die beiden Universitäten Oxford und Cambridge)

P

Paarbildung – pair formation
Paartyp – pair type
Pädagoge – education(al)ist, educator, pedagogue
Pädagogik – pedagogy
Pädagogin – education(al)ist, educator
pädagogische Sonderbetreuung (behinderter Kinder) – education of retarded / handicapped children
pädagogische Ausbildung – teacher training
Pädagogische Hochschule (PH) – Teacher Training College (T.T.C.)
Panne – hitch, slip
Papiertafel – paper board
Paradigma – paradigm
Parität – parity
paritätisch – equal
paritätisch besetzt – in equal number
Partei ergreifen – take sides
Parteienverkehr von ... bis – office hours from...to...
partielles Analphabetentum – functional illiteracy
Partizipation – participation
Partner – partner, business partner
Partnerschaft – partnership
passend – convenient, suitable
passiv – passive
passiver Wortschatz – passive vocabulary
passives Lernen – passive learning
Passivrauchen – passive smoking
patriarchalisches Führungsverhalten – patriarchal management behaviour
pauken – cram, swot; drill
Pauker – private teacher
Paukkurs – cram course, grinds
Pauschalabrechnung – lump-sum accounting
Pauschalbesteuerung – lump-sum taxation
Pauschalbewertung – group valuation
Pauschale – lump sum; flat rate; per diem (travel expenses)

pauschale Abfindung – dismissal compensation, redundancy payment
pauschale Abgeltung – dismissal compensation, redundancy payment
Pauschalenregelung – lump-sum rule
Pauschalentschädigung – lump-sum compensation
pauschalierte Beitragszahlung – lump-sum contribution payment
Pauschalierung – consolidation into a lump sum
Pauschalierung – commutation, consolidation into a lump sum
Pauschalsatz – lump-sum rate
Pauschalsatz, ermäßigter – lump-sum rate, reduced
Pauschalsatz, maßgebender – lump-sum rate, relevant
Pauschalsteuer – lump-sum tax
Pauschalvergütung – fixed allowance
Pauschalversteuerung – lump-sum payment of tax
Pauschalzahlung – lump sum
Pauschbetrag – blanket allowance
Pauschsatz – lump sum charge
Pause – work break, break, pause
Pause, bezahlte – break, paid
Pause, unbezahlte – break, unpaid
Pausenanfang – beginning of break
Pausenende – end of break
Pausenhof – playground, schoolyard
Pausenmilch – school milk
Pausenmodell – break model
Pausenraum – break-room, rest room
Pausentyp – break type
Pedant(in) – pedant
Pedanterie – pedantry
pedantisch – pedantic
Peer-Gruppe – peer group
peinlich – embarrassing
pendeln – commute
Pendelzeit – commution time
Pendler – commuter
Pendlerpauschale – commuter rate
Pendlerzuschuss – commuter allowance
Pension – pension, old-age pension, retirement benefit
Pensionär – pensioner, old-age pensioner, retired employee
pensioniert werden – be retired
Pensionierung – retirement

Pensionsalter – pensionable age, retirement age
Pensionsanspruch – pension entitlement
Pensionsanwartschaft – pension expectancy, accrued pension rights
Pensionsband – pension scale
Pensionsbeitrag – contribution to the pension scheme
pensionsberechtigt – eligible for a pension
Pensionskasse – pension fund
Pensionsleistungskonto – pension benefit account
Pensionsplan – retirement plan
Pensionsrückstellung – pension reserves
Pensionsstaffel – pension scale
Pensionszahlung – pension payment, retirement pay
Pensionszusage – employer's pension commitment, pensioning warrant, guaranteed pension payment (Zusage mit Rechtsanspruch)
Periodenarbeitszeitplan – period work schedule
Periodenparameter – period parameter
periphäre Beurteilung – 360°-feedback
Permanenzkennzeichen – grossing-up indicator
Person, anrechnungsfähige – persons entitled to imputation credit
Personal – staff, employees, personnel, human resources
Personal- und Sozialaufwand – personnel and social expenses
Personal-Controlling – personnel controlling
Personalabbau – reduction of staff, manpower cutback, labour force reduction, staff cut
Personalabbauplanung – manpower reduction planning
Personalabrechnung – personnel expense report, payroll system (AmE), payroll accounting
Personalabrechnungskreis – payroll accounting area
Personalabrechnungsparameter – payroll accounting parameter
Personalabteilung – personnel department

Personaladministration – personnel administration
Personalagentur – executive search agency
Personalakte – personnel file, personnel record
Personalakquisition – personnel recruitment
Personalanforderung – request for (additional or new) personnel (Anforderung aus einer Abteilung an die Personalabteilung nach neuem oder zusätzlichem Personal.)
Personalangelegenheiten der Firmenleitung – personnel affairs of top management
Personalanpassung – staffing adjustment
Personalarbeit – personnel work
Personalaufbau – staff increase
Personalauftrag – personnel order
Personalaufwand – total staff costs
Personalaufwendungen – personnel expenses
Personalausstattung – personnel inventory
Personalauswahl – selection of personnel
Personalausweis – identity card (ID)
Personalbasisaufwand – basic staff costs
Personalbedarf – manpower requirements, workforce requirements
Personalbedarfsplanung – manpower planning, workforce requirements planning
Personalbemessung und Trendverfahren – Personnel Assessment and Trend Procedures
Personalberater – head hunter; HR advisor
Personalbereich – personnel area
Personalbericht – personnel report
Personalberichterstattungssystem – personnel reporting system
Personalbeschaffung – recruitment, personnel recruitment
Personalbeschaffungsplanung – recruitment planning
Personalbestand – personnel inventory, headcount
Personalbestandsanpassung – adaptation of staffing level

Personalbestandsaufnahme – personnel survey, personnel inventory
Personalbestandsentwicklung – development of staffing level
Personalbestandsverzeichnis – personnel survey, personnel inventory
Personalbetreuung – personnel guardianship
Personalbeurteilung – rating of employees
Personalbewertung – rating of employees, personnel evaluation
Personalbüro – personnel office
Personalchef – head of personnel
Personaldaten – personnel data, HR data
Personaldatenbank – HR database
Personaldatenpflege – HR data maintenance
Personaldecke – personnel situation
Personaldisposition – personnel allocation, personnel disposition
Personaleinsatz – personnel placement
Personaleinsatzplanung – manpower planning
Personaleinsparung – staff cutback, staff savings
Personalentscheidungen – personnel decisions
Personalentwicklung – personnel development, staff promotion
Personalentwicklungsplan – personnel development plan
Personalentwicklungsplanung – personnel development planning
Personaler – head of personnel, HR advisor
personalfachliche Aufgaben – personnel tasks
Personalfragebogen – personnel questionnaire
Personalfreisetzung – personnel layoff
Personalführung – personnel management, human resources management
Personalführungskraft – personnel manager
Personalführungsnachwuchs(kraft) – trainee for personnel management
Personalgespräch – appraisal
Personalgewinnung – recruitment
Personalgruppe – personnel group

Personalinformationssystem – HR information system, personnel information system
personalintensiv – personnel intensive
Personalkartei – manning tables
Personalkosten – personnel costs
Personalkostenplanung – personnel cost planning
Personalkostenzuordnung – personnel cost assignment
Personalleiter – head of HR department, head of personnel
Personalmanagement – personnel management, human resources management
Personalmangel – manpower shortage
Personalmaßnahme – personnel event
Personalnummer – personnel number
Personalorganisation – personnel organization, HR organization
Personalplan – personnel plan
Personalplanung – personnel planning, human resources planning
Personalplanung und -entwicklung – manpower planning and development
Personalpolitik – personnel policy
Personalrechenschema – personnel calculation schema
Personalreferent – group personnel director, personnel officer
Personalsachbearbeiter – personnel officer, staff manager
Personalstammblatt – HR master data sheet
Personalstammdaten – HR master data
Personalstammdatenverwaltung – HR master data management
Personalstammsatz – HR master record
Personalstandreduzierung – personnel cutback, downsizing
Personalstärke – headcount
Personalstraffung – downsizing
Personalstrategie – personnel strategy
Personalstruktur – personnel structure
Personalstruktur, administrative – administrative personnel structure
Personalstruktur, organisatorische – organizational personnel structure
Personalstrukturwandel – changing personnel structure
Personalteilbereich – personnel subarea

Personalumbau – shake-up
Personalunterlagen – personnel record
Personalveränderungsmitteilung – personnel change notifications
Personalverantwortung – personnel responsibility
Personalvertretung – staff representation
Personalverwaltung – personnel administration
Personalverwaltungssatz – personnel control record
Personalverzeichnis – staff list, list of employees
Personalwechsel – staff turnover, change in staff, labour turnover, fluctuation
Personalwerbung – recruitment, personnel recruitment (Beschaffung neuer Arbeitskräfte, geht der Auswahl und Einstellung voraus.)
Personalwerbung an einer Universität/ Hochschule – college recruitment
Personalwesen – human resources department
Personalwirtschaft – personnel management, Human Resources
Personalzusatzaufwand – additional staff costs
Personalzusatzkosten – additional staff costs
personelle Weisungsbefugnis – personal authority
Personenauswahl – employee selection
personenbezogene Daten – personal data
Personengruppe – set of persons, group of persons
Personenkreis – set of persons, group of persons
Personenschaden – injury to person
Personenversicherungsschutz – insurance coverage of persons
persönlich haftender Gesellschafter (KG) – General Partner (Limited Partnership)
persönliche Bemessungsgrundlage – personal assessment basis
persönliche Eignung – personal aptitude
persönliche Habe – personal effects
persönliche Leistungsfähigkeit – individual performance and ability
persönliche Meisterschaft – personal mastery

persönliche Verantwortung – personal responsibility
persönliche Voraussetzung – personal predisposition
persönliches Anliegen – personal matter
persönliches Auftreten – demeanour, bearing
Persönlichkeit – personality
Persönlichkeitsbildung – personality formation
Persönlichkeitsentwicklung – development of personality
Persönlichkeitsmerkmal – personality characteristic
Persönlichkeitstest – personality test
Persönlichkeitstyp – personality type
Perspektive – perspective
Perspektivlosigkeit – without a perspective
petzen – tell tales, inform on someone, sneak
Petzer – telltale
pfänden – impound, seize
Pfändung – seizure, attachment, garnishment
Pfändung, bevorrechtigte – preferred seizure
Pfändung, gewöhnliche – common seizure
Pfändungsabwicklung, automatische – automatic seizure
Pfändungsart – seizure type
Pfändungsgrundlage – base amount for seizure
Pfändungsnummer – seizure number
Pfändungsschutz – garnishment exemption (USA)(CAN); court order exemption (GB)
Pflege der guten Beziehungen – public relations, PR (Maßnahmen zur Herstellung eines positiven Verhältnisses des Unternehmens zur Öffentlichkeit, zur Werbung um öffentliches Vertrauen.)
Pflege der Mustermappe – standard file maintenance
Pflegekind – a child in one's care
Pflegeurlaub – personal leave
Pflegeversicherung – nursing insurance
Pflicht – duty, obligation
Pflicht... – obligatory, compulsory

Pflichtarbeitsstunde – study period
Pflichtbeitrag – compulsory contribution
Pflichtbewusstsein – sense of duty
Pflichterfüllung – discharge of duty
Pflichtfach – compulsory subject, required course
Pflichtkasse – compulsory health insurance fund
Pflichtmitgliedschaft – statutory membership
Pflichtplatz – compulsory workplace
Pflichtschulalter – compulsory school age
Pflichtverletzung – breach of duty, lapse from one's duty
pflichtversichert – compulsorily insured
Pflichtversicherung – compulsory insurance
Pförtner – gatekeeper, porter, doorkeeper, guard
Phantasie – imagination
phantasievoll – imaginative
Phlegmatiker(in) – phlegmatic person
phobischer Mensch – phobia-ridden person
Phonotypistin – audio-typist
Physik – physics
Physiker – physicist
physiologische Bedürfnisse – physiological needs
Pilotkurs – pilot course
Pilotprogramm – pilot program
Pilotprojekt – pilot project
Pkw-Regelung – car rule
Plakatwerbung – poster advertising
Plan – plan
Planabrechnung – forecast of personnel costs
Plandatenverwaltung – administration of planning data
Planmäßigkeit – methodicalness
Planspiel – business management game, business game (Training an Simulationsmodellen für unternehmerische Entscheidungen)
Planstatus – plan status
Planstelle – position authorized (in the budget)
Planstelle, besetzte – occupied position
Planstelle, unbesetzte – unoccupied position

Planstellenbeschreibung – position description
Planstellenbesetzungsplan – staffing schedule
Planstellenhierarchie – position hierarchy
Planstelleninhaber – holder of a position
Planstellenübersicht – survey of established posts
Planstellenverzeichnis – directory of positions
Planüberwachung und Steuerung – management control
Planung – work planning, task planning
Planungsgruppe – planning group
Planungshilfe – planning aids
Planungsprämisse – planning premise
Planungstechnik – planning technique
Planungsversion – plan scenario
Planungsvorgabe – planning premise
Planvariante – plan version
Plan vorschlagen – propose change
Planzeit – planned time
Plenarsitzung – plenary session
Platz (Kurs) – place (on a course)
Platzzusage, sichere – definite confirmation of place
Plausibilitätsprüfung – validation check
plump – clumsy
Podiumsgespräch – panel discussion
Polytechnikum – polytechnic (cf. chart)
polytechnische Hochschule – polytechnic (cf. chart)
Portfolio-Planung – portfolio planning
Position – position, post
Positionsbezeichnung – position title
Positivdatenerfassung – positive time recording
Positiverfassung – positive time recording
Postgironummer – postal giro account number
Potential – potential
Potentialbeurteilung – potential assessment
Präfekt – prefect
prahlen (BW) – boast
Praktikant – probationer, trainee (AmE) (Das Wort „Praktikant" kann für fast jeden Teilnehmer an einem Lehrgang verwendet werden sowie für jeden, der

seine Ausbildung noch nicht abgeschlossen hat; Nachwuchskräfte, die nach einem bestimmten Durchlaufplan ausgebildet werden.)

Praktikantenzeugnis – school report for trainees

Praktiker – practician; field worker, practical person

Praktikum – practical training, practical, internship

praktische Erfahrung – practical experience

praktische Studieninhalte – practical subject matters

praktische Tätigkeit – practical activity

praktisches Studiensemester – term spent in industry

Prämie – award, bonus, premium

Prämie (z.B. Versicherungsprämie/ Leistungsprämie) – premium (e.g. insurance premium)

Prämie für Nachtarbeit – night-work bonus

Prämienfindung – calculation of premium

Prämienfunktion – premium function

Prämiengrenze – threshold performance

Prämienlohn – premium pay

Prämienlohnsystem – premium wage system

Prämienmodifikator – premium modifier

Prämiensatz – premium rate

Prämiensystem – bonus scheme

Präsentation – presentation

Präsident – president

Prävention – prevention

Praxis – practice, usage, custom

Praxiserfahrung – practical experience

praxisfern – impractical

Praxislernen – action learning

praxisnahe Ausbildung – practically orientated education

praxisorientierte Lehrmethode – hands-on teaching method

praxisorientierte Vorgehensweise – hands-on approach

praxisorientierter Unterricht – hands-on approach

präzise Ausdrucksweise – precise manner of expression

Präzision – accuracy

preisen – praise

Preissteigerungsklausel – escalator clause

Preisverteilung – prize distribution

Primärstufe – primary level

Prinzip – principle

private Krankenversicherung – private health insurance

private Lebensversicherung – private life insurance

private Unfallversicherung – private accident insurance

private Vorbereitungsschule für eine Public School (BrE) – preparatory school

private Vorbereitungsschule für die Hochschule (AmE) – preparatory school

privates Lehrinstitut – private college

privates Schulsystem – independent system

Privatleben – private life, privacy

Privatlehrer – private teacher

Privatperson – individual

Privatrecht – civil law

Privatschule – private school (Sie dienen dazu, das öffentliche Schulwesen zu vervollständigen und zu bereichern. Das Recht private Schulen zu errichten ist im Grundgesetz und der bayrischen Verfassung verankert.)

Privatversicherter – privately insured

Privatversicherungsbeitrag – private insurance premium

Privileg – privilege

Privilegienverlust – loss of privileges

pro Kopf – capitation

Pro-Kopf-Ausgaben – per capita expenditure

Probeabrechnung – dummy payroll run

Probearbeit – job test

Probeinterview – mock interview

Probeprüfung – mock examination

probeweise – on probation

Probezeit – probationary period

Probezeitverlängerung – extension of the probationary period

Problembereich – problem area

Problembewältigung – coping with problems

Problembewusstsein – appreciation of the difficulties
problemorientiert – problem-oriented
produktiv – productive
Produktivität – productivity
Produktivitätsanteil – productivity wage increase
Produktivitätslohnsteigerung – productivity wage increase
Produktivstunden – productive hours
Produktoptimierung – product optimization
Produktschulung – product training
professionelle Führungskraft – professional manager
professioneller Manager – professional manager
professionelles Management – professional management
Professor – professor
Profilvergleich – profile comparison
Profit Center – profit center (Organisationseinheit mit eigener Gewinn- und Verlustrechnung)
profitieren – gain, profit
Prognose – prognosis
Programm – programme, program
Programmänderung – programme change
programmatische Rede – keynote speech
Programmbudget – programme budget
Programmierer – programmer
Programmierkurs – programming course
Programmiersprache – programming language
programmierte Unterweisung – programmed instruction
Programmierung – programming
progressive Produktionsprämie – steepening incentive
Progressivlohn – progressive wage rate
Projektarbeit – project work
Projektierung – engineering
Projektmanager – project manager
Projektmitarbeiterplan – project staff plan
Projektor – projector
Prokura – procura; power of attorney

Promotion – awardance of a doctorate, conferment of a doctor's degree
Promovend – doctoral candidate
promovierter Assistentenzprofessor – lecturer
promovieren – take a doctor's degree
propädeutisch – preparatory
Prospekt – leaflet, prospectus
Protokoll – minutes
provinziell – provincial
provokant – provocative
Provision – commission
Provisionsabrechnung – calculation of commission
Provisionsgruppe – commission group
provisorisch – interim, provisional
Prozess – process
Prozessberatung – coaching
Prozessmanagement – process management
Prozessoptimierung – process optimization
Prüfer – examiner, inspector
Prüfer aus einer fremden Universität – external examiner
Prüffeld – test department
Prüfhilfe – test aid
Prüflauf – test run (AmE)
Prüfling – candidate, examinee, ratee (AmE)
Prüfliste – checklist
Prüfung – exam(ination)
Prüfung nicht bestehen – fail an exam
Prüfung wiederholen – resit an exam
Prüfungsanforderungen – examination requirements
Prüfungsangst – examination nerves, fear of exams
Prüfungsaufgabe – examination question
Prüfungsaufsicht – supervision, invigilation
Prüfungsausschuss – board of examiners
Prüfungsfach – examination subject
Prüfungsfrage – examination question
Prüfungsmaterial – test material
Prüfungssystem – examination system
Prüfungstermin – examination date
Prüfungsthema – examination topic
Prüfungszeugnis – certificate
Prügelei – fight, punch-up

Prügelstrafe – corporal punishment
psychische Belastung – psychic strain
Psychologe – psychologist
Psychologie – psychology
psychologische Arbeitsbedingungen –
psychological working conditions
psychologische Prognoseverfahren –
psychological projections
Psychoterror am Arbeitsplatz –
mobbing
Pubertät – puberty
Publikum – audience
Pult – desk
Punkte erzielen – achieve points, score
Punktebewertungssystem – point rating
system
Pünktlichkeit – punctuality
Punktsystem – credit point system

Qualität des Arbeitslebens – quality of
work life
qualitative Personalplanung –
qualitative personnel planning
Qualitätsgruppe – quality circle
Qualitätskontrolle – quality control
Qualitätsmerkmal – quality feature
Qualitätsprämie – quality bonus
Qualitätssicherung – quality assurance
Qualitätsverbesserung – quality
improvement
Quantität – quantity
Quellenabzugsverfahren –
pay-as-you-earn, PAYE
Quellensteuer – withholding tax
Querdenker – lateral thinker
Quereinsteiger – traversee
Querulant – querulous person,
troublemaker, crab
Quittung – receipt
Quorum – quorum

Q

Qualifikation – qualification, skills
Qualifikationen und Anforderungen –
qualifications and requirements
Qualifikationsdefizit – qualification
deficit
Qualifikationskatalog – qualifications
catalog
Qualifikationslücke – qualifications
loophole
Qualifikationsnetz – qualifications
network
Qualifikationspool – qualifications pool
Qualifikationsprofil – qualifications
profile
Qualifikationsstruktur – qualifications
structure
qualifizieren – qualify someone for/as
qualifizierende Berufsausbildung –
qualifying vocational training
qualifiziert – qualified
Qualifizierungsmaßnahme –
qualification measure
Qualität – quality

R

Rabatt – deduction, discount
Rabauke – bully
Radiergummi – eraser (AmE), rubber
Rahmenabkommen – skeleton
agreement
Rahmenlehrplan – framework
curriculum
Rahmenvereinbarung – framework
agreement
Rahmenzeit – skeleton time
Randbedingung – peripheral condition
Rang – rank, grade; salary level, salary
position
Rangbegriff – rank concept
Rangbezeichnung – rank designation,
ranking description
rangliche Gleichstellung – rank
equivalency
rangmäßig – according to rank
Rangordnung – hierarchy, ranking, rank
order

Rangordnungssystem – rank order system
Rangstruktur – rank structure, grade structure
Rangstufe – rank, grade (of management)
Rangsystem – ranking system
Rangvergleich – rank comparison
rar – rare
Rassen... – racial
Rassendiskriminierung – racial discrimination, discrimination on racial grounds
Rat – advice, counselling
Ratschlag – advice, suggestion
Rate der Unfallschwere – accident severity rate
raten – advise, guess
Ratensparen – instalment savings
Ratenzahlung – installment, payment on account
rationalisieren – modernize, rationalize, streamline
Rationalisierung – rationalization
Rationalität – rationality
Ratlosigkeit – helplessness, perplexity
Rauchverbot – ban on smoking, prohibition of smoking
Raufbold – bully
raufen – scrap, fight
Rauferei – scrap, rough-house
Raum – room
Raumausstattung – room equipment
Raumbelegung – room reservation
Raumbelegungsart – room reservation type
Raumbelegungsplanung – room reservations planning
Raumdaten – room data
räumlich – spatial
Rauschgiftsüchtige(r) – drug addict
Reaktion – reaction
Realeinkommen – real income
Reallohn – real wage
Realschule – Realschule, secondary modern school (Die Realschule umfasst die Jahrgangsstufen 7 mit 10. Ihr Bildungsangebot richtet sich an junge Menschen, die geistig beweglich sind und zugleich praktische Fähigkeiten und Neigungen haben. Sie vermittelt eine allgemeine und

berufsvorbereitende Bildung, die zwischen den Angeboten der Hauptschule und des Gymnasiums liegt.) (s. Graphik)
Realschüler – secondary modern pupil (BrE)
Realschullehrer – secondary modern school teacher
Realschulrektor – headmaster of a secondary modern school
Rechenkenntnis – numeracy
Rechenschema – calculation schema
Rechenzeitplanung – computer scheduling (AmE)
Rechenzentrum – computer center
Rechnen – arithmetic
Rechner – computer
Rechnungsprüfer – auditor
Rechnungswesen – accounting (department), accountancy
Rechnungszinsfuß – interest rate
Recht – law
rechtliche Verselbständigung – legal independence (Diese setzt voraus, dass vorher eine organisatorische Verselbständigung erfolgte. Sie bedeutet die Gründung einer juristisch eigenständigen Gesellschaft - oft einer GmbH - mit eigener Geschäftsführung, eigener Buchführung, eigener Steuerpersönlichkeit usw.)
rechtmäßig – legal, lawful; legitimate
Rechtsabteilung – legal department
Rechtsanspruch – legal claim, legal right, legal demand
Rechtschreibung – orthography
Rechtsfrage – legal question
Rechtsgrundlage – legal basis, statutory source
Rechtsgrundsatz – principle of law
Rechtshilfe – legal aid
Rechtskunde – law
Rechtsmittelbelehrung – instructions on right to appeal
Rechtsprechung – jurisdiction; administration of justice
Rechtsreferendar – Bachelor of Law; articled clerk
Rechtsschutz-Versicherung – legal costs insurance
Rechtsstreit – litigation, lawsuit

Rechtsvorschrift – legal provision, rule of law
Rechtsweg – legal action
Rede – speech
redegewandt – eloquent
Redewendung – figure of speech, idiom; idiomate expression
redlich – fair, honest
reduzierte Arbeitszeit vor der Pensionierung – preretirement reduced working hours
Referat – department; seminar paper
Referat Personal (RefPers) – Personnel Advisory Office
Referatsleiter – head of department
Referendarzeit – teacher on probation; traineeship; time under articles
Referent – speaker, instructor
Referenz – reference
Referenzplanung – reference planning
Referenzwert – reference value
Regel – rule
Regelfall – as a rule
regelmäßig – regular
regelmäßige Beurteilung – continuous assessment
regelmäßige Überstunden – regular overtime
regeln – settle
Regelstudienzeit – normal period of time for completion of studies
Regeltyp – rule type
Regelverstoß – breach of rules
Regionenschlüssel – region key
Registrator – registrar
registrieren – register
Regressionsanalyse – regression analysis
regulär – regular
Rehabilitation – rehabilitation
Rehabilitationsmaßnahme – rehabilitation measure
Reibung – friction
Reibung auf Grund unterschiedlicher Kulturen – culture clash
reibungslos – smooth
Reibungspunkt – point of friction
reif – mature
Reife – maturity
Reifegrad – maturity
Reifeniveau – maturity

Reifezeugnis – A levels (BrE), high-school graduation certificate (AmE)
Reihenfolge – sequence
rein – net
Reise – travel
Reise, eintägige – trip, one-day
Reise, mehrtägige – trip, several-day
Reise, untertägige – trip lasting less than one day
Reise-Abrechnungskreis – travel accounting area
Reiseantrag – trip application
Reiseart – trip type
Reiseart, (gesetzliche) – trip type, (statutory)
Reiseart, unternehmensspezifische – trip type, company-specific
Reisebereich – trip area
Reisedaten – trip data
Reisekosten – travel expenses
Reisekostenabrechnung – travel expense report
Reisekostenbestimmungen – travel expense regulations
Reisekostenbuchung – travel expense posting
Reisekostenerfassung – entry of travel expense data
Reisekostenmanagement – travel expense management
Reisekostennachweis – travel expense statement
Reisekostenprüfstelle – revision office for travel expenses
Reisekostenrichtlinien – regulations on travel expenses
Reisekostenvergütung – reimbursement of travel expenses
Reisekostenvorerfassung – preliminary entry of travel expense data
Reisekostenvorschuss – advance on travel expenses
Reiseländerschlüssel – country travel key
Reisenebenkosten – additional travel expenses
Reisenummer – trip number
Reiseprivilegien – travel privileges
Reiseschema – trip schema
Reisespesen – incidental travel expenses

Reisestatus – trip status
Reisetyp – trip type
Reisevergütung – travelling allowance
Reiseverlauf – itinerary
Reisezeit – trip duration
Reiseziel – trip destination
rekapitulieren – recapitulate
Reklamation – complaint, protest, objection, claim
Rekrutierung – recruitment
Rektor – headteacher, principal (esp. U.S.); vice-chancellor, rector (U.S.), principal (FH)
relegieren – send down
Relevanztest – relevancy test
Religion – religion
Rente – pension, old-age pension, retirement benefit
Rente beziehen – draw a pension
Rente für Familienangehörige – dependents' pension
Rentenalter – pensionable age, retirement age
Rentenanpassung – pension adjustment
Rentenanspruch – pension right
Rentenantrag – pension application
Rentenantragssteller – pension applicant
Rentenanwartschaft – pension expectancy
Rentenauskunft – information about pensions, pension advisory department
Rentenbeginn – pension as of....
Rentenberatungsstelle – pensions advisory department
Rentenberechnung – pension computation, annuity computation
Rentenbescheid – pension notice
Rentendynamik – continual pension adjustment
Rentenempfänger – pensioner, old-age pensioner, retired employee
Rentenjahr – pension year
Rentenkürzung – pension reduction
Rentensteigerungsbetrag – pension increment
Rententräger – pension-paying institution
Rentenversicherung – pension insurance fund, annuity insurance; pension scheme

Rentenversicherungsnummer – pension insurance number
Rentenversicherungspflicht – statutory pension insurance
Rentenversicherungsträger – pension insurance institution
Rentner – pensioner, old-age pensioner, retired employee
Reorganisationsmaßnahme – reorganisational measure
Repetitor – coach, crammer
Report – report
Repräsentant – representative
Repräsentantenrolle – figurehead role
repressiv – repressive
Reputation – reputation
Reservebudget – emergency budget
reservieren – book, reserve
Reserviertheit – reserve, reservation
resigniert – resigned
Resolution – resolution
Ressentiment – resentment
Ressourcen – resources
Ressourcenbelegung – resource reservation
Ressourcenplanung – resources planning
Ressourcentyp – resource type
Rest... – residual
Restlaufzeit – remaining time to maturity of a loan
Resturlaub – remaining leave
Resturlaubsanspruch – remaining leave entitlement
Resultat – consequence, result
Rettungsplan – rescue plan
Rettungsweg – rescue route
revidieren – revise
Revision – audit
Revisionsbehörde – public review board
Revisor – auditor
Rezession – recession
rezitieren – recite
richtiger Zeitpunkt – timing
Richtlinie – regulation, guideline
Richtlinien für die Festlegung der Entlohnung – compensation regulations
Richtperiode – basic period
Riesenklasse – huge class
Rigorosum – thesis defense

Ringbuch – loose-leaf book
Ringtausch der Arbeitsplätze – job rotation
Risikobereitschaft – willingness to take a risk
Risikofaktor – risk factor
Rivalität – rivalry
Rohrstock – cane
Rolle – role
Rollenspiel – role game
Rollenverständnis – understanding of one's role, role understanding
Röntgenschirmbildstelle – X-ray department
Röntgenstation – X-ray department
rotieren – rotate
Routine – routine
Routinearbeit – routine work, repetitive work
rückerstatten – reimburse, refund
Rückerstattung – repayment, reimbursement
Rückfahrt – return journey
Rückfluss – reflux
Rückgaberecht – right of return
Rückgang – decrease, decline
rückgängig machen – cancel
Rückkehr – return
Rückkaufswert – redemption value, cash surrender value
rückläufig – recessive, declining
Rückmeldeverfahren – feedback technique
Rückmeldezähler – confirmation counter
Rückmeldung – feedback
Rückrechnung – retroactive accounting
Rückrechnungsanstoß – retroactive accounting trigger
Rückrechnungsart – retroactive accounting type
Rückrechnungsdatum – retroactive accounting date
Rückrechnungsdifferenz – retroactive accounting difference
Rückrechnungsergebnis – retroactive accounting result
Rückrechnungserkennung – retroactive accounting indicator
Rückrechnungsgrenze – retroactive accounting limit

Rückrechnungskennzeichen – retroactive accounting indicator
Rückrechnungsmonat – retroactive accounting month
Rückrechnungsperiode – retroactive accounting period
Rückrechnungsperiode, tiefste – earliest retroactive accounting period
Rückrechnungspol – key date for retroactive accounting run
Rückrechnungsrelevanz – retroactive accounting relevance
Rückrechnungstiefe – earliest retroactive accounting period
Rückrechnungstyp – retroactive accounting category
Rückreise – return trip
Rücksetzkandidat – reset candidate
rücksichtslos – inconsiderate
Rückstufung im Gehalt – salary demotion
Rücktritt – employee voluntary termination notice, resignation
rückwirkend – retroactive, retrospective
rückwirkende Lohnerhöhung – retroactive wage increase
Rückzahlung – repayment, reimbursement
Rückzahlungsverpflichtung – repayment requirement
Ruf – reputation
Rufbereitschaft – on call
rufschädigend – blacken someone's name, bring someone into disrepute
Ruhegehalt – old-age pension, superannuation
Ruhegehaltsabkommen – retirement benefits agreement
Ruhegehaltsanpassung – pension adjustment
Ruhegehaltsberechtigter – employee eligible for company pension
Ruhegehaltsempfänger – pensioner, old-age pensioner, retired employee
Ruhegeld – pension, old-age pension, retirement benefit
Ruhegeldanspruch – right to pension
Ruhegeldanwartschaft – public pension expectancy
ruhegeldfähiges Einkommen – pensionable income

Ruhegeldordnung – pension scheme
Ruhegeldverpflichtung – pension
  liability
Ruhegeldzusage – promise of a pension
ruhendes Beschäftigungsverhältnis –
  dormant employment
Ruhestand – retirement
Ruhetag – day off
Ruhezeit – rest period
rühmen – praise
Rundbrief – circular
Rundschreiben – circular
ruppig – coarse, unmannerly
Rüstzeit – setup time

S

Sachbearbeiter – person in charge,
  administrator
Sachbearbeiter für Abrechnung –
  payroll administrator
Sachbezüge – non-monetary
  compensation, remuneration in kind
Sachebene – factual level
Sachgebiet – area, subject, special area
sachgerecht – adequate, factual, proper,
  appropriate
Sachkenntnis – knowledge of the facts
Sachleistung – payment in kind
Sachlichkeit – objectivity,
  matter-of-factness
Sachpfändung – seizure of personal
  property
Sachschaden – material damage, damage
  to property
Sachunterricht in der Fremdsprache –
  immersion (Der Lehrer unterrichtet
  zeitweise sein Fach in einer
  Fremdsprache)
Sachvergütung – payment in kind
Sachverstand – expertise, expert opinion
Sachverständiger – expert
Sachverzeichnis – index
Saison – season

Saisonabhängigkeit, hohe – high
  seasonality
Saisonarbeiter – seasonal worker,
  migrant worker
Sammellohnart – combined wage type
sammeln – collect
Samstagsarbeit – Saturday work
Sandwich-Lehrgang – sandwich course
  (Fortbildungslehrgang, der zwischen
  zwei Perioden praktischer Ausbildung
  eingeschaltet wird)
Sanftmut – meakness
Sanguiniker(in) – sanguine person
Sanitätsraum – sick room
Satz – sentence
Satzbettfeld – record layout field
Satzung – statute
satzungsgemäß – statutory
Säufer – drunkard, drinker
Säugling – infant
Säumniszuschlag – delay penalty
Schaden – damage, defect
Schadenersatz – damages compensation
Schadenersatzanspruch – claim for
  damages, claim for compensation
Schadenersatzklage – damage suit
Schadensabteilung – claims department
schädlicher Einfluss auf – detrimental
  effect upon
Schadstoff – harmful substance
schaffen – create
Scham – shame
Scharfsinn – acuity
scharfsinnig – astute, sharp-witted
schassen – chuck out, boot out
Schätzung – estimate
Schaubild – chart, diagram
Scheck – cheque, check (AmE)
Scheinarbeitsverhältnis – mock
  employment
Scheinselbständigkeit – fictitious
  self-employment, mock independence
Scheitern – failure
schelmisch – mischievous
schelten – scold, give out
Schema – scheme
Schenkung – gift
scheu – shy
Schicht – level
Schicht(arbeit) – shift(work)
Schichtarbeiter – shift worker

Schichtführer – shift foreman
Schichtkennzeichen – shift indicator
Schichtplan – shift schedule
Schichtplan, persönlicher – personal shift schedule
Schichtplan, rollierender – rotating shift schedule
Schichtvertretung – shift substitution
Schichtwechsel – shift rotation, changeover, change of shift
Schichtzulage – shift bonus, shift differential
Schiedsabkommen – arbitration agreement
Schiedsausschuss – arbitration board
Schiedsgericht – arbitration court
Schiedsmann – arbitrator, conciliator, mediator, referee
Schiedsrichter – arbitrator, conciliator, mediator, referee
schiedsrichterlich entscheiden – arbitrate
Schiedsspruch – arbitration award, arbitration decision
Schiedsstelle – arbitration board
Schiedsvereinbarung – arbitration agreement
Schiedsverfahren – arbitration
Schikane am Arbeitsplatz – mobbing
schikanieren – victimise, harass, bully
schimpfen – scold, give out
Schlafsaal – dormitory
schlagen – hit, slap
Schlagwort – buzz word
Schlamperkiste – lost property box, lost and found
schlanke Hierarchie – delayer
schlecht abschneiden (z.B. in einem Test) – do badly (e.g. in a test)
schlecht bezahlter Arbeitsplatz – badly paid job
schlechter Schüler – dunce
schlechtes Benehmen – misbehaviour, misconduct
schlechtes Betragen – bad behaviour
Schlechtwettergeld – bad weather pay
Schlichter – arbitrator, conciliator, mediator, referee
Schlichtung – conciliation
Schlichtungsausschuss – mediation committee

Schlichtungsvereinbarung – arbitration agreement, amicable agreement
Schlichtungsverfahren – arbitration
Schließung – closure
Schluss – conclusion, end
Schlussbesprechung – wrap-up session
Schlüsselerfahrung – key experience, central experience
Schlüsselposition – key position
Schlüsselqualifikation – key qualification
Schlüsselwort – key word
Schlussfolgerung – inference
schlüssig – coherent
Schmerzensgeld – exemplary damages, damage for pain and suffering (BrE)
Schmiergeld – bribe, kickback
Schmutzzulage – dirty money (BrE)
schnell – facile
schnell von Begriff – quick on the uptake
Schnelllesen – speed reading, rapid reading
Schnittstelle – interface
Schnittstellenkompetenz – interface competence
schöne Künste – Fine Arts
schöpferisch – creative
Schreib... – clerical
schreibbehindert – writing impediment
schreiben – write
Schreiben – letter
Schreibfähigkeit – writing ability
Schreibkraft – typist
Schreibtisch – desk
Schriftform – written form
schriftliche Bewerbung – written application
schriftliche Lernunterlagen – written learning material
schriftliche Prüfung – written examiniation
Schriftprobe – writing sample
Schriftverkehr – correspondence
schritthalten mit – keep abreast of, keep up with, keep up to date
schüchtern – shy
Schubladendenken – thinking in narrow categories
Schulabbrecher – drop out
Schulabgänger – school leaver
Schulabgangsalter – school leaving age

Schulabschluss – school leaving certificate
Schulabschlusstag – last day of school
Schulamt – supervisory school authority, school board
Schulanfänger – school beginner, 1st grader
Schularbeiten – homework
Schulart – type of school
Schularzt – school medical officer
Schulatmosphäre – school atmosphere
Schulaufgaben – homework
Schulaufsichtsbehörde – schools' inspectorate
Schulausbildung – formal education
Schulausflug – school outing
Schulausschuss – school committee
Schulbehörde – Local Education Authority (LEA)
Schulberater – school adviser, school counsellor
Schulberatungsstelle – advisory centre for education
Schulbildung – education, educational background
Schulbuch – schoolbook
Schulbus – school bus
Schulchor – school choir
Schuld – blame
schuld sein an – blame
schuldig – quilty
Schuldgefühl – sense of guilt, guilty conscience, quilty feeling
Schuldkomplex – guilt complex
schuldrechtlicher Versorgungs-ausgleich – contractual pension adjustment
Schule – school
Schule abbrechen – drop out
Schule für Behinderte und Kranke – school for the (physically) handicapped and ill
Schule mit Selbstverwaltung – self-governing school
Schulbesuch – attendance at school
Schulbube – schoolboy
Schuleinschreibung – enrolment, school registration
Schuleinzugsbereich – catchment area
Schulemblem – school badge
schulen – train

Schulentlassungsalter – school leaving age
Schüler – schoolboy, student, pupil
Schüler einer weiterführenden Schule – secondary school pupil
Schüler-Lehrer-Zahlenverhältnis – pupil-teacher ratio
Schülerbogen – report card
Schülereinsatz – employment of pupils
Schülermitverantwortung – staff-students council
Schülermitverwaltung – staff-students council
Schülermonatskarte – season ticket for schoolchildren
Schülerrückgang – diminishing number of pupils
Schülerschwund – diminishing number of pupils
Schüleruniform – school uniform
Schülerzahl – number of pupils, number on roll
Schulferien – school holidays
schulfrei – no school
schulfreier Tag – free day
Schulgarten – school garden
Schulgebet – school prayer
Schulkind – school child
Schulkindergarten – infant school
Schulgeld – school fees
Schulglocke – bell, school bell
Schulheft – exercise book
Schulhof – playground, schoolyard
Schuljahr – school year
Schuljahresbeginn – start of the school year
Schulkamerad – school friend, schoolmate
Schulleiter – headmaster
Schulleiter(in) – principal
Schulleiterin – headmistress
Schulmappe – schoolbag
Schulmeister – schoolmaster, teacher
Schulordnung – school regulations
Schulpflicht – compulsory education, compulsory schooling
schulpflichtig – of school age
Schulphobie – fear of school
Schulranzen – satchel, school bag

Schulrat – school inspector (Der Schulrat ist in Bayern für Grund- und Hauptschulen zuständig)
Schulreferat – school department
Schulspeisung – school meals
Schulsport – school sports
Schulsportanlage – school sports grounds
Schulsprecher – school spokesman
Schulschwänzer – truant
Schulung – schooling, instruction, training
Schulungspaket – training package
Schulungsprogramm – training programme, training scheme
Schulungsunterlagen – training material
Schulungszentrum – training centre
Schuluniform – uniform, school uniform
Schulversager – failure at school
Schulverwalter – bursar
Schulverwaltung – governing board of the school
Schulweg – way to school
Schulwesen – educational system
Schulwesen, privates – public schools (s. Grafik)
Schulzwang – compulsory education
schummeln – cheat
Schutz der Sozialdaten – protection of social data
Schutzanzug – protective suit
Schutzausrüstung – safety equipment
schutzbedürftig – in need of protection
Schutzbrille – protective glasses, goggles
schützen – protect
Schutzgitter – safety guard
Schutzhandschuhe – protective gloves
Schutzhaube – protective hood
Schutzhelm – helmet
Schutzkleidung – protective clothing
Schutzmantel – protective coat
Schutzmaske – face guard, protecting mask
Schutzmaßnahme – precautionary measure
Schwäche – weak point, weakness
schwacher Anreiz – weak incentive
Schwächung – debilitation, weakening
Schwamm – sponge
schwanger – pregnant
Schwangerschaft – pregnancy

Schwangerschaftsabbruch – termination of pregnancy
Schwangerschaftsbescheinigung – letter of confinement
Schwangerschaftsmeldung – notification of pregnancy
Schwankung, saisonale – seasonal fluctuation
Schwänzer – truant
Schwarzarbeit – moonlighting, illicit work, undeclared employment
Schwarzarbeiter – illicit worker, black marketeer, moonlighter (Der Moonlighter geht einer zweiten Beschäftigung nach, ohne dem Hauptarbeitgeber oder dem Finanzamt darüber eine Mitteilung zu machen; Schwarzarbeit findet außerhalb des offiziellen Systems statt und es werden weder Steuern noch Sozialabgaben abgeführt.)
Schwarzes Brett – notice board, bulletin board
schwarzes Loch – black hole
Schwarzmarkt – black market
Schweigepflicht – professional discretion
Schweißerei – welding shop
Schwelle – threshold
schwer erziehbar – maladjusted, problem child
schwer vermittelbar – hard to place
schwer von Begriff – slow on the uptake
Schwerarbeit – heavy work
schwerbehindert – severely disabled, severely handicapped
Schwerbehinderte(r) – severely disabled person, severely handicapped person, invalid
Schwerbehindertenausweis – ID for severely handicapped persons
Schwerbehindertendatei – file of severely handicapped persons
Schwerbehindertengesetz – Severely Handicapped Persons Act
Schwerbehindertenurlaub – additional vacation for severely handicapped persons
Schwerbehindertenvertretung – disabled persons' representative
Schwerbehinderung – severe handicap
Schwerbeschädigter – disabled person

schwerfällig – clumsy, awkward
schwerhörig – hard of hearing
Schwerpunkt – main focus
schwieriger Verhandlungspartner –
  tough negotiator
schwierig – difficult
Schwierigkeit – difficulty
Schwung – impetus, drive, energy
sehbehindert – partially blind
Sehkraft – eyesight, vision
Sehvermögen – strength of vision
Seiteneinsteiger – lateral entrant
Sekretariat – office
Sekretär(in) – secretary
Sekretariatspraxis – secretarial
  experience
Sekundarbereich 1 – lower secondary
  sector (Der Sekundarbereich 1 umfasst
  die Hauptschulen und die Realschulen
  und Gymnasien bis zum 16.
  Lebensjahr.) (s. Graphik)
Sekundarbereich 2 – upper secondary
  sector (Der Sekundarbereich 2 umfasst
  die gymnasiale Oberstufe, die
  Fachgymnasien, die Fachoberschule,
  die Berufsfachschule und die
  Berufsaufbauschule.) (s. Graphik)
Sekundarschule – secondary school
  (Alle Schularten, die nach
  Jahrgangsstufe 4 bzw. 6 beginnen, sind
  sogenannte weiterführende Schulen:
  Hauptschule, Realschule, Gymnasium
  und alle Schulen im beruflichen
  Bereich. Das Gymnasium baut auf der
  Grundschule auf und umfasst die
  Jahrgangsstufen 5 mit 13. Es vermittelt
  eine vertiefte allgemeine Bildung, die
  für ein Hochschulstudium
  vorausgesetzt wird; damit schafft es
  zugleich auch Voraussetzungen für eine
  berufliche Ausbildung außerhalb der
  Hochschule. Das Gymnasium schließt
  mit der Abiturprüfung ab und verleiht
  die allgemeine Hochschulreife.)
  (s. Graphik)
selbstgesteuertes Lernen – self-directed
  learning
Selbstachtung – self-esteem, self-respect
selbständig – independent, self-reliant
selbständig erwerbstätig –
  self-employed

selbständig Tätiger – self-employer
**selbständige Tätigkeit (nicht
  versichert in der gesetzlichen
  Unfallversicherung)** – independent
  activity
Selbständige(r) – self-employed
selbständiger Betrieb – independent
  plant
Selbständigkeit – independence
Selbstausbildungsverfahren – "pick-up"
  method (ohne Anleitung durch einen
  Lehrmeister; der Arbeiter muss sich die
  nötigen Kenntisse durch Beobachtung
  erfahrener Kollegen aneignen.)
Selbstbestimmung – self-determination
Selbstbeteiligung – deductible
Selbstbewusstsein – self-confidence,
  self-assurance, self-awareness
Selbstdisziplin – self-discipline
Selbsteinschätzung – self-appraisal,
  self-assessment
Selbsteinstufungstest – self-assessment
  test
Selbsterfüllung – self-fulfilment
Selbsterkenntnis – self-knowledge
Selbsthilfe – self-help
Selbsthilfegruppe – self-help group
Selbstkontrolle – self-control
Selbstlaut – vowel
Selbstlernplatz – self-tutoring station
Selbstlernzentrum – self-access centre
Selbstmanagement – management of
  one's own concerns
Selbstorganisation – self-organisation
Selbstschutz – self-protection
selbstsicher – self-confident
Selbstsicherheit – self-confidence
Selbststudium – private study
Selbstsucht – egoism
Selbstunterweisung – self instruction
selbstverständlich – natural, obvious,
  self-evident
Selbstverständlichkeit – only natural,
  taken for granted, a matter of course
Selbstverständnis – conception of
  oneself
Selbstvertrauen – self-confidence
**Selbstverwaltung der
  Sozialversicherungsträger** –
  self-government board of the social
  insurance institutions

Selbstwertgefühl – feeling of one's own value, self-esteem
Selbstzahler – self-pay patient
selten – rare
Semester – term
Semesterarbeit – term paper
Semesterende – end of term, end of semester
Semesterferien – holiday, vacation
Semesterwochenstunden – lecture hours per week
Seminar – seminar, course
Seminarablauf – course schedule
Seminarplanung – education and training planning
Seminarverwaltung – education and training administration
Senkrechtstarter – high-flyer
Sensibilisierung – sensitization
Sensibilität – sensitivity, sensitiveness
Sensitivitätstraining – sensitivity training
sensorisch – sensory
sexuelle Belästigung – sexual harassment
sich an Entscheidungen beteiligen – share in decisions
sich aneignen – appropriate
sich anstrengen – make an effort
sich auf dem Laufenden halten – keep abreast of, keep up with, keep up to date
sich äußern – express oneself
sich auszeichnen – surpass, exceed, excel
sich bekennen – out one-self
sich bemühen – make an effort
sich beraten – confer
sich bewerben um – apply for
sich dagegen entscheiden – opt out, decide against
sich einigen – agree
sich entziehen – evade
sich erinnern – remember
sich erinnern an – recall
sich erkundigen nach – inquire
sich etwas vorstellen – imagine, visualize
sich ins Gedächtnis zurückrufen – recall
sich mitteilen – communicate
sich nähern – approach
sich rühmen – boast

sich unterscheiden – differ
sich vermindern – decrease
sich verständigen – communicate
sich zusammenschließen – amalgamate
Sicherheit – safety, security
Sicherheit am Arbeitsplatz – on-the-job safety
Sicherheit des Arbeitsplatzes – job security
Sicherheitsbeauftragter – safety officer
Sicherheitsbedürfnisse – safety needs
Sicherheitsbestimmungen – safety regulations
Sicherheitseinrichtung – safety facility
Sicherheitsempfänger – warrantee
Sicherheitsfachkraft – safety expert
Sicherheitsingenieur – safety engineer
Sicherheitskontrolle – safety control
Sicherheitsschuhe – protective shoes
Sicherheitsvorkehrungen – safety precautions
Sicherheitsvorschriften – safety code
Sicherung von Arbeitsplätzen – employment protection
Silbe – syllable
Simulation – simulation
Sinekure – sinecure (einträgliches Amt ohne Arbeit)
Sinnes- – sensory
sinnloses Wort – nonsense word
Sinngebung – meaning
Situationsanalyse – commitment analysis
situationsgerecht – suitable to the situation
Sitz (eines Unternehmens) – headquarters
sitzenbleiben – repeat a year, stay down
Sitzstreik – sit-down strike
Sitzung – conference, meeting
Sitzungsniederschrift – minutes
Skala – scale
Slums – slums
Sofort-Nachfolger – immediate successor
Software-Haus – software company
Solidarität – solidarity
Solidaritätsbeitrag – solidarity contribution
Solidaritätsmaßnahme – sympathetic action

Solidaritätsstreik – sympathy strike, blacking (Ablehnung mit Material oder Produkten zu arbeiten, die von einem bestreikten Betrieb angefertigt werden)

Soll-Kopfzahl – authorized number of employees

Soll-Leistung – standard output

Sollarbeit – planned work

Sollarbeitspause – break during planned working time

Sollarbeitsstunde – planned working hour

Sollarbeitszeit – required working hours, planned working time

Sollbezahlung – planned remuneration

Sollbezüge – planned pay

Sollohnkosten – planned labor costs

Sollpaar – planned time pair

Sollpaarermittlung – determining planned time pairs

Sollplan – target plan

Sollstunde – planned hour

Sollvorgabe – target specification

Sollzeit – target time

Sommerschule – vacation school, summer school

Sommersemester – summer term

Sommertrimester – Trinity term

Sommerurlaub – summer vacation

Sonderabrechnung – special payroll run

Sonderarbeit – special work

Sonderaufgaben – special assignments

Sonderausbildung – special training

Sondererziehung – special education

Sonderausgabenpauschale – lump sum for special expenditures

Sondergenehmigung – special permission

Sondergruppe – special class

Sonderlehrgang – special course

Sonderleistungsvertrag – merit-based contract for non-salaried employees

Sonderlohnvertrag – special wage contract

Sonderprojekt – special project

Sonderrecht – privilege

Sonderregelung – special rule, special regulation

Sonderschüler – special needs students

Sonderschullehrer – teacher at a school for handicapped children

Sonderurlaub – compassionate leave

Sondervereinbarung – special agreement

Sondervergütung – special remuneration, gratuity

Sonderzahlung – bonus

Sonderzahlung, periodisch wiederkehrende – periodically recurring special payment

Sonderzahlungslauf – special payment calculations

Sonderzulage – special bonus

Sonderzuwendung – special allowance, special bonus

Sonntagsarbeit – Sunday work

sonstige Firmennebenleistungen – non-monetary compensation, remuneration in kind

Sorgenkind – problem child

Sorgfalt – thoroughness, care

sorgfältig – careful, painstaking, meticulous

souverän – superior

sozial erfolgsarm sein – socially atrophied

Sozialabgaben – social insurance contributions

Sozialamt – social services office

Sozialaufgaben – social affairs

Sozialaufwendungen – social expenditure

Sozialberater – social advisory officer

Sozialberatung – social advisory office; industrial social work

soziale Einrichtungen – welfare facilities

soziale Gesetzgebung – social legislation

soziale Intelligenz – social intelligence

soziale Isolation – social isolation

soziale Maßnahmen – social (work) measures

soziale Sicherheit – social security

soziale Vergangenheit – background

Sozialeinrichtungen – social facilities

sozialer Besitzstand – vested rights

sozialer Rückzug – social withdrawal

soziales Faulenzen – social loafing

soziales Umfeld – social environment

Sozialfürsorge – welfare

Sozialgericht – court of Social Justice

Sozialgesetzbuch – social legislation code

Sozialisierung – socialization
Sozialkompetenz – social competence
Sozialkosten – social expenses
Sozialkunde – social science
Sozialleistungen – employee benefits, social benefits
Sozialleistungsträger – social insurance agency; Department of Social Security
Sozialplan – social plan, redundancy payments scheme
Sozialpolitik – social policy
Sozialrecht – social legislation
Sozialstatistik – social statistics
Sozialversicherung – social insurance
Sozialversicherungsabkommen – social insurance agreement
Sozialversicherungsbeitrag – social insurance contribution, social security contribution (AmE)
Sozialversicherungsbrutto – gross amount for social insurance
Sozialversicherungsentgeltbescheinigung – social security statement
Sozialversicherungsgesetz – social insurance law
Sozialversicherungsgrenze – limit for social insurance
Sozialversicherungsheft – social insurance booklet
Sozialversicherungsnachweis – social insurance statement
Sozialversicherungsnummer – social insurance number
Sozialversicherungspflicht – obligation to make social insurance contribution
sozialversicherungspflichtig – liable to social security payments, subject to social insurance contribution regulations
Sozialversicherungsrecht – social insurance legislation
Sozialversicherungssystem – social security scheme
Sozialversicherungsträger – social insurance agency, Department of Social Security
Sozialwissenschaften – social sciences
sozialwissenschaftliches Gymnasium – secondary school emphasizing social sciences

Sozialzulage – social welfare supplement, welfare allowance
Sozialzulagensystem – benefit plan
Sozioökonomie – socio-economics
Spannung – tension
Spannungen in der Familie – family tensions
Sparbetrag – savings amount
Sparplan – savings plan
Sparprämie – savings premium
Sparprogramm – cost-cutting drive, austerity programme
Sparzulage – savings bonus
Späteinsteiger – late entrant
Spätentwickler – late developer
Spätschicht – night shift, late shift
Spätzünder – slow on the uptake, late starter
Speiseraum – refectory, dining-centre
Speisesaal – refectory, dining-centre
Spekulation – gossip, rumour
Sperrfrist – blocking period
Sperrvermerk – lock flag
Sperrzeit – blocked period
Spesen – incidental expenses
Spesenart – expense type
Spesenberechtigung – authorization to run an expense account
Spesenkategorie – expense category
Spesenkennzeichen – expense indicator
Spesenkonto – expense account
Spesenkürzel – expense code
Spesenrichtlinien – expense account regulations
Spezialausbildung – special training
Spezialist – specialist, skilled labour
Spezialistenausbildung – specialist training
Spezialwissen – expert knowledge
Spezifikation – specification
spezifizieren – specify
spicken – copy
Spickzettel – crib
Spiel – game
Spielgruppe – play group
Spielraum – margin
Spinoff – spin off (kleinere Betriebsteile oder Funktionen, die an das bisherige Management übertragen werden)
spitzbübisch – mischievous
Spitzen... – leading-edge

Spitzenbelastung – peak load
Spitzenbewerber – highly qualified candidate
Spitzenführungskraft – top manager
Spitzenleistung – peak performance, top efficiency
Spitzenleute – highly qualified candidates
Spitzenlohn – top wages, top wage rate, maximum wage rate
Spitzenmanager – top manager
Spitzenposition – top position
Spitzenverdiener – top earner
Spitzer – pencil sharpener
Splitt-Genauigkeit – split accuracy
Splittkennzeichen – split indicator
Splittzeitraum – split period
Sponsor – sponsor
spontan – spontaneous
Spontanbewerber – unsolicited applicant
Spontanbewerbergruppe – unsolicited applicant group
Spontanität – spontaneous
Sport – sports, physical education
Sportanlagen – sports facilities, sports grounds
Sportfest – sports-day
Sportlehrer – sports teacher
Sprachausbildung – language training
sprachbegabt – good at languages, talent for languages
Sprachbehinderung – speech impediment
Sprache – language, speech
Sprachenfolge – language sequence
Sprachentwicklung – language development
Spracherwerb – acquisition of languages
Sprachfähigkeit – language ability
Sprachgefühl – feeling for language
Sprachkenntnisse – knowledge of languages
Sprachlabor – language laboratory
sprachliche Identität – language identity
sprachliche Intelligenz – linguistic intelligence
sprachlicher Ausdruck – verbal ability
Sprachentest – language test
Sprecher des Vorstands – Chairman of the Board of Management

Sprecherausschuss – committee of spokesmen
Sprecherrolle – spokesman role
Sprechtechnik – elocution
Sprichwort – proverb
Springer – reserve pool employee, relief person, relief man
Springerschicht – relief shift
sprunghaft – volatile
Spürsinn – serendipity
staatlich finanzierte Umschulung – state sponsored retraining
staatlich geprüfter – state-certified
staatliche Ausbildungsförderung – grants for families of limited income (Aufgabe des sozialen Rechtsstaates ist es, Familien mit einem geringen Einkommen bei der Ausbildung ihrer Kinder finanziell zu unterstützen)
staatliche Schule – state school, public school (AmE) (s. Graphik)
staatliches Gymnasium – state-governed secondary grammar school (s. Graphik)
staatliches Prüfungszeugnis – official certificate
Staatliches Schulamt – government school authority, government education office
Staatsangehörigkeit – nationality
Staatsbeamte(r) – public servant, civil servant
Staatsbedienstete(r) – civil servant
Staatsdienst – civil service (BrE), public service (AmE)
Staatsexamen – Civil and Public Service Examination (CSE) (Das erste bzw. zweite Staatsexamen ist zwingend vorgeschrieben für alle, die eine Beamtenlaufbahn einschlagen wollen.)
Staatskunde – civics
Stab – staff
stabil – stable
Stabsabteilung – staff unit
Stabsfunktion – staff function
Stabskennzeichen – staff indicator
Stabsmann – staff man
Stabsstelle – staff position
Stabstellenliste – list of staff positions
städtische Schule – municipal school
Stadtschulrat – municipal schools inspector

Stammaktie – common share (BrE), common stock (AmE)
Stammhaus – parent company, head office
Stammhauslehre – head office training
Stammhauslehrling – head office trainee
Stammkostenstelle – master cost center
Stammkunde – regular customer
Stammpersonal – permanent staff
Stammstundenlohnsatz – standard hourly pay rate
Stand – level
Stand der Technik – state-of-the-art
Standard – standard
Standard-Tagestyp – standard day type
Standardwoche – standard week
ständig – regular, continual
ständige Lernbereitschaft – permanent readiness (willingness) to learn
ständiges Lernen – continual learning
Standort – location
Standortverwaltung – facility management
Standortwechsel – change of location
Stärke – strong point, strength
Starkstromelektriker – power electrician
starr – rigid
starres Budget – fixed budget
Starrheit – rigidity
starrsinnig – stubborn
Statistik – statistics
statistisches Prognoseverfahren – statistical projection
Status – status
Status eines Fellows in einem College oder wissenschaftlicher Vereinigung – fellowship
Statusänderung – change in status
Statusmerkmal – status feature
statusorientiert – status-oriented
stechen – clock in or out
Stechkarte – time card
Stechuhr – time clock
steif – rigid, stiff
Stehvermögen – stamina, staying power
steigen – rise
steigern – increase
Steigerung – increase, increment
Steigerung der Motivation – increase of motivation

Steigerungsklausel – escalator clause
Steigerungssatz – rate of increase
Stelle – position, post, job
Stellenangebot – vacancies, job offers
Stellenanzeige – job ad(vertisement), employment ad(vertisement)
Stellenausschreibung – job ad(vertisement), employment ad(vertisement)
Stellenausschreibung, interne – job advertisement, internal
Stellenbeschreibung – job description
Stellenbesetzung – filling of vacancies
Stellenbesetzungsplan – plan to fill vacancies, staffing schedule
Stellenbewerber – applicant, candidate
Stellenbezeichnung – position title
Stellendatei – job file
Stellengesuch – application
Stelleninhaber – incumbent, job holder
Stellennummer – job number
Stellenplan – plan to fill vacancies, job index
Stellensplitting – job splitting
Stellensuche – job search, job hunt
Stellensucher – job hunter, applicant
Stellenteil in Zeitungen und Zeitschriften – appointments section
Stellenvermittlung – employment bureau, employment centre, employment agency
Stellenwechselplanung – job rotation planning
Stellungnahme des Vorgesetzten – superior's statement
stellvertretender Vorsitzender des Aufsichtsrates – Deputy Chairman of the Supervisory Board
stellvertretendes Vorstandsmitglied – Senior Vice President
Stellvertreter – substitute
Stempelkarte – time card
stempeln gehen – be on the dole
Stempeluhr – time clock
Stenotypist(in) – shorthand typist
Sterbegeld – death benefit
stetig – consistent, constant, continual, steady
Steuer – tax
steuerabzugsfähig – tax-deductible
Steueramt – tax office

Steuerausgleichszahlung – tax equalization payment
Steuerbefreiung – tax exemption
steuerbegünstigt – tax-privileged
Steuerbehörden – tax authorities
Steuerberechnungsverfahren – tax calculation procedure
Steuerbescheid – tax statement
Steuerbrutto – gross tax amount
Steuereinbehaltung – tax withholding
Steuererklärung – tax return, tax declaration
Steuererlass – tax exemption
Steuererstattung – tax refund
steuerfrei – tax free, tax-exempt, non-taxable
Steuerfreibetrag – tax-free amount
Steuerklasse – tax bracket, tax group
steuern – control
Steuernachweis – statement for wage withholding tax
Steuernummer – tax payer's account number
Steuertarif – tax scale
Steuerung, betriebliche – internal control
Steuerungsbegriff – control term
Steuerungsmerkmal – control feature
Steuerzeichen – control character
Steuerzeitraum – tax period
Stichprobe – random sample
Stichwahl – final ballot, run-off (AmE)
Stichwort – key word
Stichwort geben – prompt
Stiefkind – stepchild; neglected child
Stift – pencil
Stiftung – foundation
Stillegung – closure
stiller Teilhaber – sleeping partner
Stillschweigen – silence
Stimme abgeben – vote
Stimmengleichheit – tie, parity (of votes)
Stimmrecht – right to vote, voting right
Stimmzettel – ballot, ballot paper, voting paper
Stimulus – stimulus
Stipendium – scholarship, bursary (Scottish universities), grant
Stolz – pride
stören – interrupt, interfere
Störenfried – trouble-maker

Störer – trouble-maker
störrisch – refractory
stornieren, Kurs – cancel a course
Stornierungsmitteilung – notification of cancellation of a course
Störpotential – propensity to disturb
Störquelle – source of interference
Störung – disturbance
Störzeit – malfunction period
stoßen – push
stottern – stammer, stutter
Stotterer – stammerer, stutterer
Strafarbeit – punishment exercise, lines
Strafe – punishment, penalty, fine
straffällig – culpable
strahlenarmes Gerät – low level radiation screen
Strahlenschutz – protection against radiation
Strategie – strategy
strategische Geschäftseinheit (SGE) – strategic business unit (SBU)
strategische Planung – strategic planning
Streber – swot, pusher, over-ambitious person
Streik – strike
Streik durch passiven Widerstand – go slow, work to rule
Streikaufruf – call to strike
Streikbrecher – blackleg, strikebreaker
Streikfonds – strike fund
Streikgefahr – danger of a strike
Streikgeld – strike benefit
Streikkasse – strike fund
Streikposten – picket
Streikrecht – right to strike
Streikregelung – strike control
Streikverbotsklausel – no-strike clause
Streikvergütung – strike benefits
Streikversicherung – strike insurance
Streikwache halten – picket
Streit – argument
streiten – argue
streitsüchtig – aggressive
streng – strict
Stressbewältigung – overcoming stress
Struktur – structure
Struktur des Führungskreises – management structure

strukturierter Unterricht – structured teaching
Strukturtiefe – structure depth
Strukturwandel – structural change
Stück(lohn)system – piecework system
Stückakkord – quantity-based piecerate work
Stücklohnsatz – piecerate
Stückzahl – numbers
Student – student, undergraduate
Student im ersten Studienjahr – freshman (AmE)
Student im zweiten Studienjahr – sophomore (AmE)
Studentenheim – students hostel, hall of residence, dormitory (dorm)
Studentenrat – student council
Studentenvertretung – students' representation
Studentenwohnheim – students hostel, hall of residence, dormitory (dorm)
Studentinnenvereinigung – sorority
Studiendekan – dean of studies (Der Studiendekan hat darauf zu achten, dass das Lehrangebot den jeweiligen Studien- und Prüfungsordnungen entspricht und dass das Studium in der Regelstudienzeit absolviert wird und dass Studenten ihre Professoren bewerten.)
studentische Veranstaltung – college rag (Eine karnevalistische Veranstaltung der Studenten zu Wohltätigkeitszwecken.)
Studienabbrecher – drop out
Studienanfänger – new student, freshman
Studienangebot – courses on offer
Studienassessor – secondary school teacher, grammar school teacher (Lehrer an einer weiterführenden Schule)
Studienbeihilfe – student grant
Studienberater – student adviser
Studienberatung – academic advisory service
Studienbetreuer – tutor
Studiendauer – length of studies
Studiendirektor – principal (Fachschule), deputy principal (Gymnasium) (Ein Studiendirektor

übernimmt neben der Unterrichtstätigkeit besondere Aufgaben, vornehmlich Verwaltungsaufgaben.)
Studienfach – discipline, subject
Studienfach aufgeben – drop a subject
Studiengang – course of study
Studiengebühren – university fees
Studiengewohnheiten – study habits
Studienhilfen – study aids
Studienkollege – classmate
Studienplan – curriculum, syllabus
Studienrat(rätin) – secondary school teacher, grammar school teacher (Lehrer an einer weiterführenden Schule) (s. Graphik)
Studienreferendar – student teacher (Der Studienreferendar hat das erste Staatsexamen bestanden und befindet sich in der Lehrerausbildung.) (s. Graphik)
Studienzweig – branch of study
studieren – study
Studierende(r) – scholar
Studium abbrechen – drop out
Studium auf höchster Ebene – course of advanced studies
Studium zur Erlangung eines höheren akademischen Grades – post-graduate course
Stufenausbildung – progressive training, training by stages
Stufenbonus – progressive bonus
Stufenqualifikation – up-the-ladder qualification (Neue Anforderung auf der nächst höheren Rangstufe.)
stumm – mute
Stunde – lesson, class, period
Stunde frei haben – have a free period
Stundenausfall – cancelled hours
Stundenkontingent – allocation of hours, quota of hours
Stundenlohn – hourly rate, hourly wage
Stundenlohnempfänger – hourly worker
Stundenlöhner – wage-earner, non-salaried employee
Stundenplan – time table
Stundenplangestaltung – time-table planning
Stundensatz – hourly rate
Stundentafel – roster

Stundenzettel – time report
subtil – subtle
subventionierte Schule – grant-aided
  school
Suche nach Arbeit – search for
  employment
Sucht – addiction
Suggestivfrage – leading question
Suggestopädie – suggestopaedia
summa cum laude – summa cum laude
Summenlohnart – cumulation wage type
Sündenbock – scapegoat
SV-Entgeltbescheinigung – social
  security statement
SV-Tage – days qualifying for social
  insurance coverage
Sympathiestreik – sympathy strike,
  blacking (Ablehnung mit Material oder
  Produkten zu arbeiten, die von einem
  bestreikten Betrieb angefertigt werden)
Symposion – symposium
Syndikus – head of the legal department,
  corporation lawyer, general counsel
Synergie – synergy
Synonym – synonym
Syntagma – syntactic construction
Synthese – synthesis
Synthesefähigkeit – ability to synthesize
synthetisch – synthetic
System, nachgelagertes – back-end
  system
System, vorgelagertes – front-end
  system
systematisch fördern – promote
  systematically
systematische Förderung – systematic
  promotion
systemisches Denken – systems thinking

T

Tabelle – table, scale
Tabellenwert – table value
Tabu – taboo
Tadel – rebuke, reprimand

tadeln – rebuke, reprimand
Tafel – board, blackboard
Tag – day
Tag der offenen Tür – open day
Tag der Preisverleihung – prize day
Tag, anrechenbarer – accountable day
Tag, arbeitsfreier – day off
Tagegeld – per-diem allowance
Tagelöhner – day-labourer
Tagesarbeit – a day's work
Tagesarbeitszeitplan – daily work
  schedule
Tagesgeschäft – day-to-day business
Tageshort – day care centre
Tageslichtprojektor – overhead
  projector, OHP
Tageslohn – daily wage
Tagelöhner – day-labourer
Tagesmerkmal – day feature
Tagesmodell – day model
Tagesmodifikator – day modifier
Tagesordnung – agenda
Tagesprogramm – day program
Tagessaldo – daily balance
Tagessatz – per-diem allowance, day rate
Tagesschicht – day shift
Tagesschule – day-school
Tagestyp – day type
Tagestypmodifikator – day type modifier
Tagesverarbeitung – day processing
Tageszeitsaldo – daily time balance
**tageweise Freistellung für
  Kursteilnahme** – day-release training
  (Ausbildungsverfahren, nach dem
  Beschäftigte wiederholt (meist in
  regelmäßigen Abständen) zwecks
  Teilnahme an Lehrgängen von der
  Arbeit freigestellt werden.)
tagtäglich – daily
tagtägliche Berufspraxis – daily
  experience at work
Tagträumer – day-dreamer
Tagung – conference, congress,
  convention
taktil – tactile
taktlos – tactless
taktvoll – tactful
Taktzeit – cycle time
Tarif – wage scale, pay scale
Tarifabkommen – bargaining agreement
Tarifabschluss – wage settlement

Tarifangestellte(r) – salaried employee, white collar worker, office worker, pay scale employee
Tarifart – pay scale type
Tarifausschuss – collective settlement committee
Tarifautonomie – right to free collective bargaining
Tarifbedingungen – standard agreement provisions
Tarifbewertung – pay scale valuation
Tarifbezirk – collective agreement area, collective bargaining area
Tariferhöhung – increase in wage scales
Tarifforderung – wage scale claim
Tarifgebiet – pay scale area
Tarifgehalt – salary, union rates
Tarifgrundlohn – basic pay scale salary
Tarifgruppe – salary group, pay scale group
Tarifgruppenstrukturanalyse – wage group analysis
Tarifgruppenwechsel – change in pay scale group
Tarifkennzeichen – pay scale indicator
Tarifkonflikt – trade dispute
Tarifkreis – tariff staff, non-exempt staff
tarifliche Leistungszulage – merit increase according to wage scale
tarifliche Wochenarbeitszeit – weekly working hours according to collective agreement
tarifliches Personal – non-exempt personnel
Tariflohnerhöhung – increase in wage scales
Tarifmonatslohn – standard monthly pay
Tarifpartner – partner to collective bargaining
Tarifpolitik – wage(-scale) policy
Tarifrunde – collective bargaining round
Tarifstruktur – pay scale structure
Tarifstufe – pay scale level
Tarifstufenwechsel – change in pay scale level
Tarifstundenlohn – standard hourly pay
Tarifsystem – wage scale system
Tariftabelle – pay scale table
Tarifurlaub – standard annual leave
Tarifverhandlungen – collective bargaining

Tarifverhandlungen für den gesamten Industriebereich – industry-wide bargaining
Tarifvertrag – wage/pay agreement
Tarifvertrag für den gesamten Industriebereich – industry-wide agreement
tarifvertragliche Regelungen – regulations as laid down in collective agreement
Tarifzulage – standard bonus
Taschengeld – pocket money
Taschenrechner – calculator
Tast- – tactile
Tastsinn – sense of touch
Tatendrang – energy, thirst for action
Tätigkeit – job, activity, work, occupation, employment
Tätigkeit, mitternachtsübergreifende – work extending beyond midnight
Tätigkeitsbereich – field of activity
Tätigkeitsbeschreibung – job description
Tätigkeitsbezeichnung – job title
Tätigkeitsfeld – field of activity
Tätigkeitsmerkmal – job feature, job characteristic
Tätigkeitsnachweis – activity report
Tätigkeitsprofil – job profile, activity profile
taub – deaf
Taubstummenalphabet – manual alphabet, finger alphabet, deaf and dumb alphabet
Tautologie – tautology
taxieren – assess, appraise, estimate
Taxonomie – taxonomy
Teamarbeit – group work, teamwork
Teamfähigkeit – team ability
Teamgeist – team spirit
Technik – technology
Techniker – technician
Technikerschule – advanced vocational training college (für Techniker und Meister)
technisch – technical
technisch Tätige – technical employees
technische Ausbildung – technical training
Technische Hochschule – Technical Engineering College, Technical

University, Institute of (Science and) Technology (s. Graphik)
technische Regelung – adjustment, regulation
Technische Universität (TU) – Technical Engineering College, Technical University, Institute of (Science and) Technology (s. Graphik)
technischer Aufsichtsbeamter – technical supervisor
technischer Bildungsausschuss (TBA) – committee on technical training
technischer Zeichner – draughtsman
technisches Fachwissen – specialized technical knowledge
technisches Kolleg – technical courses
technisches Zeichnen – technical drawing
Technokrat – technocrat
Technologie – process engineering
Technostruktur – technostructure
Teenageralter – teens
Teilaufgabe – subtask
Teilbeschäftigte(r) – short-time worker
teilen – divide
Teilentgelt – partial remuneration
Teilentgeltberechnung – calculation of partial remuneration
Teilgehalt – partial salary
Teilhaber – partner
Teilmonatsbetrag – partial period amount
Teilnahme – attendance, participation
Teilnahme des Arbeitnehmers – employee participation
Teilnahme, fixierbare – firmly bookable attendance
Teilnahme, offene – undetermined attendance
Teilnahmeberechtigung – eligibility
Teilnahmebestätigung – certificate of attendance
teilnehmen (an) – attend, participate in
Teilnehmer – participant
Teilnehmer von außerhalb, der am Studienort wohnt – resident participant
Teilnehmerbeurteilung – attendee appraisal
Teilnehmerliste – attendee list
Teilnehmerprofil – participant profile

Teilprozentsatz – part percentage rate
Teilschema – subschema
Teilveranstaltung – session
Teilwert – fractional value (Berechnungsgröße bei Pensionsrückstellungen)
Teilzahlung – instalment, payment on account
Teilzeit – part-time work
Teilzeitarbeit – part-time work, part-time job, part-time employment
Teilzeitarbeit für ältere Mitarbeiter – part-time work for elder co-workers
Teilzeitarbeiter – part-time worker, part timer
Teilzeitarbeitsplatz – part-time workplace
Teilzeitbeschäftigung – part-time work, part-time job, part-time employment
Teilzeitkraft – part-time worker, part timer
Teilzeitkurs – part-time course
Teilzeitlehrer – part-time teacher
Teilzeitmodell – part-time working model
Teilzeitraum – partial period
Teilzeitschulpflicht – part-time compulsory schooling
Teilzeitunterricht – part-time schooling
Telearbeit – distance working, teleworking
Telearbeiter – teleworker
Teleheimarbeiter – tele outworker
Telelernen – tele-learning
Tempo – pace, speed
Termin – appointment
Terminänderung – change in an appointment
Terminart – type of deadline date
Terminerinnerung – reminder of appointment
Terminkalender – appointments diary
Terminplan – schedule
Terminplanung – schedule planning
Terminschwierigkeit – difficulty in keeping an appointment
Termintreue – punctuality (payment, delivery)
Terminverfolgung – monitoring of dates

Tertiärbereich – higher and advanced further education sector, third-level sector

tertiäre Bildung – tertiary education, third-level education (Der tertiäre Bereich umfasst die Weiterbildung an Hochschulen, Universitäten, Fach- und Gesamthochschulen.) (s. Graphik)

Test – test

Testabwicklung – test procedure

Testergebnis – test score

Testunterlagen – test material

Teuerungszulage – cost-of-living allowance (COLA)

Thema – theme, topic

Theoretiker – abstract thinker, theorist

theoretisch – theoretical

theoretische Grundlagen – theoretical basis

theoretische Kenntnisse – theoretical knowledge

therapeutisch – therapeutic

Therapie – therapy

Tiefwassermarke – earliest possible retroactive accounting period

Tilgungsdarlehen – amortizable loan

Tilgungsrate – repayment installment, redemption

Tilgungsrhythmus – repayment pattern

Tilgungssatz – repayment rate

Timing – timing

Tinte – ink

Tintenpatrone – ink cartridge

Tischler – joiner

Tochtergesellschaft – subsidiary

Tod – death

Todesfall – death

Todesmonat – month already accounted

Toleranz – tolerance

tolerante Atmosphäre – permissive atmosphere

Toleranzspanne – tolerance interval

Tonbandgerät – tape recorder

töricht – misguided

Träger der Krankenversicherung – health insurance carrier

Trainer – instructor, apprentices' teacher, trainer

Training im Karussellverfahren – carousel training

Transport – shipping

Trendberechnung – trend extrapolation

Trendextrapolation – trend extrapolation

Trennungsentschädigung – separation allowance

Trennungsgeld – separation allowance, separation compensation

Trennungszulage – separation allowance

treten – kick

treu – loyal

Treu und Glaube – good faith

Treuhänder – trustee

Treuhandfonds – trust fund

Treuhandvermögen – trust fund

Treuhandvertrag (Urkunde) – trust deed

Triebkraft – impetus, moving power, motivating force

triftiger Grund – good reason

Trimester – trimester

Trimesterhälfte – mid term

Trimesterende – end of trimester

Trinkgeld – tip

Trunkenbold – drunkard, drinker

Trunkenheit – drunkenness, intoxication

tüchtig – able, competent

Tüftler – finicky person

Tunnelblick – narrow-minded

Turnen – gymnastics

Turnhalle – gymnasium

Turnus – rotation

Turnusteilzeit – rotated part-time work (Die Mitarbeiter arbeiten nach von der Betriebsleitung festgelegten Rhythmen, z.B. eine Woche als Vollzeitkraft, die nächste Woche haben sie frei.)

turnusmäßiger Wechsel – rotation

Tutor – tutor

tyrannisieren – bully

U

üben – practise

über den Daumen gepeilt – guesstimate

Überalterung – obsolescence

Überangebot an Akademikern – surplus
of graduates
überarbeiten – revise, overwork, rework
Überbelastung – overload
**Überbesetzung mit nicht benötigten
Arbeitskräften** – featherbedding (Eine
von Gewerkschaften geforderte
Herabsetzung der Soll-Leistung zur
Verhinderung von Arbeitslosigkeit)
überbetriebliche Ausbildungsstätte –
industry-wide training centre
Überblick vermitteln – give a survey
Überbrückungsbeihilfe – transitional
allowance
Überbrückungsgeld – transitional
allowance
Überbrückungszulage – readjustment
allowance
Überdeckung – excess
überdenken – think over, reflect,
reconsider
überdurchschnittlich – above-average
übereinstimmen – agree, be in
agreement
Übereinstimmung – agreement,
conformity
überfliegen (beim Lesen) – skim (over)
überflüssig – redundant, superfluous,
unnecessary
überfordern – overtask, ask too much of
someone
Übergangsbeihilfe – transitional
allowance
Übergangsregelung – interim
arrangement
Übergangszahlung (Pensionäre) –
transitional payment
übergeordnete Ziele – superior
objectives
überheblich – arrogant
Überlagerung – overlap
Überlastung – overstrain, overstress
überlernen – overlearning (Überlernen
tritt dann ein, wenn bereits Erlerntes bei
einem Lernvorgang unnötigerweise
wiederholt wird.)
Übermaß – excess
Übernachtung – overnight stay
Übernachtungspauschale – lump sum
per night

**Übernahme in das
Angestelltenverhältnis** – transfer to
the salary payroll
Übernahmeverpflichtung – obligation to
hire
übernehmen – take over
Überprüfung – check, review
überqualifiziert – over-qualified
überreden – persuade, convince
Überschätzung – overestimation
Überschuss – surplus
überschüssig – in excess of (e.g. luggage
in excess of 20 kilos)
Überseezulage – overseas allowance
Überseezuschlag – overseas bonus
übersetzen – translate
Übersicht – survey, overview
Überspringen (Karriere) –
leap-frogging
**überstaatliches
Sozialversicherungsrecht** –
supranational social insurance law
Überstunden – overtime, extra hours
Überstunden leisten – work overtime
Überstunden machen (arbeiten) – work
overtime
Überstundenbezahlung – overtime pay
Überstundenlohn – overtime pay
Überstundenverbot – overtime ban
Überstundenzuschlag – overtime
premium, bonus for extra hours
übertariflich – above the general pay
scale
übertarifliche Bezahlung – exempt
salary
übertarifliche Rangstufe – exempt status
übertariflicher Kreis – exempt personnel
übertariflicher Mitarbeiter – exempt
member of staff
Übertragungsplan – vesting schedule
Übertragungsregel – vesting rule
übertreffen – surpass, exceed, excel
Überwachungseinrichtung – monitoring
device
Überweisung – bank transfer
Überweisung, externe – external bank
transfer
Überweisungsmonat – transfer month
Überzeit – unapproved overtime
Überzeugung – conviction
überzeugen – convince, persuade

Überzeugungsfähigkeit – persuasiveness
Überzeugungskraft – persuasiveness
Übung – exercise
Übungsperiode – practice period
Uhrzeit-Regelung – time rule
Uhrzeitintervall – time interval
Uhrzeitfeld – time field
Uhrzeitvertretung – time substitution
Ultimatum – ultimatum
umbestellen – redirect
umbuchen – rebook
Umbuchung – rebooking
Umbuchungsbestätigung – rebooking confirmation
Umbuchungsmitteilung – rebooking notification
Umdeutung – reinterpretation
umfassend – comprehensive, across the board
umfassende Bildung – all-round education
Umgangssprache – lingua franca
Umgebung – environment
umgehen – avoid
umgruppieren – restructure, redeploy
Umgruppierung – reassignment to wage group, reclassification
Umgruppierungsart – reclassification type
umherlaufend – peripatetic
Umkehrsatz – corollary
Umkehrverknüpfung – inverse relationship
Umkleideraum – changing room, dressing room
Umlage – contribution amount
umlaufen – circulate
umlernen – retrain, change one's ideas
umorganisieren – reorganize
umrennen – knock down
Umsatz – sales, turnover
Umsatzbeteiligung – participation in turnover; commission
umschulen – retrain
Umschüler – person being retrained
Umschulung – retraining
Umschulungsmaßnahmen – retraining measures
Umsetzung (beim Personaleinsatz) – personnel re-deployment
Umstand – factor

Umstellungsprobleme – changeover problems
umstrukturieren – restructure, redeploy
Umstufung – reassignment to wage level
Umstufung, tarifliche – pay scale reclassification
Umwelt – environment
Umweltbeauftragter einer Industriefirma – hygienist (AmE)
Umwelteinfluss – environmental influence
umweltfreundlich – ecologically beneficial
Umweltingenieur – hygiene engineer (AmE)
Umweltkunde – environmental studies
umweltschonend – ecologically beneficial
Umweltschutz – environmental protection
Umweltschutzbeauftragte(r) – environment protection officer
Umwelttechniker – pollution control technician
Umweltverschmutzung – environmental pollution
umweltverträglich – environmentally compatible
umziehen – relocate, remove
Umzug – move, removal
Umzugsbeihilfe – removal costs allowance
Umzugskosten – removal costs
Umzugskostenbeihilfe – removal allowance (BrE)
unabdingbar – indispensable
Unabdingbarkeit – indispensability
unabhängig – independent
unabhängige Schule – independent school
unanfechtbar – incontestable, unchallengable
unangebracht – misguided
unangepasst – non-conformist
Unangepasstheit – non-conformity
unauffällig – unobtrusive
unausgeglichen – imbalanced, maladjusted
unbeeinflusst – business-like, objective
unbefangen – impartial

unbefristetes Arbeitsverhältnis – permanent work relationship

unbefristeter Arbeitsvertrag – permanent work contract

unbefristetes Dienstverhältnis – unlimited employment

unbegabt – untalented, ungifted

unbegründete Angst – unfounded anxiety

unbeherrscht – uncontrolled

unbelesen – unread

unbeliebt – unpopular

unberechenbar – unpredictable

unbeschränkter Zugang – flat rate

unbesetzt – unoccupied, free, vacant

unbesetzte Planstelle – vacant position

unbewusst – unconscious

unbezahlt – unpaid, without pay

unbezahlter Urlaub – leave without pay

Unbilligkeit – inequity

undeutlich – indistinct (speech)

undichte Stelle – leak

unentschlossen – indecisive

unentschuldigt abwesend – absent without leave (AWOL)

unerfahren – inexperienced

unerlässlich – sine qua non, essential, imperative

unersetzbar – irreplaceable

unfähig – incompetent

Unfähigkeit – inability, incompetence

Unfall – accident

Unfallbericht – accident report

Unfallhäufigkeitsrate – accident frequency rate

Unfallrente – accident benefit

Unfallstatistik – accident statistics

Unfallverhütung – accident prevention

Unfallverhütungsvorschriften – accident prevention regulations

Unfallversicherung – accident insurance

Unfallversicherungsgesetz – Accident Insurance Act

Unfallversicherungstarif – accident insurance tariff

unflexibel – inflexible

ungebärdig – unruly

ungebildet – uneducated

ungefähr – approximate

ungehorsam – disobedient

ungekündigt – not under notice to leave

ungelernte Arbeitskraft – unskilled worker

ungelernte Tätigkeit – unskilled work

ungelernter Arbeiter – unskilled worker

ungenau – inaccurate

ungenügend – unsatisfactory

Ungerechtigkeit – injustice

ungeschickt – clumsy

Ungezogenheit – rudeness, impertinence

Ungleichheit – inequality, disparity

Unglücksfall – accident, casualty

ungültig – invalid

ungünstig – adverse, unfavourable

Universität – university

Universitätsabschluss – university degree

Universitätsabschluss mit Auszeichnung – honours degree, pass with distinction

Universitätsabschluss ohne Auszeichnung – pass degree

Universitätsabsolvent – university graduate

Universitätsausbildung – university training

Universitätsexamen – university degree

Universitätsdozent – don

Universitätsgelände – university grounds, campus (AmE)

Universitätsverband im Nordosten der U.S.A. – Ivy League

unkontrolliert – uncontrolled

unkündbare Stellung – tenured employment

Unlust – listlessness, reluctance

unmittelbarer Nachfolger – immediate successor

Unmöglichkeit – impossibility

unmoralisch – immoral

Unmündigkeit – minority

Unordnung – disorder

unorthodoxe Denkmethode – lateral thinking

unparteiisch – impartial, unbiased

unpersönlich – impersonal

unpfändbar – unseizable

unproduktive Zeit – idle time

Unpünktlichkeit – tardiness

unreif – immature

unrichtig – incorrect, false, wrong

unschlüssig – hesitant

unschöpferisch – eclectic
unselbständig – dependent
Unsicherheit – uncertainty
unsittlich – immoral
Unstimmigkeit – discrepancy
Unterbringung – accommodation
Unterbringung eines Kindes –
  placement of a child
Unterbringungskosten –
  accommodation costs
unterbrochenes Arbeitsverhältnis –
  interrupted work relationship
Unterdeckung – deficit
unterfordern – require too little of
  someone
Untergebene(r) – subordinate
Unterhalt – alimony; maintenance
Unterhalt, laufender – regular support
  payments
unterhalten (Familie) – maintain
Unterhaltsberechtigter – dependent
Unterhaltsgeld – maintenance allowance
Unterhaltspfändung – seizure of support
  payments
Unterhaltsrückstand – support payment
  arrears
Unterkunft – accommodation
Unterkunftsabrechnung –
  accommodations accounting
Unterkunftsbeleg – receipt for overnight
  stay
Unterlagen – papers, documents,
  credentials
Unternehmen – business, business
  enterprise, company, enterprise,
  corporation
Unternehmensaufbau – corporate
  structure
Unternehmensberater – consultant,
  management consultant
Unternehmensbereich – Group
Unternehmensfaktor – company factor
Unternehmensführung – corporate
  management
Unternehmensinteresse – company
  interest
unternehmensintern – internal company
  matter
Unternehmenskultur – corporate culture
Unternehmensleiter – works manager

Unternehmensleitung – Executive
  Committee
Unternehmensorganisation – corporate
  organization
Unternehmenspolitik – corporate policy
Unternehmensspiel – business game
  (Training an Simulationsmodellen für
  unternehmerische Entscheidungen)
Unternehmensspitze – corporate
  management
Unternehmensstrategie – corporate
  strategy
Unternehmensumfeld – business
  environment
Unternehmensziel – management
  objective
Unternehmenszweck – organizational
  purpose
Unternehmer – employer, industrialist
unternehmerisches Denken – business
  thinking, thinking in business
  categories
Unterordnung – subordination
Unterricht – lessons, classes, tuition
Unterricht im Klassenzimmer –
  classroom instruction
Unterricht im Team – team teaching
unterrichten – instruct, teach
Unterrichtsboykott – school boycott
Unterrichtseinheit – period (BrE),
  teaching unit
Unterrichtserfahrung – teaching
  practice
Unterrichtsfach – subject
Unterrichtsgegenstand – subject
Unterrichtsgruppen unterschiedlicher
  Altersstufen – multigrade classroom
Unterrichtsinhalt – subject matter
Unterrichtsmaterial – teaching material
Unterrichtsmethode – teaching method
Unterrichtsplan – curriculum, syllabus
Unterrichtsraum – classroom
Unterrichtsstil – style of teaching
Unterrichtsstoff – subject matter
Unterrichtsstunde – lesson, class, period
Unterrichtsstunden – contact hours,
  teaching load
Unterrichtsstunde mit einem Tutor –
  tutorial

Unterrichtsstunden je Woche – teaching load per week; classroom hours per week

Unterschicht – lower class

Unterschied – difference

unterschiedlich behandeln – discriminate

unterschiedliche Behandlung – discrimination

unterschriftsberechtigt – authorized to sign

Unterstufe – junior high school (AmE), junior cycle, intermediate school (s. Graphik)

unterstützen – support

Unterstützungsfonds – welfare fund

Unterstützungskasse – relief fund

Unterstützungskassenrichtlinien – relief fund regulations

Unterstützungspflicht – obligation to make assistance payments

untersuchen – investigate

Untersuchung – investigation

Untersuchung der Arbeitsmoral – morale survey

Untersuchungsausschuss – commission of enquiry

Unterweisung – lessons, classes, instruction

Unterweisungsplan – plan of instruction

ununterbrochen – continuous

unvereinbar – incompatible

unverfallbar – non-forfeitable

unverfallbare Pensionsanwartschaft – vested pension right

Unverfallbarkeit (bei Pensionsanwartschaft) – non-forfeitability (of pension expectancy)

Unverfallbarkeit (z.B. von Ruhegeld) – vesting, vested right

Unverfallbarkeitsbescheinigung – non-forfeitability certificate

Unverfallbarkeitsfrist – non-forfeitability limit

Unvermögen – inability, incompetence

unverträglich – incompatible

unvoreingenommen – unbiased

unvorhergesehenes Ereignis – contingency

Unzufriedenheit – dissatisfaction

unzumutbar – unreasonable demand

unzuverlässig – unreliable

Urabstimmung – strike/no strike vote

ureigen – original, inherent, innate, one's very own

Urheberrecht – copyright

Urlaub – holiday, vacation, leave

Urlaub, abgerechneter – leave taken and deducted

Urlaubsabfindung – leave compensation

Urlaubsabgeltung – vacation allowance (AmE), leave compensation

Urlaubsabtragung – leave deduction

Urlaubsanspruch – right to holidays, leave entitlement

Urlaubsanspruchsermittlung – calculation of leave entitlement

Urlaubsart – leave type

Urlaubsaufbau – leave accrual

Urlaubsauszählung – counting of leave days

Urlaubsberechnung – calculation of vacation

Urlaubsgeld – vacation allowance, vacation bonus, holiday bonus, holiday money

Urlaubsgeldordnung – vacation pay regulations

Urlaubsgeldrückstellung – reserves for vacation bonus

Urlaubsjahr – leave year

Urlaubskalender – holiday calendar

Urlaubsklasse – leave class

Urlaubskonto – leave account

Urlaubslohnfortzahlung – continued pay during approved leave from work

Urlaubsplan – leave schedule, vacation schedule (AmE)

Urlaubsregelung – holiday regulation

Urlaubssaldo – leave balance

Urlaubssatz – leave record

Urlaubsstand – leave information, accrued and taken leave

Urlaubstag – day's leave, leave day

Urlaubstag, abgerechneter – leave days paid

Urlaubsverlängerung – longer holidays

Urlaubsvertretung – holiday replacement (BrE), vacation replacement (AmE)

Urlaubszeit – leave period

Ursache – cause
Ursächlichkeit – causality
Urteil – judgment, verdict
Urteilstendenz – appraisal tendency

V

vakant – vacant
Vakanz – vacancy
Vandale – vandal
variable Entgeltbestandteile – variable income elements
variables Budget – variable budget
Varianz – variance
Variationsbreite der Führungsstile – range in management styles
Vaterschaft – paternity
Verabredung – appointment
verändern – alter, change
Veranlassung – cause, ground, reason
Veranstalter – organizer
Veranstaltung – business event
Veranstaltung, mehrteilige – multi-session business event
Veranstaltungsbeurteilung – business event appraisal
Veranstaltungsgruppe – business event group
Veranstaltungshierarchie – business event hierarchy
Veranstaltungskatalog – business event catalog
Veranstaltungsmanagement – seminar and convention management
Veranstaltungsmarketing – business event marketing
Veranstaltungsort – business event location
Veranstaltungsplanung – business event planning
Veranstaltungstyp – business event type
Veranstaltungsumfeld – business event environment
verantwortlich – responsible

Verantwortliche(r) – person in charge, person responsible
Verantwortlichkeit – responsibility, accountability
Verantwortung – responsibility, accountability
Verantwortung für nicht delegierte Aufgaben – reserved responsibility
Verantwortung tragen – bear responsibility
Verantwortung übernehmen – assume responsibility
Verantwortung übertragen – delegate responsibility
Verantwortungsspielraum – boundaries of responsibility
verarbeiten – process
Verarbeitungsklasse – processing class
Verarbeitungstyp – processing type
Verband – association
verbale Gewalt – verbal violence
verbessern – correct, improve
Verbesserung – correction, improvement
Verbesserungsvorschlag – suggestion for improvement
Verbesserungsvorschlagswesen – suggestion scheme
Verbleibende – survivors
verbieten – forbid, prohibit
Verbindung – contact, connection; fraternity (AmE)
Verbindungsmann – liaison officer
Verbreitung von Informationen – dissemination of information
verbundenes Unternehmen – subsidiary
verbürgt – authentic
verdienen – earn
Verdienst – earnings; merit
Verdienstausfall – loss of earnings
Verdienstausfallentschädigung – earnings loss compensation; compensation for lost income
Verdienstbescheinigung – statement of earnings
Verdiensterhebung – survey of earnings
Verdienstgrad – earnings factor
Verdienstmöglichkeit – potential earnings
Verdienstnachweis – statement of earnings

Verdienstsicherung – standard wage maintenance
Verdienstübersicht – pay slip
Verdrängen – bumping, displace
Verdrängung – displacement
Verein – club, society, association, organisation
vereinbar – compatible
vereinbaren – arrange, agree, come to an understanding
Vereinbarung – agreement, settlement
vereinfachen – modernize, rationalize, streamline, simplify
Vererbung – heredity
Verfahren – method, procedure, process, technique
Verfahrenstechnik – process engineering
Verfall des Wissens – lapsing of knowledge
verfallen – expire
Verfallsfrist – expiration date
Verfassungsbeschwerde – constitutional complaint
Verfechter(in) des Gleichheitsprinzips – egalitarian
verfügbar – available
verfügbare Arbeitskräfte – labour supply
verfügbares Einkommen – disposable income
Verfügbarkeit – availability
verfügen über – have at one's disposal
Vergangenheitsprinzip – reference principle
vergessen – forget
Vergessenskurve – forgetting curve
Vergiftung – poisoning
Vergleich – comparison, compromise
vergleichbar – comparable
vergleichen – compare
Vergleichsbogen – comparison sheet
Vergrößerung – enlargement
Vergünstigungen durch die Firma – non-monetary compensation, remuneration in kind, goodies
vergüten – pay (for), remunerate
Vergütung – remuneration
Vergütung der Arbeit – remuneration of work
Vergütung von Überstunden – payment of overtime

Vergütung von Zuschlägen – payment of bonuses
Vergütungsanspruch – entitlement to remuneration
Vergütungsgruppe – entitlement group
Vergütungsmatrix – remuneration matrix
Vergütungsregelung – payment provision
Vergütungssystem – remuneration system
Vergütungsverwaltung – remuneration management
Verhalten – behaviour, bearing, demeanour
Verhalten gegenüber Außenstehenden – representative role
Verhalten in der Zusammenarbeit – teamwork performance
Verhaltensänderung – behavior modification, behavior change
verhaltensgestört – maladapted behaviour, disturbed
verhaltensgestörter Schüler – disturbed pupil
Verhaltensgitter – managerial grid
Verhaltensmuster – pattern of behaviour; behavioural pattern
Verhaltensnorm – behavioural standard
Verhaltenspsychologie – behavioural psychology, behaviourism
Verhältnis – relation
Verhältnis der Einkommen vor und nach der Pensionierung – replacement ratio
Verhältnis der Tarifparteien – industrial relations
verhandeln – negotiate
Verhandlung – negotiation
Verhandlungsgegenstand – matter for negotiation
Verhandlungsordnung – negotiation procedure
Verhandlungspartner – negotiating partner
Verhandlungsrecht – right to negotiate
Verhandlungsverfahren – negotiation procedure
verheiratet – married
verhindern – prevent

verjähren – become obsolete, become null and void
Verjährung – limitation prescription
Verjährungsfrist – term of limitation, term of prescription
Verkäufer(in) – seller, vendor; shop assistant
Verkaufsprognose – sales forecast
Verkehrssicherheit – road safety
Verkehrssprache – lingua franca ⬎
Verknüpfung – connection, linkage
verkürzte Arbeitszeit – short time
Verladung – shipping
verlängern – extend
verlängerter schuldrechtlicher Versorgungsausgleich – extended contractual pension adjustment
verlängerter Urlaub – longer holidays
verlängertes Wochenende – happy Monday
Verlängerungsantrag – request for extension
Verlangsamung der Arbeitsweise – go slow, work to rule
verlegen – adjourn
verlernen – forget
verletzen – injure
Verlierer – loser
Verlust – loss
Verlust des Arbeitsplatzes – job loss, loss of job, redundancy
Verlust des Einkommens – loss of income
Vermehrung – increase, increment
vermeiden – avoid
Vermeidung von Diskriminierung – equal employment opportunity
Vermerk – memorandum, memo, note
verminderte Erwerbsfähigkeit – impaired working ability
Verminderung – decrease
vermitteln – mediate, place
Vermittler – go-between, intermediary, placement officer, facilitator
Vermittlung – intervention, good offices
Vermittlung eines Arbeitsplatzes – work placement
Vermittlungsausschuss – conciliation board
Vermittlungsverfahren – conciliation proceedings

Vermögensanlage – investment of capital
Vermögensbestätigung – capital formation confirmation
Vermögensbildung – formation of wealth, formation of capital
Vermögensbildungsvertrag – capital formation contract
Vermögenspolitik – asset-formation policy
vermögenswirksam – asset-forming
vermögenswirksame Anlage – asset-creating capital
vermögenswirksame Leistung – asset-formation benefit, capital formation savings payment (Employer's contributions to tax-deductible savings scheme.)
vernachlässigen – neglect
Veröffentlichung – publication
Verpflegung – meals
Verpflegungsabrechnung – meals accounting
Verpflegungsaufwendungen – expenses for meals
Verpflegungsbeleg – meals receipt
Verpflegungshöchstsatz – maximum rate for meals
Verpflegungskosten – expenses incurred for meals
Verpflegungsmehraufwendung – additional expenses for meals
Verpflegungspauschale – lump sum for meals
Verpflichtung – duty, obligation
Verräter – whistle-blower, traitor (Ein Whistleblower ist ein Forscher, der Verfehlungen aus dem eigenen wissenschaftlichem Umfeld öffentlich macht.)
Verrechnungsschlüssel – compensation key
Versachlichung – objectification
versagen – fail
Versammlung – assembly
versäumen – neglect
verschieden – different, miscellaneous
verschieden sein (von) – differ (from)
Verschleißerkrankung – degenerative disease
verschmelzen – amalgamate
Verschmelzung – amalgamation

verschmitzt – mischievous
Verschuldenshaftung – fault liability
Verschwörung – conspiracy
versetzen – transfer; delegation; progress to next level
Versetzung – transfer; delegation; progress to next level
Versetzung in den Ruhestand – retirement
Versetzungsgesuch – application for transferral, transfer request
versichern – insure
Versichertenrente – pension from the social insurance
Versicherung – insurance
Versicherung auf Gegenseitigkeit – mutual insurance
Versicherungen, externe – external insurances
Versicherungsabzug – insurance deduction
Versicherungsanstalt – insurance company
Versicherungsbeitrag – insurance contribution
Versicherungsdeckung – insurance coverage
Versicherungsfall – insurance case
versicherungsfreie Beschäftigung – insurance-exempt employment
Versicherungsfreiheit – insurance exempt
Versicherungsgesellschaft – insurance company
Versicherungsindex – insurance index
Versicherungsjahr – policy year
Versicherungskonditionen – terms and conditions of insurance
Versicherungslauf – insurance progress
Versicherungsleistung – insurance benefit
versicherungsmathematische Annahmen – actuarial assumptions
versicherungsmathematische Bewertung – actuarial evaluation
Versicherungsmonat – policy month
Versicherungsnachweis – insurance records
Versicherungsnennwert – reinstatement value

Versicherungsnummer – insurance policy number
Versicherungspflicht – compulsory insurance
Versicherungspflicht, allgemeine – general obligatory insurance
Versicherungspflichtgrenze – insurance liability limit
versicherungspflichtig – subject to compulsory insurance
versicherungspflichtige Beschäftigung – contributory employment
Versicherungspflichtige(r) – person subject to compulsory insurance deductions
Versicherungsprämie – insurance premium
Versicherungsrente – retirement annuity
Versicherungsschein – insurance policy
Versicherungsschutz – insurance coverage
Versicherungssumme – amount insured
Versicherungstarif – insurance tariff
Versicherungsträger – insurer
Versicherungsverein auf Gegenseitigkeit – mutual insurance association
Versicherungsverlauf – total of all contributory and non-contributory periods to social insurance
Versicherungszeit – insurance period
Versicherungszielmonat – insurance lead month
versiert – experienced, practical
Versionsnummer – version number
versorgen – provide, supply
Versorgung – provision, supply
Versorgungsamt – state benefits office
Versorgungsanspruch – benefits claim
Versorgungsausgleich – pension adjustment
Versorgungsbezüge – pensions and related benefits
Versorgungseinrichtung – facilities, supply facilities
Versorgungsfall – insured event
Versorgungsfreibetrag – personal tax exemption for retirement benefits
Versorgungsträger – pension-paying institution

Versorgungszusage – employer's
pension commitment
Verspätung – late arrival
Versprechen – promise
Verstand – intellect, mind
Verständigung – understanding,
agreement, communication
Verständigungsproblem –
communication problem
Verständnis – ability to grasp,
comprehension, understanding
verstärken – reinforce
verständlich – understandable
Verstärkung – reinforcement
verstellen – act a part, hide one's
feelings, simulate
Verstellung – simulation
versteuerbares Einkommen – taxable
income
Verstrickung – involvement
Versuch – attempt
Versuchsanstalt – research institute
Versuchsstadium – experimental stage
versuchsweise – tentatively
vertagen – adjourn
verteilen – distribute, spread (over)
Verteilung – distribution
Verteilung auf – spread
Verteuerung der Lebenshaltung –
increase in the cost of living
Vertieftsein – absorbed
vertikale Dezentralisierung – vertical
decentralization
Vertrag – contract
vertraglich – contractual
vertragliche Arbeitszeit – contractual
working hours
Vertragsabschluss – conclusion of a
contract
Vertragsabteilung – contract department
Vertragsbedingung – contract condition
Vertragsbestimmung – contract
condition
Vertragsdaten – contract data
Vertragsdauer – contract duration
Vertragsentwurf – draft contract
Vertragsgruppe – contract group
Vertragsrecht – contract law
Vertragsstrafe – penalty for breach of
contract, forfeit

Vertragsverhandlung – contract
negotiation
Vertrauen – confidence, trust
Vertrauensarzt – medical examiner,
medical referee (AmE), health officer
Vertrauensgleitzeit – flexitime on trust
basis
Vertrauenskultur – culture of trust
Vertrauensmann – shop steward, liaison
officer
**Vertrauensmann der Schwer-
behinderten** – disabled persons'
representative
Vertrauensperson – confidant(e)
Vertrauensschüler – prefect
Vertrauensstellung – confidential
position, position of trust
vertrauenswürdig – trustworthy
vertraulich – confidential
Vertraulichkeit – confidentiality
vertreten – represent
Vertreter – substitute; representative
Vertretung – stand-in, substitution
Vertretungsstunde – substitution hour
Vertriebserfolg – sales success
Vertriebsmitarbeiter – sales staff
Verunsicherung – uncertainty
verursachen – cause
**Vervielfachung (bei
Direktversicherungsbeiträgen)** –
direct insurance contributions
verwalten – administer, handle
Verwaltung – administration
Verwaltungsakt – administrative act
Verwaltungsaufgaben – administrative
tasks
Verwaltungsrat – board of Directors
(Rat, Ausschuss, Kommission von
Direktoren, Betriebsführungsgremium
mit Befugnissen, die weiter gehen als
die eines Aufsichtsrates, fast dem
Vorstand entsprechend.)
verwarnen – reprimand, warn
Verwarnung – warning
Verwässerungsabkommen – dilution
agreement
Verweigerung – refusal
Verweigerungshaltung – attitude of
refusal
Verweis erteilen – reprimand
Verweisung – expulsion

Verwendungsverbot – prohibition of use
Verwicklung – involvement
verwirken – forfeit
verwitwet – widowed
Verzeichnis – directory
Verzicht – renunciation, waiver
verzichten auf – do without, dispense with something
verzögern – delay
Verzögerung – delay
Vetternwirtschaft – nepotism
vielseitig – versatile
Vielseitigkeit – versatility
vielsprachig – polyglot
Vier-Augen-Gespräch – personal discussion
Vier-Augen-Prinzip – double verification principle
Vier-Tage-Woche – four-day week (Die Arbeitszeit wird verkürzt, wobei die Mitarbeiter auf einen Teil des Lohns verzichten. Die Umlage von Jahressonderzahlungen auf das Monatsgehalt mindert die Einkommenseinbußen ab.)
Vierteljahresgehalt – quarterly salary
Vierzig-Stunden-Woche – forty hour week
virtuelles Unternehmen – virtual company
Vision – vision
visionär – visionary
Visitenkarte – business card, calling card, visiting card, card
visualisieren – imagine, visualize
Visualisierung – visualizing ability
Vokabeltest – vocabulary test
Vokal – vowel
Volkshochschule (VHS) – Volkshochschule; adult education centre
Volksschule – primary school (BrE), elementary school, grade school (AmE) (s. Graphik)
Volksschulpflicht – compulsory primary education
Volkswirt – graduate in national economics
Volkswirtschaft – economics, national economics, national economy

Volkswirtschaftslehre – economic science, economics
Vollbeschäftigung – full employment
Vollendung – completion
voller Leistungsanreiz – hundred per cent incentive
volljährig – of age, major
Vollmacht – authority, power, power of attorney
vollständige und dauernde Arbeitsunfähigkeit – total permanent incapacity
Vollständigkeitsvektor – completion vector
Vollversorgung (mittels einer Rente) – full superannuation
Vollzeitarbeitsplatz – full-time work place
Vollzeitausbildung – full-time training
Vollzeitbeschäftigung – full-time employment
Vollzeitentgelt – full-time payment
Vollzeitlehre – full-time apprenticeship
Vollzeitlehrer – full-time teacher
Vollzeitschulpflicht – full-time compulsory education
Vollzeittelearbeit – full-time telework
Volontär – volunteer, trainee
vom Dienst entheben (arbeitgeberseitig) – give notice
vom Lohn einbehalten – withhold from wages, deduct from pay
vom Schulleiter bestimmter Schulsprecher – headboy
von der Führung ausgeübter Druck – management pressure
von der Schule ausschließen – expel
Vorankündigung – advance notice
Voranmeldung (Kurs) – advance booking
Vorarbeit – preparatory work
Vorarbeiter – foreman, ganger
Vorarbeitgeber – previous employer
Vorausplanung – prior planning
voraussagen – predict
Vorausschau – forecast
Voraussetzung – prerequisite, precondition
voraussichtlich – presumable, prospective
Vorauswahl – preselection

Vorbehalt – reservation
vorbehaltlich – provided that, subject to
vorbereiten – prepare
Vorbereitung auf den Ruhestand – preparation for retirement
Vorbereitung auf eine Berufstätigkeit – preparation for a professional activity
Vorbereitung auf Führungsaufgaben – preparation for managerial tasks
Vorbereitungskurs – preparatory course
vorbestraft – previously convicted
Vorbild – example
Vordiplom – first diploma
voreingenommen – biased
Vorgabeleistung – standard performance
Vorgabezeit – allowed time, standard time
Vorgänger – predecessor
Vorgehen – procedure
Vorgehensweise – procedure
Vorgesetztenschulung – supervisory training
Vorgesetztenverhalten – supervisory behaviour
Vorgesetzter – superior, boss
vorgezogene Pensionierung – early retirement
vorgezogener Ruhestand – early retirement
vorhanden – available
Vorherrschen – prevalence
vorherrschend – dominant
Vorhersage – prognosis
vorhersagen – forecast, predict
Vorholzeit – compensation time
vorläufig – interim, provisional
Vorleistung – preliminary work
Vorlesung (Uni) – lecture
Vorlesungsverzeichnis – university calendar, university catalogue, lecture timetable
Vorliebe – preference, prediliciton
vormerken – prebook, reserve
Vormerkung – prebooking, reservation
Vormerkungsliste – prebooking list, reservation list
Vormerkzeitraum – desired period
Vormund – guardian
Vorname – first name, Christian name
Vorprüfung – preliminary examination
Vorrat – pool

Vorrecht der Führungskräfte – management privilege
Vorruhestand – early retirement
Vorruhestandsregulung – early retirement agreement
Vorruheständler – early retiree
Vorsatz – surname prefix; resolution
vorsätzlich – intentional, deliberate
Vorschau – forecast
vorschlagen – propose
Vorschlagskasten – suggestion box
Vorschlagskatalog – proposal catalog
Vorschlagsrecht – right of nomination
Vorschlagswesen (betriebliches) – suggestion scheme
Vorschlagswesen, innerbetriebliches – employee suggestion system
Vorschrift – regulation, guideline
vorschriftsmäßig – correct, proper
Vorschul... – preschool (s. Graphik)
Vorschulalter – pre-school age
Vorschulbildung – nursery education, pre-school level (Im Elementarbereich der Vorschulbildung können Kinder ab 3 Jahren den Vorkindergarten bzw. Kindergarten besuchen) (s. Graphik)
Vorschulerziehung – pre-school education, pre-schooling
Vorschulkind – nursery school child
Vorschuss – advance, advance payment
Vorschusserfassung – advance entry
Vorschusszahlung – advance payment
Vorsicht – caution, care
vorsichtige Schätzung – conservative estimate
Vorsitzende(r) – chairperson; president (AmE)
Vorsitzender des Aufsichtsrates – Chairman of the Supervisory Board
Vorsitzender des Bereichsvorstands – Senior Vice President, Group President
Vorsorge – provision, provisory care
Vorsorge, betriebliche – company pension plan
Vorsorgeleistung – preventive medical care
Vorsorgepauschale – provisional lump sum
Vorsorgeuntersuchung – precautionary checkup, preventive medical checkup

vorsorglich Maßnahmen ergreifen – pro-active

Vorstand – Managing Board, Board of Management

Vorstandsbereich – Managing Board, Board of Management

Vorstandsvorsitzender – Chairman of the Board

Vorstandsvorsitzender eines Unternehmens – President

Vorstandsvorsitzender, stellvertretender – Deputy Chairman of the Board

Vorstellungsdatum – date of interview

Vorstellungsgespräch – employment interview, interview, job interview

Vorstellungskraft – imagination

Vorstellungsvermögen – power of imagination

Vorstrafe – previous conviction

Vortag – previous day

Vortageskennzeichen – indicator for previous day

Vorteil – advantage

Vortrag – lecture

Vortragsweise – delivery

vorübergehende Arbeitsunfähigkeit – temporary incapacity

vorübergehende Entlassung – temporary lay-off

Vorurteil – prejudice

vorurteilsfrei – unprejudiced

Vorversicherungszeit – ante insurance period

Vorwarntermin – advance warning deadline

vorwärtsstrebend – go-ahead

Vorwegnahme – anticipation

vorzeitige Aufnahme – premature admittance

vorzeitige Invalidität – premature disability

vorzeitige Pensionierung – early retirement

vorzeitiges Altersruhegeld – early retirement benefit

vorzüglich – excellent

Vorzugsbedingungen, um einen Mitarbeiter an die Firma zu binden – golden handcuff

Vorzugsbedingungen, um einen Mitarbeiter loszuwerden – golden handshake

Vorzugspreis – preferential price

W

Wachstum – growth

Wachstum ohne neue Arbeitsplätze – jobless growth

wagemutig – courageous

wagen – hazard, risk, venture

Wagniskapital – venture capital

Wahl – selection, choice, election

Wahl der Arbeitnehmervertreter – election of employee representatives

Wahl der Wahlmänner – election of delegates

Wahlanfechtung – election contest

Wählbarkeit – eligibility

Wahlberechtigung – right to vote, voting right

Wählerliste – voters' list

Wahlfach – optional subject

wahlfreie Frage – optional question

Wahlfreiheit – freedom of choice

Wahlspruch der Schule – school motto

Wahlvorgang – electoral process

Wahlvorschlag(sliste) – nomination list

Wahlvorschriften – election regulations

Wahlvorstand – electoral board, returning board (AmE)

Wahlzettel – ballot, ballot paper, voting paper

während der Dienstzeit – in-hours

Wahrnehmung – perception

Wahrnehmungsfähigkeit – power of perception

Wahrnehmungsstörung – disturbance in the power of perception

Wahrscheinlichkeit – probability

Waisenrente – orphan's allowance

Wanderarbeitnehmer – seasonal worker, migrant worker

Wanderversicherung – itinerant insurance

Wandtafel – board, blackboard

Warenbestand – inventory

Warenvorrat – inventory

Warnstreik – token strike

Wartelistebuchung – waiting-list booking

Wartelistepriorität – waiting-list priority

Wartezeit – waiting period, qualifying period

Wartezeitfiktion – fictitious qualifying periods

Wartezeitverkürzung – reduced qualifying period

Wartung – upkeep, maintenance

Wartungspersonal – maintenance personnel

Wechselschicht – rotating shift

Wechselschichtprämie – rotating shift premium

Wechselwirkung – interaction

Wegeunfall – travel accident

Wegezeit – time travelled

Wegezeitberechnung – computation of home-to-office time

weglassen – omit

wegrationalisieren – rationalize away

Wehrdienst – national service, military service

Wehrdienstverweigerer – conscientious objector

wehrpflichtig – liable to military service

Wehrpflichtiger – draftee, conscript

Wehrübung – reserve duty training exercise

weicher Faktor (nicht fachliche Kompetenz) – soft factor (social competence)

Weihnachtsfreibetrag – tax-free Christmas allowance

Weihnachtsgeld – Christmas bonus

Weihnachtsgratifikation – Christmas bonus

weise – wise

Weisung – command, directive

weisungsabhängig – subject to directives

Weisungsbefugnis – authority to instruct, right to issue instructions, command authority

weisungsgebunden – subject to directives

Weisungsrecht – authority to instruct, right to issue instructions

weite Verbreitung – prevalence

Weiterbeschäftigung – continued employment

Weiterbezahlung – continued payment

Weiterbildung – continuing education, on-going education, further education (umfasst alle nachschulischen Angebote)

Weiterbildung i.D. – in-hours on-going education

Weiterbildung von Führungskräften – further education of management

Weiterbildungsangebot – courses on offer for further education

Weiterbildungsaufwand – costs for continuing education

Weiterbildungsbedarf – demand for further education and training

Weiterbildungskandidat – candidate for further education and training

Weiterbildungsmaßnahmen für Lehrer – in-service training for teachers

Weiterbildungsplanung – further education and training planning

Weiterentwicklung – further development

weiterführende Hochschule – graduate school

weiterführende Schule – secondary school (Alle Schularten, die nach Jahrgangsstufe 4 bzw. 6 beginnen, sind sogenannte weiterführende Schulen: Hauptschule, Realschule, Gymnasium und alle Schulen im beruflichen Bereich. Das Gymnasium baut auf der Grundschule auf und umfasst die Jahrgangsstufen 5 mit 13.) (s. Graphik)

weiterführende Studien – post-graduate studies (Studien, die nach dem ersten akademischen Grad weitergeführt werden)

Weiterversicherung – continued insurance

weitschweifig – long-winded, tedious, circumstantial, prolix

weltoffen – liberal minded, cosmopolitan

Werbemaßnahmen – advertising measures

Werbeschrift – leaflet, prospectus

Werbungskosten – income-related expenses

Werdegang – background, career

Werdegang eines Mitarbeiters – employee's background

Werk – works, plant

Werkarzt – company doctor

werksärztlicher Dienst – internal medical service

Werkleiter – works manager

Werksausweis – plant ID card

Werkschutz – plant security

Werksfeuerwehr – plant fire department

Werkschließung – plant closure

Werkskalender – factory calendar

Werkstatt – workshop

Werkstoff – material

Werkstoffkunde – material technology

Werkstofflehre – material technology

Werkstudent – industrial student, temporary student employee

Werkstudenteneinsatz – employment of industrial students

Werkswohnung – company flat

Werktag – working day

Werktagnummer – weekday number

Werktätige(r) – working man, working woman

Werktyp – plant category

Werkvertrag – work performance contract

Werkzeitschrift – house journal, plant magazine

Werkzeugmacher – toolmaker

Werkzeugmechaniker – tool mechanic

Wert – value

Wertesystem – system of values

Wertewandel – change of values

Wertung – measurement

wesentlich – essential

Wettbewerb – competition

wettbewerbsfähig – competitive

Wettbewerbsfähigkeit – competitiveness

Wettbewerbsklausel – non-competition clause

Wettbewerbsnachteil – competitive disadvantage

Wettbewerbsverbot – non-competition clause

Wettbewerbsvorteil – competitive advantage

Wichtigkeit – importance

Widerruf – cancellation, revocation

widerrufen – cancel, revoke

widerspiegeln – reflect

Widerspruch – protest, objection

Widerspruchsklage – third-party action against execution

Widerspruchsstelle – appeal committee

Widerstand – resistance

Widerstand gegen Neuerungen – resistance to change

Widerwille – aversion

wie mit Ihnen vereinbart – as per our discussions

wieder in Betrieb nehmen – reopen

wiederanstellen – reemploy, reengage

Wiederaufnahme – resumption

wiederaufnehmen – resume

Wiederbeschäftigung – reemployment

Wiedereingliederung – reintegration

Wiedereingliederungsmaßnahme – resettlement measure

Wiedereingliederungszusage – promise of reintegration

wiedereinsetzen – reinstate

Wiedereinsteiger – re-entrant

wiedereinstellen – reemploy, reengage

Wiedereinstellung – reemployment, re-hiring

Wiedereintritt – re-entry, re-hiring

Wiedererkennen – recognition

wiedereröffnen – reopen

wiedergeben – repeat, reproduce

wiederholen – repeat

Wiederholung – repetition

Wiederholungskurs – refresher course

Wiederholungslektion – review lesson

Wiederholungsprüfung – repeat exam(ination)

wilder Streik – wildcat strike

Willensstärke – willpower

wirkliches Begreifen – meaningful learning

wirksame Kommunikation – effective communication

Wirksamkeit des Trainings – effectiveness of training

Wirkung – consequence, result, effect
Wirkungsbereich – sphere of action
Wirkungslosigkeit – ineffectiveness
Wirtschaft – economy
Wirtschafts(ober)schule – commercial college
Wirtschaftsausschuss – committee for economic policies
Wirtschaftsführer – business leader
Wirtschaftsinformatik – business management (Wirtschaftsinformatik ist ein Studium der Betriebswirtschaftslehre (BWL) mit stärkerem DV-Anteil.)
Wirtschaftslage – market prospects, state of the economy
Wirtschaftsprüfer – chartered accountant (BrE), certified public accountant (AmE), CPA
Wirtschaftsrechnung – internal financial statement
Wirtschaftsschule – business school (Die Wirtschaftsschule schließt an die Jahrgangsstufe 6 oder 7 der Hauptschule an. Sie führt in vier bzw. drei Jahren zu einem mittleren Schulabschluss. Sie fördert die Allgemeinbildung und vermittelt eine berufliche Grundbildung in den Bereichen Wirtschaft und Verwaltung.)
Wirtschaftsunternehmen – business enterprise, company
Wirtschaftsverbrechen – white collar crime
Wirtschaftswachstum ohne neue Arbeitsplätze – jobless growth
Wirtschaftswissenschaft – economic science, economics
Wirtschaftswissenschaftliches Gymnasium (WWG) – secondary school emphasizing business and economics
Wirtschaftszweig – branch of industry
Wissen – knowledge
Wissensbasis – knowledge base
Wissenschaft – science
Wissenschaftler – scientist, academic expert, egg head (AmE)
wissenschaftlich – scientific, academic
wissenschaftliche Freiheit – academic freedom

wissenschaftliche Gesellschaft – scientific society
wissenschaftliche Unternehmensführung – scientific management
wissenschaftlicher Assistent – assistant lecturer
Wissenschaftlichkeit – scientific method, academic approach
Wissenschaftsfreiheit – freedom of research
Wissenserwerb – acquisition of knowledge
Wissensgesellschaft – knowledge society (eine Gesellschaft, die ihre Lebensgrundlagen aus reflektiertem und bewertetem Wissen gewinnt, die von neuen Möglichkeiten einen bewussten und lebenserleichternden, sozial nicht zerstörenden Gebrauch macht)
Wissensindustrie – knowledge industry
Wissensmanagement – knowledge management
Wissensnetze – knowledge networks
Wissensverfall – outdatedness of knowledge
Wissensvermittler – knowledge engineer
Wissensvorsprung – knowledge head start
Witwenrente – widow's pension
Witwerrente – widower's pension
Wochenarbeitszeit – working week
Wochenhilfe – maternity benefit
Wochenlohn – weekly wage
Wochentag – day of the week
wöchentliche Arbeitszeit – weekly working hours
Wohltätigkeit – charity
Wohlwollen – goodwill
Wohngeld – housing subsidy
Wohngeldbescheinigung – statement of housing subsidy
Wohngeldzuschuss – housing allowance
Wohnort – residence, location
Wohnsitz – residence
Wohnsitz, ständiger – permanent residence
Wohnsitzsteuer (USA) – residence state tax

Wohnungsbauförderung – housing support
Wohnungsdarlehen – housing loan
Wohnungspolitik – housing policy
Wohnungsreferat – housing administration
Wort – word
Wortart – part of speech
wörtliche übersetzung – literal translation
Wortschatz – vocabulary
Wortschatzübung – vocabulary drill
Wunderkind – child prodigy
Wunschberuf – dream job
Wunschprofil – preference profile
Würde – dignity

Z

zählen – count
Zahlenverhältnis – pupil-teacher ratio
Zähler – numerator
Zählkind – count child
Zählklasse – counting class
Zahlstelle – paying office, office of payments (BrE)
Zahltag – pay-day
Zahlung – payment, remuneration
Zahlungsbetrag – payment amount
zahlungsfähig – solvent
Zahlungsverbot, vorläufiges – temporary freezing of payments
Zahlungsverzug – default, delay of payment
Zahlungszeitpunkt – payment date
Zank – argument, quarel, row
Zappelphillip – fidgelty person
Zeichen – character
Zeichnen am Bildschirm – computer-aided design, CAD
Zeichnung – drawing
Zeigestock – pointer
Zeile – line, row
Zeit einteilen – manage time
Zeit planen – manage time

Zeit, abgelaufene – expiration
Zeit- und Bewegungsstudie – time and motion study
Zeitabgleich – time leveling
Zeitabrechnung – time accounting
Zeitabschlag – time deduction
Zeitabschnitt – time segment
Zeitakkord – time-based piecerate work
Zeitarbeit – job leasing, temporary employment
Zeitaufnahmebogen – time study sheet
Zeitausgleich – time offset
Zeitauswertung – time evaluation
Zeitauswertungslauf – time evaluation run
Zeitauswertungsschema – time evaluation schema
Zeitbeauftragter – time administrator
Zeitbindung – time constraint
Zeitbuchung – time posting
Zeitdaten – time data
Zeitdatenerfassung – time data entry
Zeitdatenverarbeitung – time data processing
Zeitdatenverwaltung – time data management
Zeitdruck – time pressure
Zeiteinheit – time unit
Zeiterfassung – time recording
Zeiterfassungsbeauftragter – person responsible for time recording
Zeiterfassungsbeleg – attendance form
Zeiterfassungsgerät – time recording device
Zeiterfassungssystem – time recording system
Zeiterfassungsterminal – time recording terminal
Zeitergebnis – time result
Zeitermittlung – determination of times
Zeitersparnis – time savings
Zeitfahrkarte – season ticket
Zeitgeist – spirit of the times
Zeitgrad – labor utilization rate
Zeitgruppe – time group
Zeitguthaben – positive flexitime balance, time credit
Zeitkontingent – time quota
Zeitkontrolle – timekeeping
Zeitlohn – time rate, time wage
Zeitlohnart – time wage type

Zeitmangel – lack of time
Zeitmessung – time measurement
Zeitmodell – time model
Zeitnachweisformular – time statement form
Zeitplan – schedule, time line
Zeitproblem – time problem
Zeitpunkt – point in time
Zeitraster – time periods
Zeitraum – space of time, interval
Zeitreihenanalyse – time series analysis
Zeitsaldo – time balance
Zeitsaldokorrektur – time balance correction
Zeitschulden – negative flexitime balance
Zeitspanne – time span
Zeitstempel – time stamp
Zeitstudie – time study
Zeitstudien-Beobachtungsbogen – time study observation sheet
Zeittyp – time category
Zeitunterlage – time basis
Zeitverarbeitung – time processing
Zeitvereinbarung – time agreement
Zeitvertrag – fixed-term contract
Zeitvertreib – pastime
Zeitvorgabe – time standard, time target
zeitweilige Freisetzung – temporary lay-off
zeitweilige Versetzung – secondment
Zeitwirtschaft – time management
Zeitzuschlag – time bonus
Zensur – mark, grade
zentral festgelegter Lehrplan – centrally determined curriculum
Zentralabteilung – Corporate Division
Zentralabteilung Personal – Corporate Human Resources
Zentrale – headquarters
zentrale Bildungsfragen – key educational issues
zentrale Personaldatei – central personnel file
zentrales Thema – keynote
zentrales Wohnungsreferat – corporate housing administration
Zentralisierung – centralization
Zentralismus – centralism
Zentralstelle – Central Department

Zentralstelle für die Vergabe von Studienplätzen – Central Applications Office
Zentralvorstand – Corporate Executive Committee
Zerspanungsmechaniker – cutting mechanic
Zeugnis – school report, report, credit (AmE), grade (AmE)
Zeugnis (für Angestellte) – reference
Ziel – aim, goal, objective, target
Zielbereich – target area
zielbezogen – target oriented
Zieldefinition – goal definition
Zielfestlegung – target setting
Zielgruppe – target group
Zielobjekt – target object
zielorientiert – target oriented
Zielsetzung – marking of aims
zielstrebig – determined, single-minded
Zielvorgabe – aim, goal, target
Zielvorstellung – aim, goal, target
Zins – interest
zinsloses Darlehen – interest free loan
zitieren – quote, cite
Zivildienst – civilian service
Zivilisation – civilization
zögernd – hesitant
Zögling – pupil
Zoologie – zoology
zu gegebener Zeit – at the proper time
zu leihen nehmen – borrow
zu würdigen wissen – appreciate
zuerkennen – award
zufällig – incidental, random
Zuflusslohnart – incoming wage type
Zuflussperiode – period of accrual
Zuflussprinzip – accrual principle
Zufriedenheit – satisfaction
Zugang – access
zugänglich – approachable
Zugangskontrollsystem – access control system
Zugehörigkeitsdauer – membership period
zuhören – listen
Zukunftsaspekt – aspect for the future
Zukunftsaufgabe – job for the future
Zukunftsaussichten – prospects for the future

zukunftsorientierte Bildungspolitik – progressive educational policy
Zukunftssicherung – future benefits
Zukunftssicherungsfreibetrag – future benefits exemption
Zukunftssorgen – worries about the future
Zulage – bonus, extra pay
Zulage, arbeitsplatzbezogene – work center bonus
Zulage, freiwillige – voluntary bonus
zulässig – permissible
Zulassungsanforderungen – entry qualifications
Zulassungsbeschränkung – restriction on admission
Zulassungsbestimmung – terms of admission
zumutbar – reasonable demand
Zumutbarkeit – in the bounds of reason
zumuten – demand, expect
Zunahme – increase, increment
Zuname – surname, family name
Zunft – guild
Zuordnung von Entgelt – allocation of payments
Zuordnungsrang – assignment priority
zur Sprache bringen – mention
zur Zusammenarbeit bereit – cooperative
Zurechnungsbetrag – allocated amount
Zurechnungseinkommen – attributable income
Zurechnungszeit – time credit
zurückgeblieben – retarded
zurückgesetzt – placed on waiting list
zurückgestellt aus Altersgründen – inadmissible for reasons of age
Zurückhaltung – reserve, discretion
Zurücknahme – reversal; withdrawal; retraction
zurückstellen – put on hold (eg. applicant)
Zurückstufung – downgrading (Verdrängung gelernter oder angelernter Arbeitskräfte aus ihren bisherigen Tätigkeiten in weniger qualifizierte, außerhalb ihres bisherigen Berufes liegende und schlechter bezahlte Tätigkeiten)
zurückweisen – reject

Zurückweisung – rejection
zurückzahlen – reimburse, refund
zurückziehen – withdraw
Zusage – acceptance, positive reply, committment, assent
Zusammenarbeit – cooperation
zusammenarbeiten – collaborate, work together
Zusammenballung – concentration
Zusammenbruch – collapse
Zusammenfassung – abstract, summary, résumé
Zusammengehörigkeitsgefühl – feeling of solidarity
Zusammenhalt – cohesion
Zusammenhang – context, connection
zusammenhängend – coherent
Zusammenschluss – amalgamation
Zusammensetzung des Aufsichtsrates – composition of supervisory board
Zusatzartikel zur Verfassung – amendment (AmE)
Zusatzauftrag – additional order
Zusatzausbildung – additional training
Zusatzkurs – supplementary course
Zusatzleistungen – additional benefits
zusätzliche Kompetenz – additional qualification
Zusatzqualifikation – additional qualification
Zusatzurlaub – additional leave, supplementary holiday, supplementary vacation (AmE)
Zusatzverdienst – additional income, secondary income
Zusatzversicherung – supplementary insurance
Zusatzversorgung – supplementary benefits
Zusatzversorgungseinrichtung – supplementary pension institute
Zusatzversorgungskasse – supplementary pension fund
Zuschlag – bonus
Zuschlagbezahlung – bonus payment
Zuschlagslohnart – bonus wage type
Zuschuss – allowance, subsidy
zuständige Stelle – responsible authorities
Zuständigkeit – responsibility, accountability

Zustellkosten – cost for delivery
zustimmen – agree
Zustimmung – consent, approval, sanction
Zutrauen – trust
Zutrittskontrolle, zeitliche – schedule-dependent access control
zuverlässig – reliable
Zuverlässigkeit – reliability
zuviel – in excess of
Zuwachsbedarf – growth requirements
Zuwachsrate – growth rate
zuweisen – assign, allocate
Zuwendung – benefit
zuziehen (sich) – incur
Zwang – compulsion, coercion
zwanghafter Mensch – compulsive person
Zwangsrückrechnung – forced retroactive calculation
Zwangsschlichtung – compulsory arbitration
Zwangsurlaub – compulsory holidays
Zweck – intention, purpose
Zweifel – doubt
Zweigniederlassung – branch office
zweisprachig – bilingual
zweisprachige Erziehung – bilingual education
zweisprachige Klasse – bilingual class
zweistufig – two tier
Zweiter Bildungsweg – the second way of gaining university admission through evening classes and correspondence courses
Zweiterkrankung – secondary illness
Zweitsprache – second language
Zweitstudium – second degree
Zweitwohnsitz – secondary residence
Zweiweg-Kommunikation – two-way communication
zwingend – compulsive, compelling, conclusive
Zwischen... – interim, provisional
zwischenbereichliche Versetzung – inter-departmental transfer
Zwischenbescheid – interim reply, provisional notification
zwischenbetrieblicher Arbeitsplatzwechsel – mobility within the firm

zwischenmenschlicher Bereich – human relations
Zwischenprüfung – mid-course examination
zwischenstaatliches Sozialversicherungsrecht – inter-governmental social insurance legislation
Zwischenzeugnis – term report
Zwischenziel – stopover
Zwischenzielübernachtung – stopover night
Zyklogramm – time line

Das Schulsystem in Großbritannien

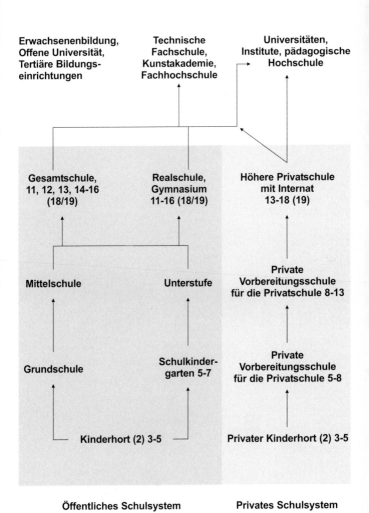

Erwachsenenbildung, Offene Universität, Tertiäre Bildungseinrichtungen

Technische Fachschule, Kunstakademie, Fachhochschule

Universitäten, Institute, pädagogische Hochschule

Gesamtschule, 11, 12, 13, 14-16 (18/19)

Realschule, Gymnasium 11-16 (18/19)

Höhere Privatschule mit Internat 13-18 (19)

Mittelschule

Unterstufe

Private Vorbereitungsschule für die Privatschule 8-13

Grundschule

Schulkindergarten 5-7

Private Vorbereitungsschule für die Privatschule 5-8

Kinderhort (2) 3-5

Privater Kinderhort (2) 3-5

Öffentliches Schulsystem

Privates Schulsystem

Das Schulsystem in Deutschland, dargestellt am Beispiel Bayern

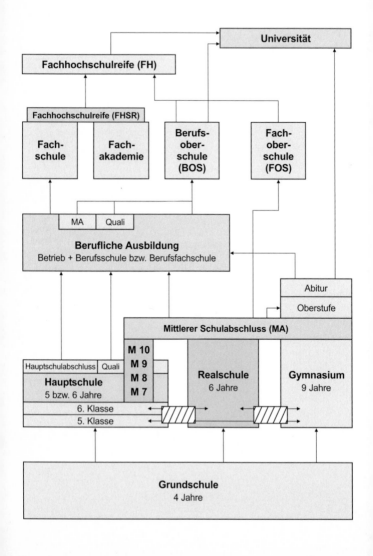

Das Schulsystem
in den Vereinigten Staaten

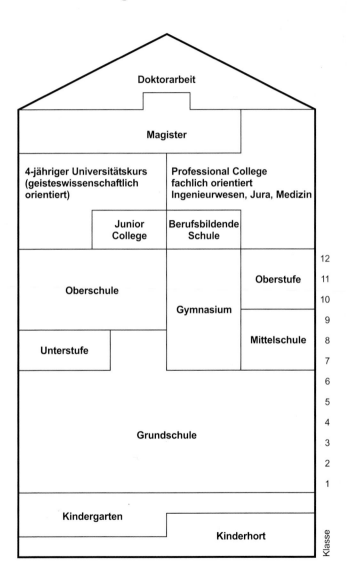

Dictionary of Personnel and Educational Terms

Part 2
English – German

A

a child in one's care – Pflegekind
a day's work – Tagesarbeit
a matter of course –
Selbstverständlichkeit
A-level – Abitur
A levels – Reifezeugnis
A-level subject – Abiturfach
Abendgymnasium – Abendgymnasium
Abendrealschule (s. Graphik) –
Abendrealschule
Abendschule (s. Graphik) –
Abendschule
ability – Befähigung; Fähigkeit; Gabe
ability category (AmE) – Kategorie;
Merkmalsgruppe
ability test – Fähigkeitstest
ability to absorb something –
Aufnahmefähigkeit
ability to accept criticism –
Kritikfähigkeit
ability to assert oneself –
Durchsetzungsfähigkeit
ability to communicate –
Kommunikationsfähigkeit
ability to cope with conflict –
Konfliktfähigkeit
ability to earn a living –
Erwerbsfähigkeit
ability to grasp – Auffassungsgabe;
Verständnis
ability to lead – Führungseigenschaft;
Führungsgabe
ability to synthesize – Synthesefähigkeit
ability to take in – Aufnahmefähigkeit
ability to work – Erwerbsfähigkeit
able – tüchtig, fähig
able to take stress – belastbar
able to work – arbeitsfähig
abnormal – abnormal
above the general pay scale –
übertariflich
above-average – überdurchschnittlich
abroad – im Ausland
absence – Abwesenheit

absence calendar –
Abwesenheitskalender;
Fehlzeitenkalender
absence category – Abwesenheitstyp
absence class – Abwesenheitsklasse
absence counting –
Abwesenheitsauszählung
absence counting class –
Abwesenheitszählklasse
absence counting rule –
Abwesenheitsauszählungsregel
absence data – Abwesenheitsinformation
absence from work – Fehlzeit
absence modifier –
Abwesenheitsmodifikator
absence period – Abwesenheitszeitraum
absence quota – Abwesenheitskontingent
absence time – Abwesenheitszeit
absence type – Abwesenheitsart
absence valuation –
Abwesenheitsbewertung
absences due to illness –
Krankheitszeiten
absent – abwesend
absent time – Fehlzeiten
absent without leave (AWOL) –
unentschuldigt abwesend
absentee – Abwesende(r)
absenteeism – Absentismus; häufiges
unentschuldigtes Fehlen
absenteeism rate – Abwesenheitsrate;
Fehlzeitenquote
absolute – absolut
absorb – aufnehmen
absorbed – Vertieftsein
abstract – Kurzfassung;
Zusammenfassung
abstract thinker – Theoretiker
abstract thinking – abstraktes Denken
academic – wissenschaftlich
academic achievement – akademische
Leistung
academic advisory service –
Studienberatung
academic approach –
Wissenschaftlichkeit
academic background – akademische
Ausbildung
academic community – Gelehrtenwelt

academic counselling – Beratung durch Akademiker, Beratung in akademischen Fragen

academic degree – akademischer Grad

academic expert – Wissenschaftler

academic finals – akademischer Abschluss

academic freedom – wissenschaftliche Freiheit

academic standard – akademisches Niveau

academic title – akademischer Titel

academic training – Hochschul(aus)bildung

academic year – akademisches Jahr

academician – Akademiemitglied

accelerated learning – beschleunigtes Lernen

acceleration – Beschleunigung

accept – annehmen

acceptance – Zusage

access – Zugang

access control system – Zugangskontrollsystem

accession – Firmenbeitritt

accident – Unfall; Unglücksfall

accident benefit – Arbeitsunfallentschädigung, Unfallrente

accident frequency rate – Unfallhäufigkeitsrate

accident insurance – Unfallversicherung

Accident Insurance Act – Unfallversicherungsgesetz

accident insurance tariff – Unfallversicherungstarif

accident prevention – Unfallverhütung

accident prevention regulations – Unfallverhütungsvorschriften

accident report – Unfallbericht

accident severity rate – Rate der Unfallschwere

accident statistics – Unfallstatistik

accommodation – Unterbringung; Unterkunft

accommodations accounting – Unterkunftsabrechnung

accommodation costs – Unterbringungskosten

accompanying measures – flankierende Maßnahmen

according to merit – leistungsbedingt

according to rank – rangmäßig

account – Konto

account assignment – Kontierungszuordnung

accountability – Fachverantwortung; Verantwortlichkeit

accountable day – Tag, anrechenbarer

accountancy – Buchführung; Buchhaltung; Rechnungswesen

accountant – Buchhalter

accounting (department) – Buchhaltung; Rechnungswesen

accounting past – Abrechnungsvergangenheit

accounting period – Abrechnungszeitraum, Abrechnungsperiode

accounting program – Abrechnungsprogramm

accounting result – Abrechnungsergebnis

accounting rule – Abrechnungsregel

accounting schema – Abrechnungsschema

accounting status – Abrechnungsstatus

accounting unit – Abrechnungseinheit

accounting variant – Abrechnungsvariante

accrual principle – Zuflussprinzip

accrued and taken leave – Urlaubsstand

accrued pension rights – Pensionsanwartschaft

accuracy – Genauigkeit; Präzision

accurate – akkurat; korrekt; genau

accuse – beschuldigen

achieve – erreichen

achieve points – Punkte erzielen

achievement – Leistung

achievement motivation – Leistungsmotivation

achievement parameter – Leistungsparameter

acknowledge – anerkennen

acknowledgement – Kenntnisnahme

acquire – aneignen

acquisition – Erwerb; Erwerbung

acquisition of knowledge – Wissenserwerb

acquisition of languages – Spracherwerb

acquisition of shares – Aktienerwerb

acquisition of skills – Aneignung von
Fertigkeiten
acribic – akribisch
across the board – global; umfassend
across the board increase – allgemeine
Lohn- und Gehaltserhöhung
act a part – verstellen
action – Maßnahme
action against wrongful dismissal –
Kündigungsschutzklage
action learning – Begreifen durch Tun,
Praxislernen, Lernen durch Tun
actionism – Aktionismus
active – aktiv
active vocabulary – aktiver Wortschatz
activity – Aktivität, Tätigkeit
activity irrelevant to the training –
ausbildungsfremde Arbeit
activity profile – Tätigkeitsprofil
activity report – Tätigkeitsnachweis
actual hours worked – Ist-Arbeitszeit
actual number of employees –
Ist-Kopfzahl
actual plan – Ist-Plan
actual salary – Ist-Gehalt
actual time – Ist-Zeit
actual time recording – Ist-Zeiterfassung
actual wage – Effektivlohn
actuarial assumptions –
versicherungsmathematische
Annahmen
actuarial evaluation –
versicherungsmathematische
Bewertung
acuity – Klugheit, Scharfsinn
ad – Anzeige; Inserat
adapt – angleichen; anpassen
adaptability – Anpassungsfähigkeit
adaptable – anpassungsfähig
adaptation period –
Anpassungszeitraum
adaptation problem –
Anpassungsschwierigkeit
add – addieren
addiction – Sucht
additional amount –
Hinzurechnungsbetrag
additional amount – Betrag,
hinzuzurechnender
additional applicant data –
Bewerbungszusatzdaten

additional benefits – Zusatzleistungen
additional burden – Mehrbelastung
additional expenses – Mehraufwand
additional expenses for meals –
Verpflegungsmehraufwendung
additional income – Hinzuverdienst;
Nebenverdienst, Zusatzverdienst
additional leave – Zusatzurlaub
additional order – Zusatzauftrag
additional qualification – außerfachliche
Kompetenz; zusätzliche Kompetenz;
Zusatzqualifikation
additional staff costs –
Personalzusatzaufwand;
Personalzusatzkosten
additional subject – Nebenfach
additional training – Zusatzausbildung
additional travel expenses –
Reisenebenkosten
**additional vacation for handicapped
persons** – Behindertenurlaub
additional vacation for older employees
– Alterszusatzurlaub
**additional vacation for severely
handicapped persons** –
Schwerbehindertenurlaub
additional wage costs –
Lohnnebenkosten
address – Adresse
address the meeting – das Wort ergreifen
address to contact in case of emergency
– Notadresse
addressee – Adressat
adequate – sachgerecht
adequate health – gesundheitliche
Eignung
adjourn – verlegen; vertagen
adjudication – Zuerkennung
adjust – angleichen; anpassen
adjustment – (technische) Regelung;
Angleichung; Anpassung
adjustment of wages – Angleichung der
Löhne; Anpassung der Löhne
administer – handhaben; verwalten
administration – Abwicklung;
Administration; Verwaltung
administration of applications –
Bewerbungsabwicklung;
Bewerbungsverfahren
administration of justice –
Rechtsprechung

administration of planning data –
Plandatenverwaltung
administrative act – Verwaltungsakt
administrative grade – höherer Dienst
administrative personnel structure –
Personalstruktur, administrative
administrative services –
Ordnungsaufgaben
administrative tasks –
Verwaltungsaufgaben
administrator – mit
Verwaltungsaufgaben befasste
Führungskraft
admiration – Bewunderung
admission – Aufnahme; Eintritt
admissions criteria – Aufnahmekriterien
admissions procedure –
Aufnahmeverfahren
admonish – ermahnen; mahnen
adolescence – Adoleszenz; jugendliches
Alter
adolescent – Jugendliche(r)
adopt – aneignen
adopted child – adoptiertes Kind
adult – Erwachsene(r)
adult education – Erwachsenenbildung
Adult Education Centre – Einrichtung
der Erwachsenenbildung,
Volkshochschule
adult learner – erwachsener Lerner
adult world – Erwachsenenwelt
advance – aufsteigen; Vorschuss
advance booking – Voranmeldung (Kurs)
advance entry – Vorschusserfassung
advance income statement –
Entgeltvorausbescheinigung
advance notice – Vorankündigung
advance on travel expenses –
Reisekostenvorschuss
advance pay – Abschlag
advance pay posting –
Abschlagsbuchung
advance pay procedure –
Abschlagsverfahren
advance payment – Abschlagszahlung;
Vorschuss; Vorschusszahlung
advance warning deadline –
Vorwarntermin
advanced – fortgeschritten
advanced course – Fortgeschrittenenkurs
advanced learner – Fortgeschrittene(r)

advanced level – Abitur
advanced level course – Leistungskurs
advanced training course – Aufbaukurs;
Fortbildungskurs
advanced vocational training college –
Technikerschule (This type of college is
reserved for technicians and master
craftsmen.)
advancement – Beförderung
advancement of science/arts –
Förderung der Wissenschaften
advantage – Vorteil
adverse – nachteilig; ungünstig
advertise – anzeigen
advertisement – Anzeige; Inserat
advertising measures –
Werbemaßnahmen
advice – Beratung; Rat, Ratschlag
advise – beraten; raten
advisory board – Beirat
advisory centre for education –
Schulberatungsstelle
advisory function – Beratungsfunktion
affective – affektiv
affective learning – affektives Lernen
affirmative – bejahend
after-hours – a.D.; außerhalb der
Dienstzeit
after-hours course – Lehrgang außerhalb
der Dienstzeit (a.D.)
after-hours on-going education –
Fortbildung a.D.
age – Alter; Lebensalter
age allowance – Altersfreibetrag
Age Discrimination Act – Gesetz gegen
Diskriminierung wegen Alters
age distribution structure –
Altersaufbau; Alterspyramide
age exemption – Altersfreibetrag
age group – Altersgruppe; Altersstufe
age limit – Altersgrenze
age pyramid – Altersaufbau;
Alterspyramide
age structure – Altersaufbau;
Altersstruktur
age-at-entry – Eintrittsalter
age-related reduction in working hours
– Altersfreizeit
agenda – Tagesordnung
aggressive – aggressiv; streitsüchtig
AGM – Jahreshauptversammlung

agree – einverstanden sein; sich einigen; zustimmen; vereinbaren; übereinstimmen
agreeable working group – Komfortzone
agreement – Einigung; Vereinbarung; Verständugung; Übereinstimmung
aggression – Aggression
aggressiveness – Aggressivität
aid to memory – Gedächtnisstütze
aim – Hauptziel; Ziel; Zielvorgabe; Zielvorstellung
aim of a promotion – Förderungsziel
air freight – Luftfracht
alcoholism – Alkoholismus
aliens office – Ausländeramt
alignment – Ausrichtung
alimony – Alimente, Unterhalt
all-day flexi leave – ganztägige Gleitzeitentnahme
all-day school – Ganztagsschule
all-round education – breite Bildung; umfassende Bildung
all-rounder – Alleskönner, Generalist
allocate – zuweisen
allocated amount – Zurechnungsbetrag
allocation of hours – Stundenkontingent
allocation of payments – Zuordnung von Entgelt
allow – gestatten
allowance – Beihilfe; Zuschuss
allowance period – Gewährungszeitraum
allowed time – Vorgabezeit
alter – ändern; verändern
alternative qualification – Ersatzqualifikation
alumnus/alumna – ehemalige(r) Schüler(in)
amalgamate – fusionieren; sich zusammenschließen
amalgamation – Fusion; Verschmelzung; Zusammenschluss
ambition – Ehrgeiz
amend – abändern; berichtigen; ergänzen
amendment – Berichtigung
amendment (AmE) – Zusatzartikel zur Verfassung
amicable agreement – Schlichtungsvereinbarung
amnesia – Amnesie
amortizable loan – Tilgungsdarlehen

amount – Betrag
amount available – Dispositionsmenge
amount cumulation – Betrags-Kumulation
amount field – Betragsfeld
amount insured – Versicherungssumme
amount of contribution – Beitragshöhe
amount of loan – Darlehenssumme
amount of remuneration – Entgelthöhe
amount paid (travel expenses) – Auszahlungsbetrag
amusing – lustig
analyse – analysieren
analysis – Analyse
analysis of unnecessary duplication of work – Doppelarbeitsanalyse
analytic – analytisch
analytical work center grading – Arbeitsplatzbewertung, analytische
anchor function – Ankerfunktion
anchor object – Ankerobjekt
anniversary – Jubilarfeier; Jubiläum, Dienstjubiläum
anniversary bonus – Jubiläumsgratifikation
announcement – Bekanntgabe; Bekanntmachung
announcement of income – Einkommensmitteilung
annoyance – Ärgernis
annual – Jahres...; jährlich; Jahrbuch
annual absence chart – Jahresfehlzeitkarte
annual adjustment – Jahresausgleich
annual contribution – Jahresbeitrag
annual contribution amount – Jahresumlage
annual cumulation – Jahreskumulation
annual earned income – Jahresarbeitslohn
annual earnings – Jahresarbeitsverdienst
annual earnings threshold – Jahresarbeitsverdienstgrenze
annual employee compensation limit – Jahresarbeitsentgeltgrenze
annual general meeting (AGM) – Jahreshauptversammlung
annual holiday – Jahresurlaub
annual income – Jahreseinkommen
annual income statement to social insurance agent – Jahresmeldung

annual income tax return –
Lohnsteuerjahresausgleich
annual income tax table –
Jahreslohnsteuertabelle
annual leave – Jahresurlaub
annual payment – Jahreszahlung
annual payroll account –
Jahreslohnkonto
annual remuneration slip –
Jahresentgeltbescheinigung
annual report – Geschäftsbericht;
Jahresmeldung
annual salary – Jahresgehalt
annual wage type development –
Lohnarten-Jahresentwicklung
annual working hours –
Jahresarbeitszeit (The prescribed
number of working hours per week is
calculated for the year. Employees are
present at work, depending on the
quantity of work and their personal
wishes. Salaries remain the same per
month, the number of working hours
per month can vary but in the end totals
the annual amount.)
annuity – Annuität; Jahreszahlung;
jährliche Zahlung
annuity computation –
Rentenberechnung
annuity insurance – Rentenversicherung
annuity payment – Annuitätenzahlung
annul – annullieren; für nichtig erklären
answer – Antwort; antworten;
beantworten
answer book – Antwortheft
ante insurance period –
Vorversicherungszeit
anticipation – Vorwegnahme
anxiety – Angst
apathetic – apathisch
apathy – Apathie
aphasie – Aphasie (a disorder of language
caused by brain damage)
apology – Entschuldigung
apparent – ersichtlich; offenbar
appeal committee – Widerspruchsstelle
(SW, BetrV.)
applicant – Anmelder; Bewerber;
Stellenbewerber
applicant action – Bewerbervorgang
applicant data – Bewerberdaten

applicant data administration –
Bewerberverwaltung
applicant file – Bewerberakte;
Bewerberkartei
applicant for an apprenticeship –
Lehrstellenbewerber
applicant group – Bewerbergruppe
applicant master data – Bewerberstamm
applicant number – Bewerbernummer
applicant profile – Bewerberprofil
applicant range – Bewerberkreis
applicant screening –
Bewerbervorauswahl
applicant selection – Bewerberauswahl
applicant test – Bewerbertest
applicant – Antragsteller
application – Bewerbung;
Bewerbunsunterlagen; Stellengesuch
application deadline – Bewerbungsfrist
application documents –
Bewerbungsunterlagen
application expenses –
Bewerbungskosten
application for a vacant position –
Bewerbung für eine bestimmte
Tätigkeit
application for transferral –
Versetzungsgesuch
application form – Antragsvordruck;
Bewerbungsbogen
application on spec – Blindbewerbung
application papers –
Bewerbungsunterlagen
application procedure –
Bewerbungsabwicklung;
Bewerbungsverfahren
application status – Antragsstatus
application strategy –
Bewerbungsstrategie
applied economics – Betriebswirtschaft
applied research – angewandte
Forschung
apply for – sich bewerben um
appoint – ernennen
appointment – Ernennung; Termin;
Verabredung, Berufung
appointments diary – Terminkalender
appointments procedure –
Berufungsverfahren
appointments section – Stellenteil in
Zeitungen und Zeitschriften

apportion – aufschlüsseln
appraisal – Beurteilung
appraisal factor – Beurteilungsmerkmal
appraisal system – Beurteilungssystem
appraisal type – Beurteilungsart
appraise – taxieren, beurteilen
appraisal tendency – Urteilstendenz
appreciate – zu würdigen wissen,
  anerkennen
appreciation of the difficulties –
  Problembewusstsein
apprehensive – ängstlich
apprentice – Auszubildende(r); Azubi;
  Lehrling
apprentice payment – Lehrlingsgehalt,
  Lehrlingsvergütung
apprenticeship agency –
  Lehrstellenvermittlung
apprentices' teacher – Ausbilder;
  Ausbildungsmeister; Lehrmeister
apprenticeship – Ausbildungsplatz;
  Ausbildungsstelle
apprenticeship contract – Lehrvertrag
apprenticeship place – Lehrstelle
apprenticeship time – Ausbildungszeit;
  Lehre; Lehrjahre; Lehrzeit
approach – herangehen; sich nähern;
  Vorgehen
approach to leadership –
  Führungsansatz
approachable – zugänglich
appropriate – angemessen; sachgerecht;
  sich aneignen
approval – Billigung; Einverständnis;
  Genehmigung; Zustimmung
approval of attendance –
  Anwesenheitsgenehmigung
approval percent –
  Genehmigungsprozentsatz
approval period –
  Genehmigungszeitraum
approve – anerkennen; billigen;
  gutheißen
approved – genehmigt
approximate – annähernd; ungefähr
approximate – annähern
**(approx. =) general sickness benefit
  fund** – Allgemeine Ortskrankenkasse;
  AOK
(approx. =) stock corporation –
  Aktiengesellschaft

aptitude – Eignung
aptitude as instructor –
  Ausbildereignung
aptitude certificate – Eignungszeugnis
aptitude interview – Eignungsgespräch
aptitude test – Arbeitserprobung;
  Eignungsprüfung
arbitrate – schiedsrichterlich entscheiden
arbitration – Schiedsverfahren;
  Schlichtungsverfahren
arbitration agreement –
  Schiedsabkommen;
  Schiedsvereinbarung;
  Schlichtungsvereinbarung
arbitration award – Schiedsspruch
arbitration board – Einigungsstelle;
  Schiedsausschuss
arbitration court – Schiedsgericht
arbitration decision – Schiedsspruch
arbitrator – Schiedsmann;
  Schiedsrichter; Schlichter
archive – Archiv
area – Bereich; Sachgebiet
area of responsibility – Arbeitsgebiet;
  Aufgabenbereich
area of training – Ausbildungszweig
area sales manager –
  Gebietsverkaufsleiter
area specific – bereichsspezifisch
**area where the regular pay scale does
  not apply** – AT-Bereich
argue – argumentieren; streiten
argument – Argument; Beweisgrund;
  Streit; Zank
arithmetic – Rechnen
arrange – vereinbaren; einrichten
arrogance – Arroganz
arrogant – überheblich
art – Kunsterziehung
articled clerk – Rechtsreferendar
articulation – Artikulation; deutliche
  Aussprache
artist – Künstler
as a rule – in der Regel; Regelfall
as if principle – Hätte-Prinzip
as per our discussions – wie mit Ihnen
  vereinbart
ask too much of someone – überfordern
aspect for the future – Zukunftsaspekt
assembly line – Fertigungsstraße,
  Fließband

aspiration – hohes Ziel
assembly – Versammlung
assembly line work – Fließbandarbeit
assent – Zusage
assess – einschätzen; taxieren
assessing needs – Bedarfsanalyse
assessment – Beurteilung; Evaluierung
assessment basis – Bemessungsgrundlage
assessment center for interpersonal skills – Assessment Center zum Erfassen sozialer Fähigkeiten
assessment system – Beurteilungssystem
assessment threshold – Bemessungsgrenze
assessor – Assessor (A German civil servant who has passed the second state examination, esp. secondary school teachers and lawyers)
asset-creating capital – vermögenswirksame Anlage
asset-formation benefit – vermögenswirksame Leistung
asset-formation policy – Vermögenspolitik
asset-forming – vermögenswirksam
assign – zuweisen; abordnen
assignment – Abordnung; Arbeitsaufgabe
assignment allowance (AmE) – Abordnungsgeld
assignment expenses – Abordnungskosten
assignment priority – Zuordnungsrang
assignment to a foreign country – Auslandsentsendung
assignment to wage group – Eingruppierung, tarifliche
assignment to wage level – Einstufung, tarifliche
assimilation – Aufnahme
assist – beistehen; helfen
assistance – Beistand, Mitarbeit
assistant – Assistent(in); Gehilfe
assistant lecturer – wissenschaftlicher Assistent
assistant manager – außer Tarif; außertarifliche(r) Mitarbeiter(in)
assistant professor (AmE) – Dozent
assisted passage – Fahrgeldzuschuss
associate – in Verbindung bringen

associate professor (AmE) – ausserordentlicher Professor
associated company – Beteiligungsgesellschaft
association – Assoziation; Gesellschaft; Verband, Verein
assume new tasks – neue Aufgaben übernehmen
assume responsibility – Verantwortung übernehmen
astute – scharfsinnig
asylum seeker – Asylbewerber
at a disadvantage – benachteiligt
at home and abroad – im In- und Ausland
at the expense of the company – auf Firmenkosten
at the proper time – zu gegebener Zeit
athlete – Athlet
atmosphere – Atmosphäre
attach – anheften
attachment – Beschlagnahmung
attachment of wages – Lohnpfändung
attain an objective – ein Ziel erreichen
attainable – erreichbar
attainment – Erreichung; Erzielung
attainment target – Leistungsziel
attempt – Versuch
attend – besuchen; teilnehmen (an)
attendance – Anwesenheit; Teilnahme
attendance at school – Schulbesuch
attendance category – Anwesenheitstyp
attendance check list – Anwesenheitskontrolliste
attendance counting rule – Anwesenheitsauszählungsregel
attendance data – Anwesenheitsinformation
attendance form – Arbeitszeitnachweis; Zeiterfassungsbeleg
attendance list – Anwesenheitsliste; Namensliste
attendance officer – Beamter einer Schulbehörde, der Fälle von Schulschwänzen untersucht
attendance quota – Anwesenheitskontingent
attendance reason – Anwesenheitsgrund
attendance survey – Arbeitszeitübersicht
attendance time – Anwesenheitszeit
attendance type – Anwesenheitsart

attendee appraisal –
Teilnehmerbeurteilung
attendee list – Teilnehmerliste
attention – Achtsamkeit;
Aufmerksamkeit
attention span –
Aufmerksamkeitsspanne
attitude – Haltung
attitude of refusal –
Verweigerungshaltung
attitude to work – Berufseinstellung
attitude – Einstellung
attributable income –
Zurechnungseinkommen
attribute – Merkmal
audience – Publikum
audio-typist – Phonotypistin
audio-vision – Audiovision
audio-visual – audiovisuell
audit – Revision
auditor – Rechnungsprüfer; Revisor
auditory perception – Hörwahrnehmung
auditorium – Hörsaal
aural learning – Lernen durch Hören
aural test – Hörtest
austerity programme – Sparprogramm
authentic – echt; verbürgt
authenticity – Authentizität; Echtheit
authoritarian management behaviour –
autoritäres Führungsverhalten
authoritarian method – autoritäre
Methode
authoritarian teacher – autoritärer
Lehrer
authoritarian leadership style –
autoritärer Führungsstil
authoritative – fachkundig; kompetent
authorities – Behörden
authority – Autorität; Befugnis;
Ermächtigung; Macht; Machtbefugnis
authority to instruct –
Weisungsbefugnis; Weisungsrecht
authorization for activities –
Berechtigung, aktionsorientierte
authorization for process organization
– Berechtigung, ablauforientierte
authorization of a loan – Bewilligung
eines Darlehens
**authorization to run an expense
account** – Spesenberechtigung

authorization vector –
Berechtigungsvektor
authorize – anweisen; berechtigen;
ermächtigen
authorized – berechtigt
authorized number of employees –
Soll-Kopfzahl
authorized representative –
Bevollmächtigter
authorized to sign –
unterschriftsberechtigt
autism – Autismus (Abnormal
withdrawal from the world of reality; a
condition which has its onset in
childhood and is marked by severly
limited responsiveness to other persons,
restricted behaviour patterns, and
usually abnormal speech development.)
automatic seizure –
Pfändungsabwicklung, automatische
automatic progression –
Bewährungsaufstieg
automation – Automatisierung
autonomous – autonom,
eigenverantwortlich;
Eigenverantwortung; eigenständig
autonomy – Eigenständigkeit
autumn half-term holidays –
Herbstferien
auxiliary plant – Nebenbetrieb
auxiliary work – Hilfsarbeit
availability – Verfügbarkeit
availability for duty –
Arbeitsbereitschaft
availability indicator –
Dispositionsmerkmal
availability time – Bereitschaftszeit
availability type – Bereitschaftsart
available – erhältlich; verfügbar;
vorhanden
available funds – liquide Mittel
average – Durchschnitt
average earnings –
Durchschnittseinkommen
average gross pay of all insured persons
– durchschnittliches
Bruttoarbeitsentgelt aller Versicherten
average marks (BrE) –
Notendurchschnitt
average pay – Durchschnittsbezahlung

average percentage –
Durchschnittsprozentsatz
average salary – Durchschnittsgehalt
average selected time – gewählter
Durchschnittszeitwert
average type – Durchschnittstyp
average wage – Durchschnittslohn
aversion – Abneigung; Abscheu;
Widerwille
avert – abwenden
avoid – umgehen; vermeiden
await – abwarten
award – Auszeichnung; gewähren;
Prämie
awardance of a doctorate – Promotion
awkward – schwerfällig

B

Bachelor of Arts (B.A.) – Bakkalaureus
Artium
Bachelor of Commerce –
Diplomkaufmann
Bachelor of Law – Rechtsreferendar
Bachelor of Science (B.Sc.) –
Bakkalaureus Scientiarum
back-end system – System,
nachgelagertes
background – Hintergrund; soziale
Vergangenheit; Werdegang
background knowledge –
Hintergrundwissen
backlog demand – Nachholbedarf
back-payment – Gehaltsnachzahlung;
Nachzahlung
back-to-back courses –
aufeinanderfolgende Kurse
bad behaviour – schlechtes Betragen
bad weather pay – Schlechtwettergeld
badly paid job – schlecht bezahlter
Arbeitsplatz
balance – Gleichgewicht
balance sheet – Bilanz
balanced – ausgeglichen; ausgewogen
balancing the accounts – Bilanzieren

ball-point pen – Kugelschreiber
ballot – Stimmzettel; Wahlzettel
ballot paper – Stimmzettel; Wahlzettel
ban on smoking – Rauchverbot
bank details – Bankverbindung
bank holiday – Bankfeiertag (Br.: 1.
January, Easter Monday, last Monday
in May, last Monday in August, 25. and
26. December, Am.: 1. January, 22.
February, 30. May, first Monday in
September)
bank transfer – Überweisung
bankruptcy – Konkurs (Insolvenz)
bar chart – Balkendiagramm
barely enough – knapp
bargaining agreement – Tarifabkommen
basal text – Basaltext (complete
representation of teaching material as a
basis for programmed instruction)
base – Basis; Grundlage
base amount for seizure –
Pfändungsgrundlage
base award – Bezugsgröße; Grundbetrag
base wage type – Basislohnart
basic amount – Bezugsgröße;
Grundbetrag
basic commercial training –
kaufmännische Grundausbildung
basic competence – Grundfertigkeiten
basic corporate policies of Siemens AG
– Grundordnung des Hauses Siemens
(SAG)
basic hourly pay – Grundstunde
basic income – Grundeinkommen
basic information – Grundinformation
basic knowledge – Grundkenntnisse;
Grundlagenwissen
basic labour protection policies –
Grundsätze des Arbeitsschutzes
basic pay – Basisbezüge
basic pay period – Basisbezugszeitraum
basic pay scale salary – Tarifgrundlohn
basic pay split – Basisbezugssplitt
basic period – Richtperiode
basic personal data – Grunddaten zur
Person
basic policies – Grundsatzfragen
basic procedure – Grundablauf
basic research – Grundlagenforschung
basic salary – Grundgehalt
basic skills – Grundfertigkeiten

basic staff costs – Personalbasisaufwand
basic studies – Grundstudium
basic time – Grundzeit
basic training – Grundausbildung
basic training course –
Grundausbildungslehrgang
basic wage – Ecklohn; Grundlohn
basic wage policy –
Entlohnungsgrundsatz
basis – Basis; Grundlage
basis for calculating average value –
Durchschnittsgrundlage
be at odds with – hadern
be at the disposal of the company – der
Firma zur Verfügung stehen
be ill – krank sein
be in agreement – übereinstimmen
be made redundant – den Arbeitsplatz
verlieren
be on the dole – stempeln gehen
be retired – in den Ruhestand versetzt
werden; pensioniert
bear responsibility – Verantwortung
tragen
bearing – persönliches Auftreten,
Verhalten
become null and void – verjähren
become obsolete – verjähren
begin – beginnen
begin time – Beginnuhrzeit
begin tolerance begin time –
Beginntoleranz-Beginn
begin tolerance end time –
Beginntoleranz-Ende
beginner – Berufsanfänger; Berufsstarter
beginners' course – Anfängerkurs
beginners' wage rate – Anlernlohnsatz
beginning of break – Pausenanfang
behavior change – Verhaltensänderung
behavior modification –
Verhaltensänderung
behaviour – Benehmen; Verhalten
behavioural pattern – Verhaltensmuster
behavioural psychology –
Verhaltenspsychologie
behavioural standard – Verhaltensnorm
behaviourism – Verhaltenspsychologie,
Behaviorismus (A system of
psychology associated with the name of
J.B. Watson. It defined psychology as a
study of behaviour and limited the data
of psychology to observable activities.)
believe – glauben
bell – Schulglocke
bend – biegen
beneficiary – Begünstigter; Nutznießer
benefit – Beihilfe; Nutzen; Zuwendung
benefit area – Leistungsbereich
benefit category – Leistungskategorie
benefit cost criteria –
Leistungskostenkomponenten
benefit period – Begünstigungszeitraum
benefit plan – Leistungsplan;
Sozialzulagensystem
benefits – Arbeitgeberleistung;
Leistungen
benefits claim – Versorgungsanspruch
benefits of a professional association –
berufsgenossenschaftliche Leistungen
benefits-related deductions – Abgaben,
soziale
Berufsaufbauschule –
Berufsaufbauschule (College of further
education) (cf. chart)
Berufsoberschule – Berufsoberschule (a
type of Technical College) (cf. chart)
biased – voreingenommen; befangen
biennial incremental increase –
Biennalsprung
big industries – Großindustrie
bilingual – zweisprachig
bilingual class – zweisprachige Klasse
bilingual education – zweisprachige
Erziehung
billing – Berechnung
biodata – Lebenslauf
biography – Biographie
biology – Biologie
biro – Kugelschreiber
birth – Geburt
birth certificate – Geburtsurkunde
birthday list – Geburtstagsliste
black hole – schwarzes Loch (a condition
where one or more managers fail to
fulfill their sponsor responsibilities e.g.
by withholding or distorting
information)
black market – Schwarzmarkt
black marketeer – Schwarzarbeiter
blackboard – Tafel; Wandtafel

blacken someone's name –
rufschädigend
blacking – Solidaritätsstreik;
Sympathiestreik
blackleg – Streikbrecher
blame – beschuldigen; Schuld; schuld
sein an
blanket agreement – Flächentarifvertrag
blanket allowance – Pauschbetrag
blind application – Blindbewerbung
blind person – Blinde(r)
block release instruction –
Blockunterricht
blocked period – Sperrzeit
blocking period – Sperrfrist
blood alcohol – Blutalkohol
blood circulation training –
Kreislauftrainingskur
blue collar worker – Arbeiter
board – Tafel; Wandtafel
board of Directors – Verwaltungsrat (Im
angelsächsischen Raum ist die BoD die
einstufige Unternehmungsleitung einer
Kapitalgesellschaft, die von den
Aktionären gewählte Aufsichts- und
Verwaltungsinstanz.)
board of examiners – Prüfungsausschuss
Board of Management – Vorstand;
Vorstandsbereich
boarder – Interne(r)
boarding school – Internat
boast – angeben; prahlen; sich rühmen
body corporate – Körperschaft des
öffentlichen Rechts
body language – Körpersprache
bold – kühn
bona fide – in gutem Glauben
bond – Bindung
bond index – Bindungsindex
bonus – Sonderzahlung; Zulage;
Zuschlag; Prämie
bonus calculation – Bonusermittlung
bonus calculation procedure –
Bonusverfahren
bonus due date – Bonusstichtag
bonus for extra hours –
Überstundenzuschlag
bonus for hazardous or unpleasant
work – Erschwerniszulage
bonus for years of service –
Dienstjubiläum

bonus paid after birth of child –
Geburtszulage
bonus payment – Zuschlagbezahlung
bonus percentage – Bonusprozentsatz
bonus scheme – Prämiensystem
bonus settlement – Bonusabrechnung
bonus wage type – Zuschlagslohnart
book – Buchen
book knowledge – Buchwissen
book learning – Bücherweisheit
booking – Belegung; Buchung
booking note – Buchungsbeleg
boost – erhöhen
boot out – schassen
bootlicking – kriecherisch
booted out – geschasst
border crossing – Grenzübertritt;
Grenzübertritt, Hinreise
border value – Grenznutzen (that which
one can do better than another)
border worker – Grenzarbeitnehmer;
Grenzgänger
borderline case – Grenzfall
bored – gelangweilt
boredom – Langeweile
boring – langweilig
borrow – borgen; leihen; zu leihen
nehmen
boss – Chef, Vorgesetzter
bother – belästigen
bottleneck – Engpass
boundaries of responsibility –
Verantwortungsspielraum
bound by law – Gesetzespflicht
boycott – Boykott; boykottieren
brain – Gehirn
brain damage – Hirnschaden
brain drain – Abwanderung von
Wissenschaftlern
brainstorming – Brainstorming (A
method of finding answers to problems
in which all members in a group think
very quickly of as many ideas as they
can)
brainwave – Geistesblitz; guter Einfall
branch of business – Geschäftszweig
branch of industry – Industriezweig;
Wirtschaftszweig
branch of study – Studienzweig
branch office – Zweigniederlassung
breach of duty – Pflichtverletzung

breach of rules – Regelverstoß
breadwinner – Ernährer
break – Arbeitspause; Pause, bezahlte
break during planned working time – Sollarbeitspause
break model – Pausenmodell
break type – Pausentyp
break-room – Pausenraum
breakdown – Aufgliederung; Aufteilung
breakdown of compensation for mileage – Kfz-Verrechnung
breakeven analysis – Breakeven-Analyse; Ermittlung der Ertragsschwelle
bribe – Schmiergeld
bridge day – Brückentag
briefing – Einsatzbesprechung
bright – aufgeweckt; gescheit; intelligent
bring someone into disrepute – rufschädigend
bring to a close – erledigen
bring up – aufziehen; großziehen
broad band model – Breitenbandmodell (Employees can repeatedly re-define their contract working hours within a defined bandwith of e.g. between 15 and 40 hours per week. Salaries increase or decrease respectively.)
broken – ge(zer)brochen
brush up – auffrischen
brutality – Brutalität
buddy – Kumpel
buddy system – Betreuung neuer Mitarbeiter
budget – Budget; Etat
budget planning – Budget-Planung
building and loan association – Bausparkasse
building loan – Baudarlehen
building trade – Baugewerbe
bulletin board – Anschlagtafel; Schwarzes Brett
bully – schikanieren; tyrannisieren; Rabauke; Raufbold
bumping – Verdrängen
bureaucracy – bürokratische Organisation; Bürokratie
bureaucratic organization – bürokratische Organisation; bürokratische Organisationsstruktur
burn-out – Burn-out

burnt out – ausgebrannt sein
bursar – Finanzverwalter; Schulverwalter
bursary (Scottish universities) – Stipendium
business – Geschäft(e); Geschäftsbetrieb; Unternehmen; kaufmännisch
business administration – Betriebswirtschaft
business card – Karte; Visitenkarte
business development – Geschäftsentwicklung
business development and strategy – Geschäftsfeldplanung
business economics – Betriebswirtschaft
business employer – Unternehmer
business enterprise – Unternehmen; Wirtschaftsunternehmen
business environment – Unternehmensumfeld
business event – Veranstaltung
business event appraisal – Veranstaltungsbeurteilung
business event catalog – Veranstaltungskatalog
business event environment – Veranstaltungsumfeld
business event group – Veranstaltungsgruppe
business event hierarchy – Veranstaltungshierarchie
business event location – Veranstaltungsort
business event marketing – Veranstaltungsmarketing
business event planning – Veranstaltungsplanung
business event type – Veranstaltungstyp
business expenses – Betriebsausgaben
business game – Planspiel; Unternehmensspiel
business leader – Wirtschaftsführer
business management – Wirtschaftsinformatik
business management game – Planspiel
businessman – Geschäftsmann
business partner – Geschäftspartner
business planning – Geschäftsplanung
business process – Geschäftsprozess
business school – Wirtschaftsschule

business school graduate –
Diplomkaufmann
business strategy – Geschäftsstrategie
business thinking – unternehmerisches
Denken
business trip – Dienstreise;
Geschäftsreise
business unit – organisatorisch
selbständige Einheit
business year – Geschäftsjahr
business-like – objektiv; unbeeinflusst
bussing (AmE) – Beförderung von
Schülern in Bussen
busy – beschäftigt
buyer – Einkäufer
buzz word – Schlagwort

C

CAD – computerunterstütztes
Konstruieren; Zeichnen am Bildschirm
CAI – computerunterstützte
Unterweisung
CAL – computerunterstütztes Lernen
calculate – berechnen; errechnen
calculated minimum – kalkulierte
Mindestleistung
calculation – Berechnung; Kalkulation
calculation of averages –
Durchschnittsberechnung
calculation of commission –
Provisionsabrechnung
calculation of contribution amount –
Beitragsermittlung
calculation of cut-off dates –
Fristenberechnung
calculation of entitlement –
Anspruchermittlung
calculation of gross amount –
Bruttoermittlung
calculation of leave entitlement –
Urlaubsanspruchsermittlung
calculation of overtime –
Mehrarbeitsberechnung

calculation of partial remuneration –
Teilentgeltberechnung
calculation of premium –
Prämienfindung
calculation of remuneration –
Entgeltberechnung
calculation of vacation –
Urlaubsberechnung
calculation of wages – Lohnberechnung
calculation schema – Rechenschema
calculator – Taschenrechner
calendar day – Kalendertag
call – Anruf; Aufruf
call to strike – Streikaufruf
call-in pay – Mindestlohn für
Aufforderung und Erscheinen am
Arbeitsplatz (selbst wenn nicht
gearbeitet wird) (payment guaranteed
to workers who report for work even if
there is no work for them to do)
calling card – Karte; Visitenkarte
calm – gelassen
campus (AmE) – Universitätsgelände
cancel – absagen; für nichtig erklären;
rückgängig machen
cancel a course – Kurs stornieren
cancellation – Widerruf
cancellation clause – Kündigungsklausel
cancellation contract –
Aufhebungsvertrag
cancelled hours – Stundenausfall
candidate – Anwärter; Kandidat;
Prüfling; Stellenbewerber
candidate for a doctor's degree –
Doktorand
**candidate for further education and
training** – Weiterbildungskandidat
cane – Rohrstock
canteen – Kantine, Mensa
canteen food – Kantinenessen
capability – Befähigung; Fähigkeit;
Leistungsfähigkeit
capable – fähig
capable of learning – lernfähig
capacity – Kapazität
capacity of management trainees –
Nachwuchspotential
capacity to work – Arbeitsfähigkeit
capital expenditure budget –
Investitionsbudget

capital formation confirmation –
Vermögensbestätigung
capital formation contract –
Vermögensbildungsvertrag
capital formation savings payment –
Leistung, vermögenswirksame
capital increase – Kapitalerhöhung
capital payment – Kapitalzahlung
capital-gains tax – Kapitalertragssteuer
capitation – pro Kopf
capping – Kappung
car loan – Autodarlehen
car rule – Pkw-Regelung
carcinogenic agent – krebserzeugender
Arbeitsstoff
card – Karte; Visitenkarte
care – Vorsicht; Sorgfalt
care of the employees – Betreuung der
Mitarbeiter
career – beruflicher Werdegang;
Berufslaufbahn, Werdegang
career advancement – Aufstieg,
beruflicher
career adviser – Berufsberater
career advisory service –
Berufsberatung
career anchor assessment – Instrumente
zur Selbsteinschätzung des Mitarbeiters
Career and Succession Planning –
Karriere- und Nachfolgeplanung
career break – Berufspause
career candidate – Karrierekandidat
career development – berufliche
Entwicklung; Berufsentwicklung
career expectations – Karriereerwartung
career for specialists – Fachlaufbahn
career goal – Karriereziel
career interruption – Karriereknick
career ladder – Karriereleiter
career model – Laufbahnmodell
career opportunities –
Aufstiegsmöglichkeiten, berufliche
career planning – Berufsplanung;
Berufswegplanung
career preparation – Berufsvorbereitung
career prospects –
Aufstiegsmöglichkeiten;
Karrieremöglichkeiten
careful – sorgfältig
caretaker – Hausmeister

carousel training – Training im
Karussellverfahren
carry out research work – in der
Forschung tätig sein
carry-on luggage – Handgepäck
case of sickness – Krankheitsfall
case study – Fallstudie
cash – Bargeld
cash budget – Einnahmen- und
Ausgabenbudget
cash flow – Cash Flow
cash office – Kasse
cash payment – Barzahlung
cash remuneration – Barentgelt
cash surrender value – Rückkaufswert
cash value – Barwert;
Berechnungsgrundlage bei
Pensionsrückstellungen
cash wages – bar gezahlter Lohn;
Barlohn; Geldlohn
casual work – Gelegenheitsarbeit
casual worker – Aushilfskraft;
Gelegenheitsarbeiter
casualty – Unglücksfall
catastrophic employment situation –
Beschäftigungskatastrophe
catch-up allowance –
Ausgleichsvergütung
catchment area – Einzugsgebiet;
Schuleinzugsbereich
category – Kategorie; Merkmalsgruppe
causality – Kausalität; Ursächlichkeit
cause – Ursache; verursachen;
Veranlassung
caution – Vorsicht
CBL (computer-based learning) –
(computerunterstütztes Lernen)
CBT – Lernen mit einem interaktiven
Lernprogramm
center of excellence – Center of
excellence
Central Applications Office –
Zentralstelle für die Vergabe von
Studienplätzen
central department – Hauptabteilung;
Zentralstelle
central experience – Schlüsselerfahrung
central personnel file – zentrale
Personaldatei
central works council –
Gesamtbetriebsrat

centralism – Zentralismus
centralization – Zentralisierung
centralized decision making –
zentralisierte Entscheidungsfindung
centrally determined curriculum –
zentral festgelegter Lehrplan
cerebral – Gehirn...
certificate – Berufsnachweis;
Bildungsabschluss; Prüfungszeugnis
certificate of apprenticeship – Lehrbrief
certificate of aptitude for higher
education – Hochschulreife
certificate of attendance –
Teilnahmebestätigung
certificate of disability –
Arbeitsunfähigkeitsbescheinigung
certificate of discharge –
Entlassungszeugnis
certificate of employment –
Arbeitszeugnis;
Beschäftigungsnachweis
certificate of proficiency –
Facharbeiterbrief
certificate of qualification –
Befähigungsnachweis
certification – Bescheinigung
certified public accountant, CPA (AmE)
– Wirtschaftsprüfer
certified unfit for work due to illness –
krankgeschrieben
certify – bescheinigen
cession of wages – Abtretung,
Gehaltsabtretung
chain employment contract –
Kettenarbeitsvertrag
chain of command –
Autoritätshierarchie; Hierarchie der
Weisungsbefugnis
chair – Lehrstuhl; den Vorsitz führen;
Vorsitzende(r); Lehrstuhlinhaber
Chairman of the Board –
Vorstandsvorsitzender
Chairman of the Board of Management
– Sprecher des Vorstands
Chairman of the Parents' Council –
Elternbeiratsvorsitzender
Chairman of the Supervisory Board –
Aufsichtsratsvorsitzender; Vorsitzender
des Aufsichtsrats
chairperson – Vorsitzende(r)
chalk – Kreide

challenge – herausfordern;
Herausforderung
challenging – herausfordernd
Chamber of Commerce –
Handelskammer
Chamber of Industry and Commerce –
Industrie- und Handelskammer
chance of promotion –
Beförderungschance
chancellor – Kanzler (Univ.) (A titular
head of a university, now usually an
honorary office, the actual duties being
performed by a vice-chancellor.)
change – ändern; verändern
change document – Änderungsbeleg
change in an appointment –
Terminänderung
change in basic pay –
Basisbezugsänderung
change in pay – Änderung der Bezüge
change in pay scale group –
Tarifgruppenwechsel
change in pay scale level –
Tarifstufenwechsel
change in staff – Personalwechsel
change in status – Statusänderung
change of awareness –
Bewusstseinsänderung
change of employer –
Arbeitgeberwechsel
change of job – Arbeitsplatzwechsel
change of locality – Ortswechsel
change of location – Standortwechsel
change of pay – Bezugsänderung
change of pay group – Änderung der
Eingruppierung
change of shift – Schichtwechsel
change of values – Wertewandel
change of working hours –
Arbeitszeitänderung
change one's ideas – umlernen
change in salary – Gehaltsänderung
changeover – Schichtwechsel
changeover problems –
Umstellungsprobleme
changing personnel structure –
Personalstrukturwandel
changing room – Umkleideraum
channel – Kanal
chaotic – chaotisch
chapter – Kapitel

character – Charakter; Zeichen
character trait – Charaktereigenschaft
characteristic – Merkmal
charisma – Charisma, Ausstrahlung
charitable – gemeinnützig
charity – Wohltätigkeit
charming – liebenswürdig
chart – Diagramm; graphische Darstellung; Schaubild
chartered accountant (BrE) – Wirtschaftsprüfer
cheap – billig
cheap labourer – Billigarbeiter
cheat – mogeln; schummeln
check – Kontrolle; kontrollieren; Überprüfung
check (AmE) – Scheck
checklist – Kontrollliste; Prüfliste
check necessity of an adjustment – Anpassungsprüfung (bei Betriebsrente)
checkoff system (AmE) – Lohnabzugsverfahren
chemical engineer – Chemie-Ingenieur
chemistry – Chemie
cheque – Scheck
child – Kind
child allowance – Kindergeld, Kinderzulage, Kinderfreibetrag
child benefit – Kindergeld
child benefit fund – Kindergeldkasse
child benefits – Erziehungsrente
child care-leave – Erziehungsurlaub
child category – Kinderart
child education – Kindererziehung
child number – Kindnummer
child prodigy – Wunderkind
childhood – Kindheit
childish – kindisch
children's home – Kinderheim
children's allowance – Kindergeld
chlorinic hydrocarbon – Chlorkohlenwasserstoff
choice – Auswahl; Wahl
choice of profession – Berufswahl
choice of subjects – Fächerwahl
choleric person – Choleriker(in)
choose – aussuchen; Auswahl treffen; auswählen
Christian name – Vorname
Christmas bonus – Weihnachtsgeld; Weihnachtsgratifikation

chronic shortage of educational facilities – Bildungsnotstand
chuck out – schassen
chucked out – geschasst
chum – Kumpel
church tax – Kirchensteuer (BRD)
church tax area – Kirchensteuergebiet
church tax rate – Kirchensteuerhebesatz
church tax reduction – Kirchensteuerkappung
circular – Rundbrief; Rundschreiben
circulate – umlaufen
circumstantial – weitschweifig
cite – zitieren
citizenship – Nationalität
City and Guilds of London Institute – Institut der Stadt und der Zünfte von London
civics – Staatskunde
Civil and Public Service Examination – Staatsexamen (the CSE is compulsory for lawyers, doctors, teachers and all who wish to become (senior) civil and public servants)
civil engineer – Bauingenieur
civil engineering – Bautechnik; Hoch- und Tiefbau
civil law – bürgerliches Recht; Privatrecht
civil liability insurance – Haftpflichtversicherung
civil servant – Beamte(r); Staatsbeamte(r); Staatsbedienstete(r)
civil service – Dienst, öffentlicher Staatsdienst
civilian service – Zivildienst
civilization – Zivilisation
civilized nation – Kulturvolk
claim – Anspruch; Forderung; Reklamation
claim benefit – beanspruchen
claim benefits – Leistungen beanspruchen
claim for compensation – Schadenersatzanspruch
claim for damages – Schadenersatzanspruch
claim to power – Machtanspruch
claimant – Kläger, Antragsteller
claims department – Schadensabteilung
clash – Konflikt

class – Stunde; Unterrichtsstunde; Unterricht; Unterweisung
class diary – Klassenbuch
class numbers – Klassenstärke
class of risk – Gefahrenklasse
class prefect – Klassensprecher
class size – Klassengröße
class spokesman – Klassensprecher
class teacher – Klassenlehrer
class test – Klassenarbeit
class-book – Klassenbuch
classical education – humanistische Bildung
classical organisation theory – klassische Organisationstheorie
classics – Altphilologie
classification – Eingruppierung; Einstufung
classification criterion – Klassifikationskriterium
classification figure – Kennzahl
classify – einteilen; klassifizieren
classmate – Mitschüler; Studienkollege
classroom – Klassenzimmer; Unterrichtsraum
classroom hours per week – Unterrichtsstunden je Woche
classroom instruction – Unterricht im Klassenzimmer
clause – Klausel
clear – anschaulich, klar
clear differences – Konflikte bereinigen, Konflikte lösen
clearance card – Abgangszeugnis (eines(r) Angestellten)
clerical – Büro...; Schreib...
clerical assistant – Bürogehilfe; Bürogehilfin
clerical grade – mittlerer Dienst
clerical staff – Büropersonal
clerk – Büroangestellter
clique – Klüngel
clock in – einstempeln
clock in or out – stechen
clock off – ausstempeln
clock on – einstempeln
clock out – ausstempeln
clock-in entry – Kommenbuchung
clock-in/clock-out times – Kommt-Geht-Zeiten
clock-out entry – Gehenbuchung

close down – Betriebsstillegung
closed circuit television (C.C.T.V.) – Fernsehen für eine geschlossene Gruppe, Industriefernsehen
closed shop – gewerkschaftspflichtiger Betrieb (system where an organisation agrees to employ only union members in certain occupations or at certain places of work)
closing date – Abschlussdatum
closing time – Geschäftsschluss
closure – Schließung; Stillegung
club – Klub; Verein
club house – Klubhaus
clumsy – plump; ungeschickt, schwerfällig
cluster structure – Cluster-Struktur
co-determination – Mitbestimmung
Co-determination in the Mining Industry – Montan-Mitbestimmungsgesetz
co-determination right – Mitbestimmungsrecht
coeducation – Gemeinschaftserziehung; Koedukation
co-owner – Geschäftsteilhaber
co-worker – Arbeitskollege; Kollege; Mitarbeiter
coach – Coach; Repetitor
coaching – Lernbetreuung; Prozessberatung
coaching and counselling – Förderung und Beratung durch Gespräche
coarse – ruppig
code of commerce – Handelsgesetzbuch
coercion – Gewalt; Zwang
coercive power – Macht durch Einschüchterung
coffee corner – Kaffeeküche
cognitive – kognitiv
cognitive ability – kognitive Fähigkeit
cognitive disorder – kognitive Störung
cognitive learning – kognitives Lernen
cognizance – Kenntnisnahme
coherent – schlüssig; zusammenhängend
cohesion – Zusammenhalt
cold reasoning – kalter Verstand
collaborate – mitarbeiten; zusammenarbeiten
collaboration – Mitarbeit
collapse – Zusammenbruch

colleague – Arbeitskollege; Kollege; Mitarbeiter
collect – sammeln
collected – gelassen
collecting agency – Einzugsstelle
collecting of premiums – Einzug der Beiträge
collective agreement – Kollektivvertrag; Manteltarifvertrag
collective agreement area – Tarifbezirk
collective bargaining – Tarifverhandlungen
collective bargaining area – Tarifbezirk
collective bargaining round – Tarifrunde
collective contract – Kollektivvertrag
collective settlement committee – Tarifausschuss
college – Bildungsanstalt; Hochschule; Lehranstalt
College of Advanced Vocational Studies – Berufsakademie
College of Art – Kunsthochschule, Kunstakademie
College of Further Education – Fachoberschule
college rag – studentische Veranstaltung
college recruitment – Hochschulrekrutierung; Personalwerbung an einer Universität/ Hochschule
collegial management behaviour – kollegiales Führungsverhalten
collision check – Kollisionsprüfung
colloquialism – Ausdruck der Umgangssprache
collusion – betrügerisches Einverständnis; Kollusion
colo(u)r blind – farbenblind
color legend – Farblegende
combat fluctuation – Fluktuation bekämpfen
combine works council – Konzernbetriebsrat
combined wage type – Sammellohnart
come into force – In Kraft treten
come to an understanding – vereinbaren
comfort – Behaglichkeit
comfort zone – Komfortzone
comfortable zone – Komfortzone
command – Anweisung; Weisung
command authority – Weisungsbefugnis

commence – beginnen; einführen; ins Leben rufen
comment – Äußerung; Kommentar
commerce – Handel
commercial – kaufmännisch
commercial agent – Handelsvertreter
commercial college – Wirtschafts(ober)schule
commercial law – Handelsrecht
commercial occupation – kaufmännischer Beruf
commercial power of attorney – Handlungsvollmacht
commercial representative – Handelsvertreter
commercial training – kaufmännische Ausbildung
commission – Provision; Umsatzbeteiligung
commission group – Provisionsgruppe
commission of enquiry – Ermittlungsausschuss; Untersuchungsausschuss
Commission of the European Communities – Europäische Kommission
commissioner for women's issues – Frauenbeauftragte(r)
commitment – Zusage, Hingabe
commitment analysis – Situationsanalyse
committed – engagiert
committee – Ausschuss; Kuratorium
committee for economic policies – Wirtschaftsausschuss
committee for labour protection – Arbeitsschutzausschuss
committee of spokesmen – Sprecherausschuss
committee of the Central Works Council – GBR-Ausschuss; Gesamtbetriebsratsausschuss
committee of the Supervisory Board – Ausschuss des Aufsichtsrats
committee on commercial training – Kaufmännischer Bildungsausschuss (KBA) (SAG)
committee on technical training – technischer Bildungsausschuss (TBA) (SAG)

common – allgemein; gemeinsam; gemeinschaftlich
common seizure – Pfändung, gewöhnliche
common law – Gewohnheitsrecht
common room – Lehrerzimmer
common sense – gesunder Menschenverstand
common share (BrE) – Stammaktie
common stock (AmE) – Stammaktie
communicate – kommunizieren; sich mitteilen; sich verständigen
communication – Kommunikation; Verständigung
communication problem – Verständigungsproblem
communications electronics technician – Kommunikationselektroniker
communications equipment installer – Fernmeldemonteur
communicative skills – kommunikative Fähigkeiten
community – Gemeinschaft
community work – Bürgerarbeit (Community work on an honorary basis. It is unpaid but may lead to an improvement in pension rights and be compensated for by public honours.)
commutation – Pauschalierung
commute – pendeln
commuter – Pendler
commuter allowance – Pendlerzuschuss
commuter rate – Pendlerpauschale
commution time – Pendelzeit
company – Betrieb; Gesellschaft; Unternehmen
company administration guideline – Firmenrichtlinie
company bonus – Betriebszulage
company booking (without names) – Firmenbuchung (ohne Namensangabe)
company car – Dienstwagen
company code / plant combination – Buchungskreis-Werk-Kombination
company deduction – Abzug, firmeninterner
company department – Betriebsabteilung
company doctor – Betriebsarzt; Werkarzt
company factor – Unternehmensfaktor
company flat – Werkswohnung

company health insurance fund – Betriebskrankenkasse
company holidays – Betriebsferien
company interest – Unternehmensinteresse
company magazine – Firmenzeitschrift, Hauszeitung
company medical officer – Betriebsarzt
company organization structure – Aufbauorganisation
company pension – Betriebsrente; Firmenpension
company pension plan – Vorsorge, betriebliche
company pension scheme – betriebliche Altersversorgung
company pensioner – Firmenrentner
company reference number – Geschäftszeichen
company restaurant – Kasino
company social policy – betriebliche Sozialpolitik
company sports grounds – Betriebssportplatz
company training centre – betriebliche Ausbildungsstätte
company welfare plans – freiwillige Sozialleistungen der Firma
company-facilitated sports activities – Betriebssport
company-specific trip type – Reiseart, unternehmensspezifische
company-sponsored day care – Betriebskindergarten
company-sponsored education – betriebsverbundenes Studium (BVS)
comparable – vergleichbar
compare – vergleichen
comparison – Vergleich
comparison (jobseekers/ vacancies) – Abgleich
comparison sheet – Vergleichsbogen
compassionate leave – Sonderurlaub
compatible – im Einklang mit; kompatibel; vereinbar
compelling – zwingend
compendium – Kompendium
compensate – abgelten; ausgleichen; entschädigen
compensation – Abgeltung; Lohn

compensation comparison –
Einkommensvergleich
compensation development –
Einkommensentwicklung
compensation evaluation –
Einkommensbemessung
compensation for absence –
Abwesenheitsvergütung
compensation for accumulated flextime
– Gleitzeitausgleich
**compensation for competitive
restriction** – Karenzentschädigung
compensation for rendered services –
Leistungsausgleich
compensation for wage deficiencies –
Lohnausgleich
compensation in money –
Geldentschädigung
compensation key –
Verrechnungsschlüssel
compensation of personnel –
Entlohnung der Mitarbeiter
compensation payment – Entschädigung
compensation regulations – Richtlinien
für die Festlegung der Entlohnung
compensation survey –
Einkommensübersicht
compensation system –
Einkommenssystem
compensation time – Vorholzeit
compensation type – Entlohnungsart
compensatory – ausgleichend;
Entschädigungs...
compete – im Wettbewerb stehen;
konkurrieren
competence – Fähigkeiten; Fertigkeiten;
Kompetenz; Können
competence area – Kompetenzbereich
competency development –
Kompetenzentwicklung
competent – fachkundig; kompetent,
fähig, tüchtig
competition – Konkurrenzdruck;
Wettbewerb
competition clause – Konkurrenzklausel
competitive – auf Wettbewerb eingestellt;
konkurrenzfähig
competitive advantage –
Wettbewerbsvorteil
competitive disadvantage –
Wettbewerbsnachteil

competitiveness – Wettbewerbsfähigkeit
competitor – Mitbewerber; Konkurrent
complaint – Beschwerde; Reklamation
complaints book – Beschwerdebuch
complete – absolvieren; vervollständigen;
vollständig
completed – erledigt
completion – Erledigung, Vollendung
completion vector –
Vollständigkeitsvektor
comply (with) – einhalten; entsprechen
component – Bestandteil
composition – Aufsatz
composition of supervisory board –
Zusammensetzung des Aufsichtsrates
comprehension – Auffassungsgabe;
Verständnis
comprehensive – global; umfassend
comprehensive school – Gesamtschule
comprehensive university –
Gesamthochschule
compromise – Vergleich
compulsion – Gewalt; Zwang
compulsive – zwingend
compulsive person – zwanghafter
Mensch
compulsorily insured – pflichtversichert
compulsory – obligatorisch; Pflicht...
compulsory arbitration –
Zwangsschlichtung
compulsory contribution –
Pflichtbeitrag
compulsory education – allgemeine
Schulpflicht; Schulpflicht, Schulzwang
compulsory health insurance fund –
Pflichtkasse
compulsory holidays – Zwangsurlaub
compulsory insurance –
Pflichtversicherung;
Versicherungspflicht
compulsory primary education –
Volksschulpflicht
compulsory school age –
Pflichtschulalter
compulsory schooling – allgemeine
Schulpflicht; Schulpflicht
compulsory subject – Pflichtfach
compulsory vocational schooling –
Berufsschulpflicht
compulsory workplace – Pflichtplatz

computation of home-to-office time –
Wegezeitberechnung
computer – Computer; DV-Anlage;
Rechner
computer center – Rechenzentrum
computer expert – Computer-Experte
computer illiterate – EDV-Laie
computer literacy – Computerwissen
computer manufacturer –
Computerhersteller
computer scheduling (AmE) –
Rechenzeitplanung
computer science – Informatik
computer science instruction –
Informatik-Unterricht
computer scientist – Informatiker
computer studies – Computer-Unterricht
computer-aided design –
computerunterstütztes Konstruieren;
Zeichnen am Bildschirm
computer-aided instruction –
computerunterstützte Unterweisung
computer-aided learning –
computerunterstütztes Lernen
computer-based training – Lernen mit
einem interaktiven Lernprogramm
concealment – Geheimhaltung
conceit – Eitelkeit
concentration – Häufung; Konzentration;
Zusammenballung
concept – Konzept
conception of oneself – Selbstverständnis
conceptional – konzeptionell
concern (as a whole) –
Gesamtunternehmen
concession period –
Begünstigungszeitraum
conciliation – Schlichtung
conciliation board – Einigungsstelle;
Vermittlungsausschuss
conciliation proceedings –
Vermittlungsverfahren
conciliator – Schiedsmann;
Schiedsrichter; Schlichter
conclusion – Ende; Schluss
conclusion of a contract –
Vertragsabschluss
conclusive – zwingend
concrete thinking – gegenständliches
Denken
condition – Bedingung

condition precedent – aufschiebende
Bedingung
conduct – führen; leiten
conduct of business – Geschäftsführung
confer – konferieren; sich beraten
conference – Konferenz; Sitzung; Tagung
conferment of a doctor's degree –
Promotion
conferring of a degree – Erteilung eines
akademischen Grades
confidant(e) – Vertrauensperson
confidence – Vertrauen
confidential – vertraulich
confidential data – vertrauliche Daten
confidential position –
Vertrauensstellung
confidentiality – Vertraulichkeit
confinement – Entbindung
confirm – bestätigen
confirmation – Bestätigung,
Bescheinigung
confirmation counter –
Rückmeldezähler
confirmation of course booking –
Buchungsbestätigung
confirmation of course registration –
Anmeldebestätigung
confirmation of receipt –
Eingangsbestätigung
confiscate – beschlagnahmen
conflict – Konflikt
conform – übereinstimmen
conformance – Übereinstimmung
congenital – angeboren
congratulate – gratulieren
congress – Kongress; Tagung
connection – Kontakt; Verbindung;
Verknüpfung; Zusammenhang
connection power – Macht durch
Beziehungen
connotation – Assoziation; Konnotation
conscience – Gewissen
conscientious – gewissenhaft
conscientious objector –
Wehrdienstverweigerer
conscript – Wehrpflichtiger
consensus – Konsens
consensus on educational issues –
bildungspolitischer Konsens
consent – Einverständnis; Zustimmung;
Billigung; billigen

consequence – Folge; Konsequenz; Resultat; Wirkung
conservative – konservativ
conservative estimate – vorsichtige Schätzung
consistent – folgerichtig; gleichmäßig; stetig
consolidate – festigen; konsolidieren
consolidated financial statement – Konzernabschluss
consolidated subsidiaries and associated Companies – Beteiligungen
consolidation into a lump sum – Pauschalierung
consonant – Konsonant
conspiracy – Komplott; Verschwörung
constancy – konstant
constant – stetig
constituent (part) – Bestandteil
constitutional complaint – Verfassungsbeschwerde
construction job – Bauberuf
construction sector – Baubranche
constructive – aufbauend; konstruktiv
constructive criticism – konstruktive Kritik
constructive feedback – konstruktive Rückmeldung
consult – jemanden um Rat fragen
consultancy contract – Beratervertrag
consultant – Berater; Unternehmensberater
consultation by specialists – Fachberatung
consultancy – Beratungsunternehmen
consumer price index – Lebenshaltungsindex
consumption of alcohol – Alkoholkonsum
contact – Kontakt; Verbindung; Ansprechpartner
contact hours – Unterrichtsstunden
contact person – Ansprechpartner
contestable – anfechtbar
context – Zusammenhang
contingency – Eventualfall; unvorhergesehenes Ereignis
contingency planning – Eventualitätsplanung; Kontingenzplanung

contingency theory – Kontingenztheorie
continual – stetig, ständig
continual learning – ständiges Lernen
continual pension adjustment – Rentendynamik
continue – fortdauern; fortfahren mit; fortsetzen
continued employment – Weiterbeschäftigung
continued existence of a company – Firmenfortführung
continued insurance – Weiterversicherung
continued pay – Lohnfortzahlung
continued pay during approved leave from work – Urlaubslohnfortzahlung
continued payment – Weiterbezahlung
continued payment of salary – Gehaltsfortzahlung
continued remuneration – Entgeltfortzahlung
Continued Remuneration Act – Entgeltfortzahlungsgesetz
continuing education – Fortbildung; nachschulische Angebote in der Erwachsenenbildung
continuous – fortlaufend; ununterbrochen, laufend
continuous assessment – regelmäßige Beurteilung
continuous reading method – kontinuierliche Zeitmessung; kontinuierliche Zeitnahme
contract – Vertrag
contract condition – Vertragsbedingung; Vertragsbestimmung
contract data – Vertragsdaten
contract department – Vertragsabteilung
contract duration – Vertragsdauer
contract group – Vertragsgruppe
contract law – Vertragsrecht
contract negotiation – Vertragsverhandlung
contract out – ausschreiben
contract wage payment – Akkordsatz
contracting out – Auftragsvergabe
contractual – vertraglich
contractual pension adjustment – schuldrechtlicher Versorgungsausgleich

contractual power –
Abschlussvollmacht;
Handlungsvollmacht
contractual working hours –
vertragliche Arbeitszeit
contribute – Beitrag leisten; beitragen
contribution – Beitrag
contribution accounting –
Beitragsabrechnung
contribution allowance –
Beitragszuschuss
contribution amount – Umlage
contribution group – Beitragsgruppe
contribution linked – beitragsorientiert
contribution rate – Beitragssatz (SW)
contribution rate table – Beitragstabelle
contribution scale – Beitragsklasse
**contribution to a professional
association** –
berufsgenossenschaftlicher Beitrag
contribution to employee association –
Kammerumlage
contribution to the pension scheme –
Pensionsbeitrag
**contributions according to actual
earnings** – Beiträge nach dem
wirklichen Arbeitsverdienst
contributions account – Beitragskonto
contributions in arrears –
Beitragsrückstände
contributions to the old-age pension –
Beiträge zur
Arbeiterrentenversicherung
contributory employment –
versicherungspflichtige Beschäftigung
contributory period – Beitragszeiten
control – steuern; kontrollieren;
Kontrolle
control by exception – Kontrolle durch
Planabweichung
control by inspection – Kontrolle durch
Aufsicht
control character – Steuerzeichen
control feature – Steuerungsmerkmal
control group – Kontrollgruppe
control term – Steuerungsbegriff
controlled – kontrolliert
controlled reading – gesteuertes Lesen
controlled vocabulary – ausgewählter
Wortschatz
controller – Controller

convalescence home – Erholungsheim
convalescent leave – Genesungsurlaub
convene (meeting) – einberufen (eine
Besprechung)
convenient – günstig; passend
convention – Kongress; Tagung
conventional – konventionell
converge – annähern
conversation – Gespräch
conviction – Überzeugung
convince – überreden, überzeugen
cool – gelassen
cooling-off period –
Überdenkungsperiode
cooperate – mitarbeiten
cooperation – Zusammenarbeit,
Mitarbeit
cooperative – zur Zusammenarbeit bereit
cooperative management – kollegiale
Leitung
cooperative management behaviour –
kooperatives Führungsverhalten
cooperative mutual pension fund –
Genossenschaftskasse auf
Gegenseitigkeit
cooperativeness – Kollegialität
coopetition – kooperative Konkurrenz
(cooperation and competition combine)
coordinate – koordinieren
coordination – Koordination
coping with problems –
Problembewältigung
copy – abschreiben; Durchschlag;
Exemplar; Kopie; spicken
copyright – Urheberrecht
Copyright law – Gesetz über Copyright
core competence – Kernkompetenz
core job – Arbeitsstelle für die Elite
(relatively few top jobs and only for top
people)
core night work – Kernnachtarbeit
core skills curriculum – Lehrplan für die
Hauptfächer
core subject – Hauptfach
core team of employees –
Kernbelegschaft
core time – Kernzeit
core vocabulary – Grundwortschatz
core working hours – Kernzeit;
Kernarbeitszeit
corollary – Folgesatz; Umkehrsatz

corporal punishment – Prügelstrafe
corporate culture – Unternehmenskultur (The beliefs, behaviours, and assumptions shared by individuals within an organisation. It includes such things as procedures, values, and unspoken norms.)
Corporate Division – Zentralabteilung
Corporate Executive Committee – Zentralvorstand (SAG)
corporate housing administration – zentrales Wohnungsreferat
Corporate Human Resources – Zentralabteilung Personal
corporate identity – Firmenidentität
corporate management – Firmenleitung; oberster Führungskreis
corporate organization – Unternehmensorganisation
corporate policy – Unternehmenspolitik
corporate strategy – Unternehmensstrategie
corporate structure – Unternehmensaufbau
corporate university – betriebsinterne Weiterbildung
corporation – Aktiengesellschaft; Gesellschaft
corporation lawyer – Justitiar; Leiter der Rechtsabteilung
correct – korrigieren; verbessern; vorschriftsmäßig; korrekt
correction – Verbesserung
correction document – Korrekturbeleg
correction period – Korrekturperiode
correction run – Korrekturlauf
correction sheet – Korrekturbeleg
corrective training – Entziehungskur
correspondence – Korrespondenz; Schriftverkehr
correspondence course – Fernlehrgang; Fernstudium
cosmopolitan – weltoffen
costs – Aufwand; Auslagen; Kosten; Kostenbetrag
cost accounting – Betriebsbuchhaltung; Kostenrechnung
cost center – Kostenstelle
cost center for fixed costs – Fixkostenstelle

cost consciousness – Kostenbewusstsein; Kostendenken
cost criteria group – Gruppe der Kostenkriterien
cost for delivery – Zustellkosten
cost of living – Lebenshaltungskosten
cost of on-going training – Aufwendungen für Weiterbildung
cost of vocational training – Aufwendungen für Berufsausbildung
cost planning – Kostenplanung
cost reduction – Kostenabbau
cost transfer – Kostenübernahme
cost-conscious – kostenbewusst
cost-cutting drive – Sparprogramm
cost-of-living adjustment – Ausgleichzulage
cost-of-living allowance (COLA) – Lebenshaltungskosten-Ausgleich, Teuerungszulage, Ortszuschlag
costs for continuing education – Weiterbildungsaufwand
costs projection – Kostenplan
Council of Honorary Members – Ehrenmitglieder des Aufsichtsrates
counselling – Beratung; Rat
count – zählen
count child – Zählkind
counting class – Zählklasse
counting of leave days – Urlaubsauszählung
country group – Ländergruppe
country group key – Ländergruppenschlüssel
country ID – Nationalitätskennzeichen
country indicator – Länderkennzeichen
country modifier – Länder-Modifikator; Ländermodifikator
country of birth – Geburtsland
country of origin – Herkunftsland
country report – Länderreport
country travel key – Reiseländerschlüssel
country version – Länderversion
country-dependent – länderabhängig
courageous – wagemutig
course – Lehrveranstaltung; Seminar; Ausbildungsgang; Ausbildungslehrgang; Lehrgang
course booking – Kursbuchung
course brochure – Kursbroschüre

course cancellation – Kursabsage
course content – Lehrinhalt; Lehrstoff
course evaluation – Kursauswertung;
  Kursbewertung
course fee – Kursgebühr
course fees – Kursgebühren
course group – Kursgruppe
course location – Kursort
course notification – Kursmitteilung
course of advanced studies – Studium
  auf höchster Ebene
course of education – Bildungsgang
course of study – Studiengang
course offer – Kursangebot
course prerequisite – Kursvoraussetzung
course procedure – Kursverfahren
course registration – Kursbelegung
course schedule – Kursablauf;
  Seminarablauf
course type – Kurstyp
courses on offer – Studienangebot
courses on offer for further education –
  Weiterbildungsangebot
coursework – Mitarbeit (Coursework is
  work completed throughout the course
  and may vary from course to course. It
  contributes by at least 20% towards the
  final marks.)
court costs – Gerichtskosten
court of Social Justice – Sozialgericht
court order – Lohnpfändungsbeschluss
court order exemption –
  Pfändungsschutz
courteous – höflich
courtesy – Höflichkeit
cover – Deckung
coverage – Deckung
covering letter – Begleitbrief
covering note – Begleitbrief
crab – Nörgler; Querulant
crackdown (AmE) – Maßregelung
craft – Gewerbe; handwerklicher Beruf
craft trades – Handwerk
craft union – Fachgewerkschaft
Crafts Code – Handwerksordnung
craftsman – Facharbeiter
craftsmanship – handwerkliches Können
cram – büffeln; pauken
cram course – Paukkurs
crammer – Repetitor
crash course – Intensivkurs

create – hervorbringen; schaffen
creative – kreativ; schöpferisch
creativity – Kreativität
creativity technique –
  Kreativitätstechnik
crèche – Ganztagshort; Kinderhort, Hort,
  Kinderkrippe, Kindertagesstätte
credentials – Unterlagen
credibility – Glaubwürdigkeit
credit (AmE) – Jahreszeugnis; Zeugnis
credit factor – Anrechnungsfaktor
credit point – Bonuspunkt
credit point system – Punktsystem
crediting – Anrechnung
crediting regulation –
  Anrechnungsregelung
credulous – gutgläubig
crib – Spickzettel
crisis – Krise
criteria base – Kriterienbank
criterion – Kriterium
critical – kritisch
critical path method – Methode des
  kritischen Pfads
critical performance analysis –
  Aufgabenanalyse
critical value – Grenzwert
cross-border employee – Grenzgänger
cross-cultural experience –
  interkulturelle Erfahrung
cross-cultural study – interkulturelle
  Studie
cross-cultural training – interkulturelles
  Training
cross-national – länderübergreifend
culpable – schuldig
cultural change – kulturelle Veränderung
cultural differences – kulturelle
  Unterschiede
cultural heritage – kulturelles Erbe
culturally deprived – kulturell
  Benachteiligte
culture – Kultur
culture clash – Reibung auf Grund
  unterschiedlicher Kulturen
culture of error discussion –
  Fehlerkultur
culture of trust – Vertrauenskultur
cum laude – cum laude; mit Lob
cumulated gratuity basis –
  Gratifikationsbasis, kumulierte

cumulated wage type – Lohnart,
   summierte
cumulation – Kumulation
cumulation indicator –
   Kumulationskennzeichen
cumulation of gross amount –
   Bruttozusammenfassung
cumulation of net amount –
   Nettozusammenfassung
cumulation rule – Kumulationsregel
cumulation wage type –
   Kumulationslohnart; Summenlohnart
curatorship – Kuratorium
cure – Kur
cure home – Kurheim
curiosity – Neugier
current – gegenwärtig; laufend
current administration – laufende
   Verwaltungsgeschäfte
current benefits – laufende Leistungen
current employee compensation –
   laufendes Arbeitsentgelt
current job – ausgeübter Beruf
current payroll –
   Abrechnungsgegenwart
curriculum – Lehrplan; Studienplan;
   Unterrichtsplan
curriculum vitae (cv) – Lebenslauf
cursory – flüchtig; oberflächlich
custom – Praxis
customary – (betriebs)üblich
customary law – Gewohnheitsrecht
customer – Kunde
customer file – Kundenkartei
customer loyalty – Kundenloyalität
customer orientation –
   Kundenorientierung
customer satisfaction –
   Kundenzufriedenheit
customer specific – kundenspezifisch
customizing – Erfüllung individueller
   Wünsche des Kunden
customizing option –
   Anpassungsmöglichkeit
cut – Herabsetzung; Kürzung
cut in salary – Gehaltskürzung
cut-off rate – Kappungssatz
cutback – Arbeitsplatzabbau;
   Herabsetzung; Kürzung
cutting mechanic –
   Zerspanungsmechaniker

cutting of wages – Herabsetzung der
   Löhne; Lohnabbau
cybernetics – Kybernetik
cycle time – Taktzeit

D

daily – tagtäglich, täglich
daily balance – Tagessaldo
daily experience at work – tagtägliche
   Berufspraxis
daily round – Alltagstrott
daily time balance – Tageszeitsaldo
daily wage – Tageslohn
daily work schedule –
   Tagesarbeitszeitplan
daily working hours – Dienstzeit
daily office routine – Berufsalltag
damage – beschädigen; Schaden
damage for pain and suffering (BrE) –
   Schmerzensgeld
damage suit – Schadenersatzklage
damage to property – Sachschaden
damages compensation – Schadenersatz
danger – Gefahr
danger analysis – Gefährdungsanalyse
danger money – Gefahrenzulage
danger of a strike – Streikgefahr
danger of explosion – Explosionsgefahr
dangerous – gefährlich
dangerous materials – Gefahrenstoffe;
   gefährliche Arbeitsstoffe
daring – gewagt
data – Daten
data bank – Datenbasis
data bank system – Datenbanksystem
data capture – Datenerfassung
data communications regulation –
   Datenübermittlungsverordnung
data conversion – Datenumsetzung
data input regulation –
   Dateneingabeverordnung (DEVO)
data preparation – Datenaufbereitung
data privacy measures –
   Datenschutzmaßnahmen

data processing – Datenverarbeitung
data processing assistant –
datentechnische(r) Assistent(in)
data processing system –
Datenverarbeitungssystem
data processing technology –
DV-Technologie
data protection – Datenschutz
data protection officer –
Datenschutzbeauftragter
data system – Datensystem
data transmission regulation –
Datenübertragungsverordnung
(DUVO)
database – Datenbasis
date – Datum
date abbreviation – Datumsabkürzung
date of birth – Geburtsdatum
date of confinement –
Entbindungsdatum
date of delivery (USA) (CAN) –
Entbindungsdatum
date of entrance – Eintrittsdatum;
Eintrittstermin
date of hire – Eintrittstag
date of interview – Vorstellungsdatum
date of joining the company –
Firmeneintritt
date of leaving – Austrittsdatum
date of notice of dismissal –
Kündigungsdatum
date of origin – Entstehungsdatum
date specification – Datumsangabe
day – Tag
day boarder – Externe(r) eines Internats
day care centre – Tageshort
day feature – Tagesmerkmal
day labourer – Tagelöhner
day model – Tagesmodell
day modifier – Tagesmodifikator
day nursery – Kinderkrippe;
Kindertagesstätte
day of absence – Abwesenheitstag
day of the week – Wochentag
day off – arbeitsfreier Tag; freier Tag;
Ruhetag
day processing – Tagesverarbeitung
day program – Tagesprogramm
day program for flexitime –
Gleitzeittagesprogramm
day pupil – Externe(r) eines Internats

day rate – Tagessatz
day shift – Tagesschicht
day type – Tagestyp
day type modifier – Tagestypmodifikator
day's leave – Urlaubstag
day-dreamer – Tagträumer
day-release – Befreiung; Freisetzung;
Freistellung
day-release training – tageweise
Freistellung für Kursteilnahme
day-school – Tagesschule
day-to-day business – Tagesgeschäft
day-to-day business matters – laufende
Geschäftsangelegenheiten
days of absence – Fehltage
days of service – Diensttage
days qualifying for social insurance –
SV-Tage
deadline – frist
deadline for application – Antragsfrist
deaf – taub; gehörlos
deaf-mute sign language –
Gebärdensprache
dean – Dekan
dean of studies – Studiendekan
deanship – Dekanat
death – Ableben; Tod; Todesfall
death benefit – Sterbegeld
debilitation – Entkräftung; Schwächung
debiting contributions –
Beitragsverrechnung
debriefing – Erfahrungsberichterstattung
deburocratize – entbürokratisieren
decay meter – Halbwertszeit
decentralization – Dezentralisierung
decide – entscheiden
decide against – sich dagegen
entscheiden
decision – Entscheidung; Schied(s)spruch
decision analysis –
Entscheidungsanalyse
decision maker – Entscheidungsträger
decision making – Entscheidungsfindung
decision operation –
Entscheidungsoperation
decision-making ability –
Entscheidungsfähigkeit
decisive – maßgebend; entschlussfreudig
declaration made by third party debtor
– Drittschuldnererklärung
decline – Rückgang

decline in performance – Leistungsabfall
declining – rückläufig
decrease – Abnahme; abnehmen; Rückgang; sich vermindern
decrease in efficiency – Leistungsverlust
dedication – Engagement, Hingabe
deduct – abziehen, abführen
deduct from pay – vom Lohn einbehalten
deductible – Selbstbeteiligung
deductible amount – Absetzbetrag
deduction – Abzug; Rabatt
deduction amount – Abzugsbetrag
deduction due to entertainment – Abzug wegen Bewirtung
deduction method – Abzugsmethode
deduction of contribution amount – Beitragsabführung
deduction principle – Abzugsprinzip
deduction rate – Abzugssatz
deduction type – Abzugsart
deductive – deduktiv
deductive thinking – deduktives Denken
default – Zahlungsverzug
defect – Defekt, Fehler, Schaden
defendant – Beklagte(r)
defence mechanism – Abwehrmechanismus
defensive communication – defensive Kommunikation
deferment of decisions – Aussetzen von Beschlüssen
deficiency – Mangel
deficiency analysis – Mängelanalyse
deficiency compensation – Nachteilsausgleich
deficit – Unterdeckung; Defizit (persönliche)
defined benefit plan – feste (Renten-)Leistung
defined contribution plan – beitragsbasierter Pensionsplan
definite confirmation of place – Platzzusage, sichere
deformity – Entstellung; Missbildung
defraudation of contributions – Beitragshinterziehung
degenerative disease – Verschleißerkrankung
degree – akademischer Grad; Grad

delay – aufschieben; Aufschub; verzögern; Verzögerung
delay of payment – Zahlungsverzug
delay penalty – Säumniszuschlag
delayer – schlanke Hierarchie
delegate – abordnen; delegieren
delegate responsibility – Verantwortung übertragen
delegate tasks – Aufgaben übertragen
delegated personnel responsibility – delegierte Personalverantwortung
delegation – Versetzung; Abordnung
delegation of employees – Entsendung von Arbeitnehmern
deliberate – vorsätzlich
delimitation – Abgrenzung
delimitation date – Abgrenzdatum
delinquent – Delinquent
delivery – Vortragsweise
delivery (USA) – Entbindung
demand – Bedarf; Forderung; Anfordernis; zumuten
demand for education and training – Aus- und Fortbildungsbedarf
demand for executives – Führungskräftebedarf
demand for further education and – Weiterbildungsbedarf
demands made on personnel policies – Anforderungen an die Personalpolitik
demanding – anspruchsvoll
demarcation – Abgrenzung
demeanour – persönliches Auftreten, Verhalten
demonstration bench – Experimentierplatz
demonstration place – Experimentierplatz
denominational high school – konfessionelle Oberschule
denominational school – konfessionelle Schule
denominator – Nenner
department – Abteilung; Dienststelle; Referat
department manager – Abteilungsleiter
department meeting – Abteilungsversammlung
Department of Education and Science – Kultusministerium

Department of Social Security –
Sozialversicherungsträger,
Sozialleistungsträger
department office – Bereichssekretariat
department specialized in ... –
Fachbereich
departmental interest –
Bereichsinteresse
departure – Abfahrt
dependent – Abhängige(r); unselbständig
dependent on age – abhängig vom
Lebensalter
dependents' benefits – Leistungen an
Unterhaltsberechtigte
dependents' pension – Rente für
Familienangehörige
deployment – Einsatz
deployment of employees –
Mitarbeitereinsatz
depreciation – Abschreibung
depressive person – depressiver Mensch
deprivation – Entbehrung; Entziehung;
Entzug
deprived child – Kind, das an
Liebesentzug leidet
Deputy Chairman of the Board –
Vorstandsvorsitzender, stellvertretender
**Deputy Chairman of the Supervisory
Board** – stellvertretender Vorsitzender
des Aufsichtsrates
deputy headmaster – Konrektor
deputy member – Ersatzmitglied
deputy principal (Gymnasium) –
Studiendirektor
deregistration – Abmeldung
derived wage type – Lohnart, abgeleitete
describe – beschreiben
description of sickness –
Krankheitsbeschreibung
deserve acknowledgement –
Anerkennung verdienen
design – Konstruktionslehre (Lehrfach)
design engineer – Konstrukteur
designate – benennen; bezeichnen
desired period – Vormerkzeitraum
desired salary – Gehaltsvorstellung
desk – Pult; Schreibtisch
despondent – mutlos
detail maintenance – Detailpflege
detailed – eingehend
details – Einzelheiten

detection – Aufdeckung; Entdeckung
detention – Festnahme; Haft
detention centre (BrE) –
Jugendstrafanstalt
detention home (AmE) –
Jugendstrafanstalt
determination – Entschlossenheit
determination of times – Zeitermittlung
determine – bestimmen; festsetzen
determine needs – Bedürfnisse
bestimmen
determined – entschlossen; zielstrebig
determining planned time pairs –
Sollpaarermittlung
detrimental effect upon – schädlicher
Einfluss auf
develop – ausbauen; entwickeln
development – Entwicklung; Entfaltung
development aim – Entwicklungsziel
development goal – Entwicklungsziel
development level – Entwicklungsgrad
development margin –
Entwicklungsspielraum
development measures –
Entwicklungsmaßnahmen
**development of high-potential
employees** – Entwicklung qualifizierter
Nachwuchskräfte
development of personality –
Persönlichkeitsentwicklung
development of staffing level –
Personalbestandsentwicklung
development process –
Entwicklungsprozess
developmental age – Entwicklungsalter
deviation – Abweichung
dexterity – Geschicklichkeit;
Gewandtheit
diagonal communication –
Kommunikation zwischen
verschiedenen Führungsebenen
diagram – Diagramm; graphische
Darstellung; Schaubild
dialect – Dialekt
dialog wage type – Dialoglohnart
dictation – Diktat
diction – Ausdrucksweise; Diktion
didactics – Didaktik
differ – sich unterscheiden; verschieden
sein (von)
difference – Unterschied

difference hours – Differenzstunden
difference table – Differenztabelle
different – verschieden
different payment – Bezahlung,
abweichende
differential piecerate –
Differentialstücklohn
difficult – schwierig
difficulty – Schwierigkeit
difficulty in keeping an appointment –
Terminschwierigkeit
dignity – Würde
digression – Abschweifung
diligence – Fleiß
dilution agreement –
Verwässerungsabkommen
diminishing number of pupils –
Schülerrückgang; Schülerschwund
diminution of the earning capacity –
Minderung der Erwerbsfähigkeit
dining-centre – Mensa; Speiseraum;
Speisesaal
diploma – Diplom
direct – anordnen
direct insurance – Direktversicherung
direct insurance contributions –
Vervielfachung (bei
Direktversicherungsbeiträgen)
direct labour costs – direkte
Arbeitskosten; Fertigungslohn
direct method – direkte Methode
directive – Anweisung; Weisung
director of industrial relations –
Arbeitsdirektor
director of training – Ausbildungsleiter
directory – Verzeichnis
directory of positions –
Planstellenverzeichnis
dirty money (BrE) – Schmutzzulage
disability benefit –
Arbeitsunfähigkeitsentschädigung
disability fund – Invalidenkasse
disability insurance –
Invaliditätsversicherung
disability pension –
Berufsunfähigkeitsrente; Invalidenrente
disabled – arbeitsunfähig; behindert
disabled person – Behinderter; Invalide;
Schwerbeschädigter

disabled persons' representative –
Schwerbehindertenvertretung;
Vertrauensmann
disadvantage – Benachteiligung;
Nachteil
disallowance of ... (e.g. old-age pension
– Aberkennen von ... (z.B.
Altersversorgung)
discharge – Dienstenthebung;
Entlassung; freisetzen
discharge of duty – Pflichterfüllung
discharge of labor – Freisetzung von
Arbeitskräften
disciplinarian – autoritärer Lehrer
disciplinary punishment – Maßregelung
discipline – Disziplin; Studienfach
discontinuous work – diskontinuierliche
Arbeit
discount – Rabatt
discounting – Abzinsung
discouraged – mutlos
discover – entdecken
discovery – Aufdeckung; Entdeckung
discovery method – Lernen durch
entdecken
discrepancy – Diskrepanz;
Unstimmigkeit
discretion – Belieben; Diskretion;
Ermessen
discretionary decision –
Ermessensentscheidung
discretionary pension – Ermessensrente
discriminate – diskriminieren;
unterschiedlich behandeln
discrimination – Benachteiligung;
Diskriminierung
discrimination on racial grounds –
Rassendiskriminierung
discuss – beraten über; diskutieren
discussion – Gespräch
discussion group – Diskussionsgruppe
discussion method –
Diskussionsmethode
disheartened – mutlos
disgrace – blamieren
dismemberment grading – Gliedertaxe
(Unfallversicherung)
dismiss – entlassen; feuern; hinauswerfen
dismissal – Dienstenthebung; Entlassung;
Kündigung durch die Firma

dismissal compensation – Abfindung;
Entlassungsentschädigung

dismissal for variation of contract –
Änderungskündigung

dismissmal with notice – ordentliche
Kündigung

disobedience – Gehorsamsverweigerung

disobedient – ungehorsam

disorder – Unordnung

disparity – Ungleichheit

dispassionate – leidenschaftsarm

dispense with something – verzichten
auf

displace – freisetzen; verdrängen

displacement – Verdrängung

disposable income – verfügbares
Einkommen

dispose – anordnen

dispute – Auseinandersetzung; bestreiten;
in Zweifel ziehen

disregard – Missachtung

disrespect – Missachtung

dissatisfaction – Unzufriedenheit

dissemination of information –
Verbreitung von Informationen

dissertation – Dissertation; Doktorarbeit

distance in kilometers –
Entfernungskilometer

distance teaching – Fernunterricht

distance working – Telearbeit

distant – distanziert

distant education – Fernschulung

distraction – Ablenkung

distractor – Distraktor

distribute – verteilen

distribution – Verteilung

distrust – Misstrauen

disturbance – Störung

disturbance in the power of perception
– Wahrnehmungsstörung

disturbed – verhaltensgestört

disturbed pupil – verhaltensgestörter
Schüler

divide – teilen

Division – Geschäftsgebiet

divorce – Ehescheidung

divorced – geschieden

do badly (e.g. in a test) – schlecht
abschneiden (z.B. in einem Test)

do research work – forschen

do well – gut abschneiden

do without – verzichten auf

doctor's appointment – Arztbesuch

doctor's degree – Doktor(titel);
Doktorwürde

doctoral candidate – Doktorand,
Promovend

doctorate – Doktor(titel); Doktorwürde

document of thanks – Dankesurkunde

document – Beleg

documentary list – Nachweisliste

**documentation of contributory and
non-contributory periods** –
Kontenklärung

documents – Unterlagen

documents on marital status –
Familienstandsnachweis

domain – Aufgabengebiet

domestic – inländisch

domestic help – Haushaltsgehilfe

domestic income – Inlandseinkommen

domestic matters – Inlandsaufgaben

Domestic Siemens Company –
Siemens-Gesellschaft im Inland

domestic science – Hauswirtschaftslehre

domestic tasks – Inlandsaufgaben

domestic trip – Inlandsreise

dominancy – Dominanz

dominant – dominierend; vorherrschend

don – Universitätsdozent (At British
universities, esp. Oxford and
Cambridge: a head, fellow, or tutor of
a college, a member of the teaching
staff.)

done – erledigt

doorkeeper – Pförtner

dormant employment – ruhendes
Beschäftigungsverhältnis

dormitory – Schlafsaal

dormitory (dorm) – Studentenheim;
Studentenwohnheim

double income household –
Doppelverdienerhaushalt

double income rate –
Doppelverdienersatz

double qualification of skilled workers
– Doppelqualifikation von
Facharbeitern

double taxation convention –
Doppelbesteuerungsabkommen

double verification principle –
Vier-Augen-Prinzip

doubly occupied position – doppelt besetzte Planstelle
double first (honours) – beide Fächer mit "sehr gut" bestanden
doubt – Zweifel
down time – Leerlaufzeit
downgrading – Abgruppierung; Herabstufung; Zurückstufung; Downgrading
downsizing – Gesundschrumpfen; Personalstraffung; Arbeitsplatzabbau; Personalstandreduzierung
draft contract – Vertragsentwurf
draftee – Wehrpflichtiger
draughtsman – Bauzeichner; technischer Zeichner
draw a pension – Rente beziehen
draw benefits – Leistungen beziehen
draw wages – Lohn beziehen
drawing – Zeichnung
dream job – Wunschberuf
dressing room – Umkleideraum
drinker – Säufer; Trunkenbold
drive – Elan; Energie, Schwung
drop a subject – Studienfach aufgeben
drop out – Schulabbrecher; Schule abbrechen
drop-out rate – Abbruchquote; Ausfallrate
drug – Droge
drug addict – Drogenabhängige(r); Rauschgiftsüchtige(r)
drunkard – Säufer; Trunkenbold
drunkenness – Trunkenheit
dry-out – Entziehungskur
dual course of studies – dualer Studiengang
dual system of vocational training – duales System der Berufsausbildung
dual-career marriage – Familie und Karriere
due notice – ordnungsgemäße Kündigung
dull – schwerfällig (von Begriff)
dummy payroll run – Probeabrechnung
dunce – Dummkopf; schlechter Schüler
duplication of work – Doppelarbeit
duration – Dauer
duration of delegation – Gesamtdauer der Abordnung
duration of education – Ausbildungszeit

duty – Pflicht; Verpflichtung
duty rota – Dienstplan
duty to disclose information – Auskunftspflicht
dynamics of innovation – Innovationsdynamik
dysphasia – Aphasie

E

E&T Planning – Aus- und Weiterbildungsplanung
eagerness to learn – Lernfreude
ear – Ohr
earliest retroactive accounting period – Rückrechnungstiefe, Rückrechnungsperiode, tiefste
early retiree – Vorruheständler
early retirement – Frührente; vorgezogene Pensionierung; Frühverrentung
early retirement agreement – Vorruhestandsregulung
early retirement benefit – vorzeitiges Altersruhegeld
early shift – Frühschicht
earn – verdienen
earned income – Arbeitseinkommen; Arbeitsentgelt
earned-income allowance – Arbeitnehmerfreibetrag
earnings – Verdienst
earnings factor – Verdienstgrad
earnings gap – Lohnschere
earnings loss compensation; compensation – Verdienstausfallentschädigung
earnings-related benefit – Arbeitslosengeld
earnings replacement benefit – Lohnersatzleistung
earphones – Kopfhörer
ease – Leichtigkeit
eclectic – eklektisch; nachahmend; nicht eigenständig; unschöpferisch

ecologically beneficial –
umweltfreundlich; umweltschonend
economic science – Volkswirtschaftslehre
economic value added (EVA) –
Geschäftswertbeitrag
economics – Volkswirtschaft,
Volkswirtschaftslehre,
Wirtschaftswissenschaft
economy – Wirtschaft
EDP – EDV; elektronische
Datenverarbeitung
EDP experience – EDV-Erfahrung
EDP organization – EDV-Organisation
EDP school – EDV-Bildungseinrichtung
EDP specialist – EDV-Fachmann
EDP-implementation –
EDV-Realisierung
educated – gebildet
educated classes – die Gebildeten
education – Bildung; Erziehungswesen;
Schulbildung
education and training – Aus- und
Fortbildung
education and training administration
– Seminarverwaltung
education and training planning –
Seminarplanung
education centre – Bildungszentrum
education laws – Bildungsgesetze
education of retarded/ handicapped –
pädagogische Sonderbetreuung
(behinderter Kinder)
education(al)ist – Pädagoge; Pädagogin
educational – lehrreich
educational and training grant –
Erziehungs- und Ausbildungsbeihilfen
educational background – Schulbildung
educational boom – Bildungsboom
educational channel – Bildungsweg
educational consultant –
Bildungsberater
educational counselling –
Ausbildungsberatung
educational establishment –
Bildungsanstalt
educational event –
Bildungsveranstaltung
educational facility –
Bildungseinrichtung
educational goal – Bildungsziel

educational guidance –
Ausbildungsberatung
educational institution –
Bildungseinrichtung
educator – Pädagoge
educational issues – Fragen der
(Schul)bildung
educational leave – Bildungsurlaub
educational level – Bildungsniveau;
Bildungsstand
educational needs – Bildungsbedarf
educational objective – Bildungsziel
educational planning – Bildungsplanung
educational policy – Bildungspolitik
educational possibilities –
Bildungsmöglichkeiten
educational psychologist –
Erziehungsberater
educational resources – Bildungsmittel
educational services – Bildungsangebot
educational standard – Bildungsniveau;
Bildungsstand
educational structure – Bildungsstruktur
educational system – Bildungswesen;
Erziehungswesen; Schulwesen
educationally handicapped –
lernbehindert
educator Erzieher(in); Pädagoge,
Pädagogin
EFA-meeting – EFA-Gespräch
EFA-staff dialogue – EFA-Gespräch
effect – Wirkung
effective – effektiv
effective communication – wirksame
Kommunikation
effectiveness – Effektivität
effectiveness of training – Wirksamkeit
des Trainings
efficiency – Arbeitsleistung; Effizienz
efficiency contest – Leistungswettbewerb
efficiency payment – Leistungsentgelt
efficient – leistungsfähig
effort – Anstrengung; Bemühen;
Bemühung
egalitarian – egalitär; Verfechter(in) des
Gleichheitsprinzips
egg head (AmE) – Hochgebildete(r);
Wissenschaftler
egocentric – egozentrisch
egoism – Egoismus; Selbstsucht
egoist – Egoist

egotism – Eigendünkel
eidetic – anschaulich; eidetisch
elaboration of objectives – Erarbeiten
  von Zielen
elbow power – Durchsetzungsvermögen
election – Auswahl; Wahl
election contest – Wahlanfechtung
election of delegates – Wahl der
  Wahlmänner
election of employee representatives –
  Wahl der Arbeitnehmervertreter
election regulations – Wahlvorschriften
electoral board – Wahlvorstand
electoral process – Wahlvorgang
electrical engineer – Elektroingenieur
electrical engineering – Elektrotechnik
electrical engineering assistant –
  elektrotechnische(r) Assistent(in)
electrical machine fitter –
  Elektromaschinenmonteur
electrician – Elektriker;
  Elektroinstallateur
electronic data processing – EDV;
  elektronische Datenverarbeitung
electronic learning – Lernen am
  Computer
electronics – Elektronik
electronics technician – Elektroniker
elementary education –
  Grundschulbildung
elementary knowledge – Basiswissen
elementary school – Grundschule;
  Volksschule
eleven plus examination (BrE) –
  Aufnahmeprüfung in eine
  weiterführende Schule
elicit – entlocken
eligibility – Wählbarkeit;
  Teilnahmeberechtigung
eligible for a pension –
  pensionsberechtigt
eliminate – ausschalten; beseitigen
elite – Elite; Auslese
elitist – elitär
elocution – Sprechtechnik
eloquent – redegewandt
embarrassing – peinlich
emergency budget – Notbudget;
  Reservebudget
emergency exit – Notausgang
emergency lighting – Notbeleuchtung

emergency relief – Notstandsbeihilfe
emergency service – Bereitschaftsdienst
emoluments – Bezüge; Einkommen
emotion – Emotion; Gemütsbewegung
emotional block – emotionale Sperre
emotional disorder – emotionale
  Störung
emotional intelligence – emotionale
  Intelligenz
emotional sales intelligence –
  emotionale Vertriebsintelligenz
empathy – Einfühlungsvermögen;
  Empathie
empiric(al) – empirisch
empirical knowledge – empirisches
  Wissen
empirical method – empirische Methode
employ – beschäftigen; einstellen
employed – berufstätig
employee – Arbeitnehmer;
  Belegschaftsmitglied, Mitarbeiter;
  Betriebsangehöriger
employee association – Arbeiterkammer
employee attitude survey –
  Mitarbeiterbefragung
employee benefits – Sozialleistungen
employee compensation –
  beitragspflichtiges Arbeitsentgelt
employee counselling –
  Arbeitnehmerberatung
employee count – Mitarbeiterzahl
employee eligible for company pension
  – Ruhegehaltsberechtigter
employee group – Mitarbeitergruppe
employee handbook –
  Arbeitnehmerhandbuch
employee I.D. card – Firmenausweis
employee master file –
  Mitarbeiter-Stammdatei
employee participation – Teilnahme des
  Arbeitnehmers
**employee participation in asset
  formation** – Arbeitnehmerbeteiligung
employee pension scheme – betriebliche
  Altersversorgung
employee remuneration information –
  Entgeltbeleg
employee remuneration information –
  Entgeltnachweis

employee representative –
Arbeitnehmervertreter;
Belegschaftsvertreter
employee savings bonus –
Arbeitnehmersparzulage
employee selection – Personenauswahl
employee suggestion system –
Vorschlagswesen, innerbetriebliches
employee turnover – Fluktuation
employee voluntary termination notice
– eigene Kündigung; Kündigung durch
den Mitarbeiter
employee's background – Werdegang
eines Mitarbeiters
**employee's contribution (to
benefits-related payments)** –
Arbeitnehmerbeitrag
employee's customer/vendor account –
Mitarbeiterkontokorrentkonto
employee's planned working time –
Mitarbeitersollarbeitszeit
employee's shares – Belegschaftsaktien
**employee's share of sum to buy into
retirement fund** – Alterseinkaufsgeld
(Arbeitnehmer)
employee's withholding exemption –
Lohnsteuerfreibetrag
employee's contribution –
Arbeitnehmeranteil
employees – Belegschaft;
Mitarbeiterstab; Personal
employer – Arbeitgeber; Jobanbieter
employer number – Betriebsnummer
(SW)
employer's allowance –
Arbeitgeberzuschuss
employer's association –
Arbeitgeberverband
employer's contribution –
Arbeitgeberanteil
**employer's contribution (to
benefits-related payments)** –
Arbeitgeberbeitrag
employer's liability – Haftpflicht des
Arbeitgebers
employer's liability insurance –
Berufsgenossenschaft
employer's loan – Arbeitgeberdarlehen
employer's obligation to supply benefits
– Arbeitgeberfürsorgepflichten

employer's pension commitment –
Direktzusage; Pensionszusage;
employer's representative –
Arbeitgebervertreter
**employer's share of sum to buy into
retirement fund** – Alterseinkaufsgeld
(Arbeitgeber)
employer-employee relationship –
Arbeitgeber-Arbeitnehmer-Verhältnis
employer/apprentice relationship –
Berufsausbildungsverhältnis
employment – Anstellung;
Arbeitsverhältnis; Beschäftigung;
Tätigkeit
employment (at a company) – Arbeit,
nichtselbständige
employment ad(vertisement) –
Stellenanzeige; Stellenausschreibung
employment agency – Arbeitsamt;
Arbeitsvermittlung, Stellenvermittlung
employment bureau –
Stellenvermittlung
employment centre – Stellenvermittlung
employment certificate –
Arbeitsbescheinigung
employment conditions –
Beschäftigungsbedingungen
employment contract –
Anstellungsvertrag; Arbeitsvertrag
employment contract for foremen –
Meister-Dienstvertrag
employment contract law –
Dienstvertragsrecht
employment counselling –
Arbeitsberatung
employment exchange – Arbeitsamt
employment history – Entwicklung,
berufliche
employment interview –
Einstellungsgespräch;
Einstellungsinterview
employment level – Beschäftigungsgrad
employment limit –
Beschäftigungsbeschränkung
employment of industrial students –
Werkstudenteneinsatz
employment of pupils – Schülereinsatz
employment office – Arbeitsamt
employment opportunities for teachers
– Lehrerarbeitsmarkt

employment percentage –
  Arbeitszeitanteil;
  Beschäftigungsprozentsatz
Employment Promotion Law –
  Arbeitsförderungsgesetz
employment prospects –
  Arbeitsmarktaussichten
employment protection –
  Beschäftigungsschutz; Sicherung von
  Arbeitsplätzen
employment relationship – Anstellung;
  Arbeitsverhältnis; Beschäftigung
employment status – Dienstverhältnis
empower – berechtigen; ermächtigen;
  genehmigen
empowerment – delegative Führung
emulate – nacheifern
enclose – beilegen, beifügen
enclosed conditions – beigefügte
  Bedingungen
enclosure – Anlage
encourage – ermutigen
encouragement – Ermutigung
end – Ende; Schluss
end of break – Pausenende
end of semester – Semesterende
end of term – Semesterende
end of trimester – Trimesterende
end of working hours – Arbeitszeitende
end time – Endeuhrzeit
end tolerance begin time –
  Endetoleranz-Beginn
end tolerance end time –
  Endetoleranz-Ende
endeavo(u)r – Anstrengung; Bemühen;
  Bemühung
endostructure – Endostruktur
endurance – Ausdauer;
  Durchhaltefähigkeit
energy – Elan; Energie, Schwung
energy – Tatendrang
enforce – durchsetzen; erzwingen
enforcement – Geltendmachung
engage – einstellen
engagement – Anstellungsvertrag;
  Arbeitsvertrag
engineer – Diplomingenieur; graduierter
  Ingenieur
engineering – Engineering; Projektierung
engineering assistant –
  Ingenieurassistent(in)

enjoyment of learning – Lernfreude
enlargement – Erweiterung;
  Vergrößerung
enrichment – Anreicherung;
  Bereicherung
enrol – immatrikulieren
enrolment – Einschreibung;
  Immatrikulation
enterprise – Betrieb; Initiative;
  Unternehmen
enterprise strategy – Firmenstrategie
entertaining – kurzweilig
entertainment – Bewirtung
entertainment allowance –
  Bewirtungsspesen
entertainment expenses –
  Bewirtungskosten
enthusiasm – Begeisterung
entice employees – Mitarbeiter abwerben
enticement – Mitarbeiterabwerbung
entitle – berechtigen; ermächtigen;
  genehmigen
entitled – berechtigt
entitlement – Anspruchsberechtigung;
  Berechtigung, Anspruch
entitlement group – Vergütungsgruppe
entitlement to continued pay –
  Lohnfortzahlungsanspruch
entitlement to education and training –
  Ausbildungsanspruch
entitlement to remuneration –
  Vergütungsanspruch
entrance – Aufnahme; Eintritt
entrance exam(ination) –
  Aufnahmeprüfung
entrance examination –
  Eingangsprüfung
**entrance examination for talented
  students** – Begabtenprüfung (BRD)
entrance qualification –
  Aufnahmebedingung
entrance rate – Anfangslohn;
  Einstell-Lohn
entrant – Beitretende(r); Berufsanfänger;
  neues Mitglied
entrust – anvertrauen (jdm. etw.);
  betrauen (mit)
entry – Eintritt
entry age – Eintrittsalter
entry date – Eintrittsdatum

entry of travel expense data –
Reisekostenerfassung
entry qualifications –
Zulassungsanforderungen
entry test – Aufnahmeprüfung
environment – Umgebung; Umwelt
environment protection officer –
Umweltschutzbeauftragte(r)
environmental influence –
Umwelteinfluss
environmental pollution –
Umweltverschmutzung
environmental protection –
Umweltschutz
environmental studies – Umweltkunde
environmentally compatible –
umweltverträglich
equable – ausgeglichen
equal – gleich; gleichberechtigt
equal employment opportunity (EEO)
– Vermeidung von Diskriminierung,
Gleichberechtigung am Arbeitsplatz
(EEO ist der Inhalt eines Gesetzes in
den U.S.A., das die Arbeitgeber
verpflichtet, bei Neueinstellungen so zu
verfahren, dass die Struktur ihrer
Belegschaft (Männer, Frauen,
Schwerbehinderte, rassische und
religiöse Minderheiten, ältere
Mitarbeiter) möglichst der Struktur des
Ortes entspricht, in dem der Betrieb
liegt.)
equal opportunity – Chancengleichheit
equal rights – Gleichberechtigung
equality of wages (for men and women)
– Lohngleichheit (für Mann und Frau)
equalization – Gleichstellung
equalization board – Ausgleichsstelle
equalization fund – Ausgleichskasse
equalization levy – Ausgleichsabgabe
equalization limit – Ausgleichsgrenze
equipment pool – Gerätepark
equivalence – Gleichwertigkeit
equivalent – Gegenwert; gleicher Wert
equivalent qualifying periods –
Ersatzzeiten
eraser (AmE) – Radiergummi
ergonomic – arbeitswissenschaftlich
ergonomics – Arbeitswissenschaft;
Ergonomie
erect – errichten

erroneous assessment –
Fehleinschätzung
error – Fehler
error handling – Nachbearbeitung
error in judgement – Beurteilungsfehler
error – Irrtum
escalator clause – Lohngleitklausel;
Preissteigerungsklausel;
Steigerungsklausel
escape and rescue plan – Flucht- und
Rettungsplan
escape route – Fluchtweg
esprit de corps – Gemeinschaftssinn
essay – Aufsatz
essential – unerlässlich, wesentlich
essential course booking –
Muss-Buchung
establish – einrichten; gründen;
feststellen; fixieren
establish clear objectives – klare Ziele
setzen
establishing one's livelihood –
Existenzgründung
estimate – Schätzung; schätzen, taxieren
ethical – ethisch
ethnic minority pupils – Kursteilnehmer
aus ethnischen Minderheiten
ethnical minority – ethnische Minderheit
EU – Europäische Union; EU
EU-member states – EU-Staaten
European Council – Europäischer Rat
European Monetary System (EMS) –
Europäisches Währungssystem (EWS)
European Parliament – Europäisches
Parlament
European Union – Europäische Union;
EU
evade – ausweichen; sich entziehen
evaluate – abschätzen; auswerten;
bewerten
evaluation – Auswertung; Bewertung;
Evaluierung
evaluation class – Auswertungsklasse
evaluation of absent time –
Fehlzeitenauswertung
evaluation path – Auswertungsweg
evaluation sheet – Evaluierungsbogen
evening course – Abendkurs
evening school – Abendschule
**evening school leading to leaving
certificate** – Abendgymnasium

event – Ereignis
everyday managerial life –
  Führungsalltag
evidence – Beweis; Beweismaterial;
  Beweismittel
evident – ersichtlich; offenbar; klar
ex officio – kraft Gesetzes (von Amts
  wegen)
exact – akkurat, genau
exam(ination) – Prüfung; Klausur
examination date – Prüfungstermin
examination nerves – Prüfungsangst
examination question – Prüfungsfrage,
  Prüfungsaufgabe
examination requirements –
  Prüfungsanforderungen
examination subject – Prüfungsfach
examination system – Prüfungssystem
examination topic – Prüfungsthema
examinee – Kandidat; Prüfling
examiner – Prüfer
example – Vorbild
exceed – hervorragen; sich auszeichnen;
  übertreffen
excel – hervorragen; sich auszeichnen;
  übertreffen
excellent – ausgezeichnet; vorzüglich
except – ausgenommen; außer; es sei
  denn, dass
exception – Ausnahme
exceptional – außergewöhnlich
excess – Mehr...; Überdeckung; Übermaß
excess costs – Mehrkosten
excess labour supply –
  Arbeitskräfteüberschuss
exchange of experience –
  Erfahrungsaustausch
exchange of ideas – Gedankenaustausch
exchange programme –
  Austauschprogramm
exchange teacher – Austauschlehrer
exchangeability – Austauschbarkeit
excitement – Aufregung
exciting – aufregend
exclude – ausschließen
exclusion – Ausgrenzung;
  Ausschließung; Ausschluss
exclusive – ausschließlich; exklusiv
excuse – Entschuldigung
execute – ausführen; durchführen

execution – Ausführung; Durchführung;
  Erledigung
executive – Führungskraft
Executive Committee – Hauptausschuss
  eines Verbands
executive duty – Führungsaufgabe
executive employee – Angestellter,
  leitender
executive floor – Chefetage;
  Führungsetage
executive grade – gehobener Dienst
executive search agency –
  Personalagentur
Executive Vice President – Leiter eines
  Zentralbereiches
executives' appraisal – Beurteilung der
  Führungskräfte
exemplary damages – Schmerzensgeld
exempt – ausgenommen (von); außer
  Tarif; befreien
exempt life insurance – befreiende
  Lebensversicherung
exempt member of staff – übertariflicher
  Mitarbeiter
exempt personnel – außertarifliches
  Personal; übertariflicher Mitarbeiter
exempt salary – übertarifliche Bezahlung
exempt status – übertarifliche Rangstufe
exemption – Befreiung; Freisetzung;
  Freistellung
exemption from compulsory insurance
  – Befreiung von der
  Versicherungspflicht
exemption from contributions –
  Beitragsfreiheit
exemption from liability –
  Haftungsbefreiung
exemption limit – Freigrenze
exercise – Ausübung; Übung
exercise book – Schulheft
exercise of power – Machtausübung
exhaust – erschöpfen
exhaustion – Erschöpfung
exhibitionism – Imponiergehabe
exit document – Abgangserklärung
exit interview (AmE) –
  Abgangsinterview (interview with an
  employee when he or she is leaving an
  organization to find out his or her views
  on how the organization is run and
  reasons for leaving)

exit statement – Exit-Anweisung
expand – auffächern
expansion report – Auffächerungsreport
expatriate – ein ins Ausland versetzter
    Mitarbeiter
expect – erwarten; zumuten
expectant mother – Mutter, werdende
expectations – Erwartungen
expectations from work –
    Arbeitserwartung
expel – von der Schule ausschließen
expense account – Spesenkonto
expense account regulations –
    Spesenrichtlinien
expense audit clerk (AmE) – Abrechner
expense budget – Absatzbudget
expense category – Spesenkategorie
expense code – Spesenkürzel
expense indicator – Spesenkennzeichen
expense type – Spesenart
expenses for meals –
    Verpflegungsaufwendungen
expenses incurred for meals –
    Verpflegungskosten
experience – Erfahrung; Lebenserfahrung
experience abroad – Auslandserfahrung
experience of life – Lebenserfahrung
experienced – versiert
experiment – Experiment;
    experimentieren
experimental learning – Lernen durch
    Erfahrung
experimental stage –
    Entwicklungsstadium;
    Versuchsstadium
expert – Experte; Fachmann; Kapazität
expert knowledge – Spezialwissen
expert opinion – Gutachten eines
    Sachverständigen
expert power – Macht durch
    Sachkompetenz
expertise – Fachkompetenz; Gutachten
    eines Sachverständigen
experts – Fachleute
expiration – Ablauf; Erlöschen durch
    Zeitablauf; Zeit, abgelaufene
expiration of work permit – Ablauf der
    Arbeitserlaubnis
expire – erlöschen, verfallen
expiration date – Verfallsfrist
explain – erklären

explanation – Erklärung
exploit – ausbeuten; ausnutzen;
    ausschlachten
explore – erforschen; erkunden
export business – Auslandsgeschäft
export trade – Auslandsgeschäft
express oneself – sich äußern
expression – Ausdruck
expressive – ausdrucksvoll
expulsion – Verweisung
extend – verlängern
**extended contractual pension
    adjustment** – verlängerter
    schuldrechtlicher Versorgungsausgleich
extending across different areas –
    bereichsübergreifend
extending across disciplines –
    fachübergreifend
extension of the probationary period –
    Probezeitverlängerung
extensive listening – extensives Hören
extensive reading – extensives Lesen
external audit – Betriebsprüfung
external bank transfer – Überweisung,
    externe
external duty – Außendienst
external examiner – Prüfer aus einer
    fremden Universität
external insurances – Versicherungen,
    externe
external measures – externe Maßnahmen
external recruitment – Beschaffung,
    externe
External Training Administration –
    Aus- und Weiterbildungsverwaltung
extra hours – Überstunden
extra load – Mehrbelastung
extra pay – Aufgeld; Lohnzuschlag;
    Zulage; Lohnzulage
extra pay indicator –
    Aufgeldkennzeichen
extract from – exzerpieren
extramural courses – Hochschulkurse
    für nicht eingeschriebene Studenten
extracurricular activity – außerhalb des
    Studienplans liegende Tätigkeit
extraordinary notice of dismissal –
    Kündigung, außerordentliche
extrinsic – äußerlich
extrovert – Extravertierte(r)
eye – Auge

eye contact – Augenkontakt
eyesight – Sehkraft

F

face guard – Schutzmaske
Fachabitur – Fachabitur
Fachhochschule – Fachhochschule (BRD)
Fachoberschule – Fachoberschule (BRD)
facile – leicht; schnell
facilitate – erleichtern; fördern
facilitator – Vermittler
facilities – Anlagen; Einrichtungen
facility – Leichtigkeit; Einrichtung
facility management – Standortverwaltung
factor – Faktor; Umstand
factory – Betrieb; Fabrik
factory calendar – Fabrikkalender; Werkskalender
factory worker – Fabrikarbeiter; Industriearbeiter
factual – sachgerecht
factual level – Sachebene
faculty – Fachbereich; Fakultät
fail – durchfallen; versagen
fail an exam – bei einer Prüfung durchfallen; Prüfung nicht bestehen
failure – Misslingen; Scheitern
failure at school – Schulversager
failure rate – Durchfallrate
fair – aufrichtig; fair; gerecht; redlich
fair labor practice (AmE) – Arbeitsschutz
false – falsch; unrichtig
false beginner – falsche(r) Anfänger(in)
false friends – falsche Freunde
familiarization – Einarbeitung
family – Familie
family allowance – Familienbeihilfe; Familienzulage; Familienzuschlag; Familienzuschuss; Kindergeldkasse

family allowance card – Familienbeihilfekarte
family assistance – Familienhilfe
family burdens equalization fund – Familienlastenausgleichsfonds
family environment – häusliches Milieu
family equalization fund – Familienausgleichskasse
family friendly index – Index der Familienfreundlichkeit
family insurance – Familienversicherung
family member – Familienangehöriger
family name – Familienname; Zuname
family separation – Familientrennung
family tensions – Spannungen in der Familie
family type – Familienart
family values – Familienwerte
family work – Familienarbeit
fatigue – Ermüdung
fault-finder – Nörgler
fault liability – Verschuldenshaftung
fear – Angst
fear of exams – Prüfungsangst
fear of school – Schulphobie
feasibility study – Durchführbarkeitsstudie
feasible – durchführbar
feather structure – Federstruktur
featherbedding – Überbesetzung mit nicht benötigten Arbeitskräften (employing more staff than necessary, usually as a result of union pressure)
feature – Merkmal
Federal Benefits Authority – Bundesversicherungsamt
Federal Educational Promotion Act – Bundesausbildungsförderungsgesetz
Federal Law on Financial Support for Education and Training – (Bundesausbildungsförderungsgesetz) BAFöG
Federal Office – Bundesamt
Federal Social Insurance Office for Salaried Employees – Bundesversicherungsanstalt für Angestellte
Federal Social Security Act – Bundessozialhilfegesetz
federal state – Bundesland
federation – Bund

fee – Honorar (freie Berufe)
feeble – kurzatmig
feedback – Feedback; Rückmeldung
feedback session – Feedbackgespräch
feedback technique – Rückmeldeverfahren
feel – Gespür
feeling – Gefühl; Gespür
feeling for language – Sprachgefühl
feeling of one's own value – Selbstwertgefühl
feeling of solidarity – Zusammengehörigkeitsgefühl
fellow – Mitglied einer wissenschaftlichen Vereinigung; Mitglied eines Colleges (BrE)
fellow student – Kommilitone
fellow worker – Arbeitskollege; Kollege; Mitarbeiter
fellowship – Forschungsstipendium; Gemeinschaft der Fellows in einem College oder Universität; Status eines Fellows in einem College oder wissenschaftlicher Vereinigung
felt-tipped pen – Filzstift
female reentrant – Berufsrückkehrerin
fictitious – fiktiv
fictitious age limit – fiktive Altersgrenze
fictitious qualifying periods – Ersatzzeiten; Wartezeitfiktion
fictitious self-employment – Scheinselbständigkeit
fidgelty person– Zappelphillip
field of action – Aufgabengebiet
field of activity – Tätigkeitsbereich; Tätigkeitsfeld
field of research – Forschungsgebiet
field of training – Ausbildungsbereich; Ausbildungsrichtung
field study – Feldforschung
field test – Feldtest; Feldversuch
field work – Feldforschung
field worker – Praktiker
fight – Prügelei; raufen
figure of speech – Redewendung
figurehead role – Repräsentantenrolle
file – Aktenordner; Datei; Ordner; ablegen
file an application – einen Antrag stellen
file of severely handicapped persons – Schwerbehindertendatei

fill a vacancy – eine Stelle besetzen
fill time – Füllzeit
filling a vacancy – Stellenbesetzung
filling of vacancies – Besetzen von Stellen
final ballot – Stichwahl
final certificate – Abschlusszeugnis
final exam(ination)s – Abschlussprüfung
final interview – Abgangsinterview (interview with an employee when he or she is leaving an organization to find out his or her views on how the organization is run and reasons for leaving)
final mark – Endnote
final processing – Endeverarbeitung
final processing of averages – Durchschnittsendeverarbeitung
final qualification – Abschluss
final school year – letztes Schuljahr
final secondary school examination – Abitur
final year project – Diplomarbeit
finals – Abschlussprüfung
financial accounts – Finanzkonto
financial support for talented students – Begabtenförderung
financially strong – kapitalstark
financing – Finanzierung
find fault with a contribution – beanstanden von Beiträgen
fine – Bußgeld (bei Verstößen); Strafe, Geldstrafe
Fine Arts – schöne Künste
fine mechanics – Feinmechanik
finger alphabet – Taubstummenalphabet
finicky person – Tüftler
finish – beendigen; absolvieren
fire – entlassen; feuern; hinauswerfen
fire category – Brandkategorie
fire prevention – Brandschutz
fire-extinguisher – Feuerlöscher
fire-extinguishing installation – Feuerlöscheinrichtung
fired on the spot – fristlose Entlassung
firm – Betrieb; Gesellschaft
firm where apprentices are trained – Lehrbetrieb
firmly book – fixieren
firmly bookable attendance – Teilnahme, fixierbare

185 **forfeit**

first aid – Erste Hilfe
first class – Eingangsklasse
first diploma – Vordiplom
1st grader – Schulanfänger
first name – Vorname
first sickness – Ersterkrankung
first survey – Ersterhebung
first-hand information – Information aus erster Hand
fiscal year – Geschäftsjahr
fisticuffs – Handgreiflichkeit
fit for work – arbeitsfähig
fixed allowance – Pauschalvergütung
fixed budget – starres Budget
fixed gratuity – Gratifikation, feste
fixed holiday – Feiertag, unbeweglicher
fixed wage – Festlohn
fixed-term contract – Zeitvertrag
fixed-term employment contract – befristeter Arbeitsvertrag
fixing of (statutory) minimum wages – Festsetzung von (gesetzlichen) Mindestlöhnen
flagging attention – nachlassende Aufmerksamkeit
flash card – Leselernkarte
flat rate – Pauschale; unbeschränkter Zugang
flat rate for transportation – Fahrtkostenpauschalsatz
flexi timespan – Gleitzeitspanne
flexibility – Beweglichkeit; Anpassungsfähigkeit
flexibility in thinking – geistige Flexibilität
flexible – anpassungsfähig; flexibel
flexible learning – flexibles Lernen
flexible partial retirement – flexible Altersteilzeit
flexible pension-age approach – gleitende Altersruhe (Employees can "pay in" hours into a long-term account over a longer period of time. When approaching pension age they can "claim" these and reduce the number of their working hours – without forfeiting pension rights.)
flexible pensionable age – flexible Altersgrenze
flexible retirement – flexible Altersgrenze

flexible retirement benefit – flexibles Altersruhegeld
flexible working hours – flexible Arbeitszeit; gleitende Arbeitszeit
flexitime – Gleitzeit
flexitime regulation – Gleitzeitregelung
flexitime – Gleitzeit
flexitime balance – Gleitzeitsaldo
flexitime deficit – Gleitzeitunterschreitung
flexitime excess – Gleitzeitüberschuss
flexitime model – Gleitzeitmodell
flip chart – Flipchart
floating holiday – Feiertag, beweglicher
flouting – Missachtung
flow of information – Informationsfluss
fluctuation – Fluktuation; Personalwechsel
fluctutation – Arbeitsplatzwechsel
fluent – fließend
flunk (AmE) – durchfallen
fluorine hydrocarbon – Fluorchlorkohlenwasserstoff (FCKW)
follow-up action – Folgevorgang
follow-up course – Folgekurs
follow-up examination – Nachuntersuchung
for-period – Für-Periode
forbid – verbieten
force – durchsetzen; erzwingen
forced retroactive calculation – Zwangsrückrechnung
forecast – Vorausschau; vorhersagen; Vorschau
forecast of personnel costs – Planabrechnung
foreign company – Auslandsgesellschaft
foreign delegation – Auslandsabordnung; Auslandseinsatz
foreign employee – ausländischer Arbeitnehmer
foreign experience – Auslandserfahrung
foreign family – Ausländerfamilie
foreign language – Fremdsprache
foreign language skills – Fremdsprachenkenntnisse
foreign worker – ausländischer Arbeitnehmer
foreigner – Ausländer
foreman – Vorarbeiter
forfeit – Geldstrafe; Vertragsstrafe

forget – vergessen, verlernen
forgetting curve – Vergessenskurve
form – Formular; Jahrgangsstufe
form master – Klassenlehrer
form of address – Anrede
formal – formal
formal authority – formelle
 Weisungsbefugnis
formal communication – formelle
 Kommunikation
formal education – Schulausbildung
formal insurance – Formalversicherung
formal organization – formelle
 Organisation; geregelte
 Kompetenzordnung
formal policy – formelle Direktive
formal step – Formalstufe
formalistic – formalistisch
formality – Formsache
formation of capital –
 Vermögensbildung
formation of wealth –
 Vermögensbildung
formative years – entscheidende Jahre
former pupil – ehemalige(r) Schüler(in)
forty hour week –
 Vierzig-Stunden-Woche
found – gründen
foundation – Stiftung
foundation course – Grundkurs
foundation subject – Grundfach
founding one's own company –
 Existenzgründung
fountain pen – Füllfederhalter
four-day week – Vier-Tage-Woche
four-directional management – Führung
 in vier Richtungen
fractal company – fraktales Untenehmen
fractional value – Teilwert
framework agreement –
 Rahmenvereinbarung
framework curriculum –
 Rahmenlehrplan
fraternity – Verbindung
free collective bargaining – Lohnkampf
free day – schulfreier Tag
**free provision of schoolbooks and
 equipment** – Lernmittelfreiheit
free time – Freizeit
freedom of choice – Wahlfreiheit

freedom of research –
 Wissenschaftsfreiheit
freelance – freiberuflich
freelancer – als freie(r) Mitarbeiter(in)
 tätig sein
freeze – einfrieren
frequent – häufig
freshman – Studienanfänger
freshman (AmE) – Student im ersten
 Studienjahr
friction – Reibung
friendly – freundlich
fringe benefit – Gehaltszusatz
fringe benefit remuneration –
 Ersatzleistung
fringe jobs – Arbeitsstellen für die
 Massen (large numbers of
 interchangeable jobs for the poorly
 qualified)
front-end system – System,
 vorgelagertes
frontier worker – Grenzarbeitnehmer;
 Grenzgänger
frustrate – frustrieren
frustration – Frustration
fulfil(l)ment – Erfüllung
fulfilment of demand – Bedarfsdeckung
full employment – Vollbeschäftigung
full liberty of action –
 Bewegungsfreiheit
full superannuation – Vollversorgung
 (mittels einer Rente)
full-day nursery – Ganztagshort;
 Kinderhort
full-time – hauptamtlich; hauptberuflich
full-time apprenticeship – Vollzeitlehre
full-time basic vocational school –
 Berufsgrundschule
full-time compulsory education –
 Vollzeitschulpflicht
full-time employment –
 Ganztagsbeschäftigung;
 Vollzeitbeschäftigung
full-time job – Ganztagsarbeit
full-time payment – Vollzeitentgelt
full-time specialized vocational school –
 Berufsfachschule (BRD)
full-time teacher – Vollzeitlehrer
full-time telework – Vollzeittelearbeit
full-time training – Vollzeitausbildung

full-time work place –
Vollzeitarbeitsplatz
function – Funktion
function assessment –
Funktionsbewertung
function character – Funktionscharakter
function level – Funktionsstufe
function value – Funktionswert
function-based – funktionsbezogen
function-related training –
funktionsbezogene Weiterbildung
functional – funktionsmäßig
functional area – Funktionsbereich
functional authority – funktionale
Weisungsbefugnis
functional illiteracy – partielles
Analphabetentum
functional structure –
Funktionsordnung; Funktionsstruktur
functional unit – Funktionseinheit
fund – Fonds
furniture – Möbel
further compensation – Nachvergütung
further development –
Weiterentwicklung
further education – Fortbildung;
nachschulische Angebote in der
Erwachsenenbildung
**further education and training
planning** – Weiterbildungsplanung
further education courses for graduates
– Kontaktstudium
further education of management –
Weiterbildung von Führungskräften
furtherance of gifted students –
Hochbegabtenförderung
future benefits – Zukunftssicherung
future benefits exemption –
Zukunftssicherungsfreibetrag

G

gain – erlangen; Gewinn; gewinnen;
profitieren
gainful employment – Erwerbsarbeit

gainfully employed – erwerbstätig
game – Spiel
gang – Bande; Clique; Horde
ganger – Vorarbeiter
garnishment – Pfändung
garnishment exemption (USA) –
Pfändungsschutz
garnishment of wages – Lohnpfändung
garnishment ruling (USA) –
Lohnpfändungsbeschluss
gate control – Kontrolle (an der Pforte)
gatekeeper – Pförtner
general – allgemein
**General Certificate of Secondary
Education** – Mittlerer Schulabschluss
**general committee for labour
protection** –
Hauptarbeitsschutz-Kommission
general contribution rate – Beitragssatz
(allgemein)
general counsel – Justitiar; Leiter der
Rechtsabteilung
general disability – Erwerbsunfähigkeit
general education – Allgemeinbildung
General Executive Manager –
Hauptgeschäftsführer (GmbH &
Co.KG)
general knowledge – Allgemeinwissen
general loan – Darlehen, allgemeines
general loan repayment –
Darlehenstilgung, allgemeine
general obligatory insurance –
Versicherungspflicht, allgemeine
General Partner (Limited Partnership)
– Komplementär (KG); persönlich
haftender Gesellschafter (KG)
general personnel policy – allgemeine
Personalpolitik
general school – allgemeinbildende
Schule
general sickness benefit fund (approx.)
– Allgemeine Ortskrankenkasse
general training syllabus –
Ausbildungsrahmenplan
general valuation basis – allgemeine
Bemessungsgrundlage
generating rule – Generierungsregel
generation gap – Generationskonflikt
generous – großzügig
genius – Genie
geometry – Geometrie

German commercial code – Handelsgesetzbuch

German Corporation Act – Aktiengesetz

German Data Entry Regulation – Datenerfassungsverordnung

German Data Protection Act – Bundesdatenschutzgesetz

German Federal Labor Court – Bundesarbeitsgericht

German Federal Labor Office – Bundesanstalt für Arbeit

get up – aufstehen

getting on in one's career – berufliches Fortkommen

gift – Gabe; Schenkung

gifted – begabt

girl – Mädchen

girls' secondary school – Mädchengymnasium

give a survey – Überblick vermitteln

give instructions – Anweisungen geben

give notice – Dienstenthebung; entlassen; firmenseitige Kündigung

give orders – Anweisungen geben

give out – schelten; schimpfen

global – global

go slow – Dienst nach Vorschrift; Streik durch passiven Widerstand

go slow day – Bummeltag

go slow strike – Bummelstreik

go-ahead – vorwärtsstrebend

go-between – Vermittler

goal – Hauptziel; Ziel; Zielvorgabe; Zielvorstellung

goal definition – Zieldefinition

goggles – Schutzbrille

gold-collar worker – Goldkragen-Mitarbeiter

golden handcuff – Vorzugsbedingungen, um einen Mitarbeiter an die Firma zu binden

golden handshake – Vorzugsbedingungen, um einen Mitarbeiter loszuwerden

good at languages – sprachbegabt

good faith – Treu und Glaube

good offices – Vermittlung

good reason – triftiger Grund

goodies – Vergünstigungen durch die Firma, Nebenleistungen

goodwill – Wohlwollen; Goodwill

gossip – Gerücht; Spekulation

governing board of the school – Schulverwaltung

government education office – Staatliches Schulamt

government school authority – Staatliches Schulamt

grade – Dienstgrad; Jahrgangsstufe; Rang, Zensur

grade (AmE) – Jahreszeugnis; Zeugnis

grade (of management) – Rangstufe

grade point averages (AmE) – Notendurchschnitt

grade school (AmE) (s. Graphik) – Grundschule; Volksschule

grade structure – Rangstruktur

grant-aided school – subventionierte Schule

grading system – Benotungssystem

gradual – allmählich

graduate – Absolvent

graduate from (AmE) – absolvieren

graduate in business administration – Betriebswirt

graduate in national economics – Volkswirt

graduate school – weiterführende Hochschule

graduates per annum – "Ausstoß" der Hochschulen

graduation (AmE) – Erteilung eines akademischen Grades

grammar – Grammatik

grammar school – Gymnasium; Mittelschule

grammar school emphasizing Latin and Greek – humanistisches Gymnasium (HG)

grammar school emphasizing modern languages – Neusprachliches Gymnasium (NG)

grammar school teacher – Studienassessor; Studienrat(rätin)

grandfathering – Besitzstand erhalten

grant – bewilligen; gewähren; Stipendium

grants for families of limited income – staatliche Ausbildungsförderung

grapevine – Gerücheküche

gratitude – Dankbarkeit

gratuitous meal – Bewirtung, unentgeltliche
gratuity – Gratifikation; Sondervergütung
green card – Grüne Karte
grievance – Beschwerde(grund); Grund zur Klage
grievance committee – Beschwerdestelle
grievance procedure – Beschwerdeverfahren
grinds – Nachhilfeunterricht während der Ferien
gross – Brutto...
gross amount for social insurance – Sozialversicherungsbrutto
gross annual amount – Jahresbrutto
gross bucket – Bruttotopf
gross calculation of pay – Bruttoteil
gross income – Bruttoeinkommen
gross indicator – Bruttokennzeichen
gross level of satisfaction – Bruttoversorgungsgrad
gross pay – Bruttoentgelt
gross payroll accounting – Bruttoabrechnung
gross retroactive accounting – Bruttorückrechnung
gross salary – Bruttogehalt
gross tax amount – Steuerbrutto
gross wage calculation – Bruttolohnfindung
gross wage evaluation – Bruttolohn-Auswertung
gross wages – Bruttolohn
grossing-up indicator – Permanenzkennzeichen
ground – Beweggrund; Veranlassung
Group – Bereich; Unternehmensbereich
group – Gruppe
group activity – Gruppenaktivität
group bonus system – Gruppen-Prämiensystem
group communication – Kommunikation innerhalb einer Gruppe
group decision – Gruppenentscheidung
group discussion – Gruppendiskussion
group dynamics – Gruppendynamik
Group Executive Management – Bereichsvorstand
group exercise – Gruppenübung

group incentive – Gruppen-Leistungsprämie
group incentive wage – Gruppenleistungslohn
group leader – Gruppenführer; Gruppenleiter
group norm – Gruppennorm
group of affiliated companies (BrE) – Konzern
group of employees – Beschäftigtengruppe
group of persons – Personenkreis, Personengruppe
group payment – Gruppenlohn
group personnel director – Personalreferent
group piecework – Gruppenakkord
group piecework rate – Gruppenakkordlohn
Group President – Vorsitzender des Bereichsvorstands
group process – Gruppenprozess
group specific – bereichsspezifisch
group valuation – Pauschalbewertung
group work – Gruppenarbeit; Teamarbeit
group works council – Konzernbetriebsrat
grousing – Nörgelei
grovelling – kriecherisch
growth – Wachstum
growth rate – Zuwachsrate
growth requirements – Zuwachsbedarf
grumbler – Nörgler
grumbling – Nörgelei
guarantee of a slot in a course – Kurszusage
guaranteed minimum rate – garantiertes Minimum (bei Akkordentlohnung)
guaranteed minimum wage – garantierter Mindestlohn
guaranteed net amount – Nettozusage
guaranteed pension payment – Pensionszusage
guaranteed value – Garantiewert
guaranteed wage type – Garantielohnart
guard – Pförtner
guardian – Vormund
guess – raten
guesstimate – über den Daumen gepeilt
guest – Gasthörer
guest lecturer – Gastdozent

guest speaker – Gastsprecher
guidance – Anleitung
guide – Handbuch; Leitfaden
guide – Anleitung
guideline – Bestimmung; Richtlinie;
  Vorschrift
guidelines for management – Leitsätze
  für Führungskräfte
guiding idea – Leitgedanke
guiding principle – Leitsatz
guiding principles for managerial staff
  – Leitsätze für Führungskräfte
guild – Gilde; Innung; Zunft
guild health insurance fund –
  Innungskrankenkasse
guilt complex – Schuldkomplex
guilty – schuldig
guilty conscience – Schuldgefühl
guilty feeling – Schuldgefühl
gullible – einfältig; leichtgläubig
guru – Guru
gymnasium – Turnhalle
gymnastics – Turnen

H

habit – Gewohnheit
habituation – Gewöhnung
half timer – Halbtagskraft
half-cover – Halbbelegung
half-dynamic employer's pension –
  halbdynamische Versorgungszusage
half-orphan benefit – Halbwaisenrente
half-term – Ferien in der Mitte des
  Trimesters
half-time – halbtags
half-time worker – Halbtagskraft
hall of residence – Studentenheim;
  Studentenwohnheim
handbook – Handbuch; Leitfaden
handicap – Erschwernis
handicap group – Behindertengruppe
handicap type – Behinderungsart
handicapped – behindert
handicapped child – behindertes Kind

handicapped person – Behinderter
handicraft – Kunsthandwerk
handle – handhaben; verwalten
hands-on approach – praxisorientierte
  Vorgehensweise
hands-on teaching method –
  praxisorientierte Lehrmethode
handwriting – Handschrift
happy Monday – verlängertes
  Wochenende
harass – belästigen, schikanieren
hard of hearing – schwerhörig
hard to place – schwer vermittelbar
harmful substance – Schadstoff
have a free period – Stunde frei haben
have at one's disposed – verfügen über
have the care of – betreuen
hazard – aufs Spiel setzen; wagen
hazardous – gefährlich; gewagt
hazardous duty pay – Gefahrenzulage
head hunter – Personalberater
Head of ... – Leiter einer Hauptabteilung
  (SAG)
head of department – Abteilungsleiter;
  Bereichsleiter; Leiter
head of department unit –
  Dienststellenleiter
head of HR department – Personalleiter
head of operational sector –
  Geschäftsgebietsleiter
head of personnel – Leiter des
  Personalwesens; Personalchef;
  Personaler
head of the family – Familienvorstand
head of the legal department – Justitiar;
  Leiter der Rechtsabteilung
head of the personnel department –
  Leiter der Personalabteilung
head of training – Ausbildungsleiter
head office – Stammhaus
head office trainee – Stammhauslehrling
head office training – Stammhauslehre
headboy – vom Schulleiter bestimmter
  Schulsprecher
headcount – Personalbestand;
  Personalstärke
headcount budget (AmE) – Kopfzahletat
headcount saving – Einsparung von
  Arbeitskräften
headhunter – Headhunter

headmaster – Schulleiter,
  Oberstudiendirektor(in)
**headmaster of a secondary modern
  school** – Realschulrektor
headmistress – Schulleiterin
headquarters – Hauptverwaltung; Sitz
  (eines Unternehmens)
headstrong – eigensinnig
headteacher – Rektor
health – Gesundheit
health exclusion – Handicap
health hazard – Gesundheitsgefährdung;
  Gesundheitsrisiko
health insurance – Krankenversicherung
health insurance carrier – Träger der
  Krankenversicherung
health insurance company –
  Krankenversicherungsgesellschaft
health insurance for old age pensioners
  – Krankenversicherung der Rentner
health insurance for students –
  Krankenversicherung der Studenten
health insurance for the disabled –
  Krankenversicherung der Behinderten
health insurance fund – Krankenkasse
health insurance premium –
  Krankenversicherungsbeitrag
health officer – Vertrauensarzt
health protection – Gesundheitsschutz
hearing deficit – Gehördefekt
heavy work – Schwerarbeit
heir – Erbe
heiress – Erbin
helmet – Schutzhelm
helpful – hilfsbereit
helplessness – Ratlosigkeit
hereditary – erblich
heredity – Vererbung
heresy – Irrglauben
hesitant – unschlüssig; zögernd
heuristic – heuristisch
hide one's feelings – verstellen
hierarchism – Hierarchiedenken
hierarchy – Hierarchie; Rangordnung
hierarchy of needs – Hierarchie der
  Bedürfnisse
high school (AmE) – Gymnasium;
  höhere staatliche Schule
high school graduate – Abiturient
high seasonality – Saisonabhängigkeit,
  hohe

high-flier – Senkrechtstarter, Erfolgstyp
high-school graduation certificate –
  Reifezeugnis
higher and advanced further education
  – Tertiärbereich
higher contribution rate – Beitragssatz
  (erhöht)
higher education –
  Hochschul(aus)bildung
higher-education diploma –
  Hochschuldiplom
highly gifted – Hochbegabte(r)
highly intelligent – Hochbegabte(r)
highly qualified candidate –
  Spitzenbewerber(in)
highly qualified candidates –
  Spitzenleute
highly qualified employee –
  hochqualifizierte Kraft
Hilary term (BrE) – Frühjahrstrimester
hindrance – Hindernis
hire – einstellen
hired help – Leiharbeitskraft
hirer – Entleiher
hiring – Einstellung
hiring conditions –
  Anstellungsbedingungen
hiring transaction –
  Einstellungstransaktion
hiring-out of labour –
  Arbeitnehmerüberlassung
history – Geschichte
hit – schlagen
hitch – Panne
holder – Inhaber
holder of a position – Planstelleninhaber
holidays – Ferien; Semesterferien;
  Urlaub
holiday bonus – Feiertagszuschlag;
  Urlaubsgeld
holiday calendar – Urlaubskalender
holiday money – Urlaubsgeld
holiday pay – Feiertagslohn
holiday regulation – Urlaubsregelung
holiday replacement (BrE) –
  Urlaubsvertretung
holistic – ganzheitlich
home – Heim; Elternhaus
home economics –
  Hauswirtschaft(slehre)

home address in country of birth –
Heimatanschrift

home country – Heimatland

homogeneous – homogen

homosexual – Gleichgeschlechtlich

homework – Schulaufgaben;
Hausaufgabe; Schularbeiten

honest – ehrlich, redlich

honesty – Ehrlichkeit

**Honorary Chairman of the Board of
Directors** – Ehrenvorsitzender
(Verband)

**Honorary Chairman of the
Supervisory Board** –
Ehrenvorsitzender des Aufsichtsrates
(SAG)

honorary degree – ehrenhalber
verliehener akademischer Grad

honorary doctor – Ehrendoktor

honorary member – Ehrenmitglied

honorary post – Ehrenamt

Honorary President – Ehrenpräsident

honours degree – Universitätsabschluss
mit Auszeichnung

hopelessness – Hoffnungslosigkeit

horizontal decentralization –
horizontale Dezentralisierung

horizontal information – horizontale
Information

hospital – Krankenhaus

hospital per diem allowance –
Krankenhaustagegeld

host country – Gastland

hot skills – gefragter Beruf; gefragter
Experte

hostility towards foreigners –
Ausländerfeindlichkeit

hotel receipt – Hotelbeleg

hour of absence – Abwesenheitsstunde

hour of travel – Fahrtstunde

hourly rate – Stundenlohn; Stundensatz

hourly wage – Stundenlohn

hourly worker – Stundenlohnempfänger

hours of overtime – Mehrarbeitsstunden

house journal – Hauszeitung;
Werkzeitschrift

house master (BrE) – Lehrer, der für ein
Gruppenhaus zuständig ist

household allowance – Haushaltszulage

household day – Hausarbeitstag

hausekeeper – Haushälterin

housewife – Hausfrau

housing administration –
Wohnungsreferat

housing allowance – Mietzuschuss;
Wohngeldzuschuss

housing loan – Wohnungsdarlehen

housing policy – Wohnungspolitik

housing subsidy – Wohngeld

housing support –
Wohnungsbauförderung

howler – grober Schnitzer

HR advisor – Personalberater; Personaler

HR data – Personaldaten

HR data maintenance –
Personaldatenpflege

HR database – Personaldatenbank

HR information system –
Personalinformationssystem

HR master data – Personalstammdaten

HR master data management –
Personalstammdatenverwaltung

HR master data sheet –
Personalstammblatt

HR master record – Personalstammsatz

HR organisation – Personalorganisation

HR-culture – Führungskultur

huge class – Riesenklasse

human capital – Humankapital

human factors engineering –
Arbeitsgestaltung;
arbeitswissenschaftliche
Organisationsgestaltung

human relations –
zwischenmenschlicher Bereich

human resources – Belegschaft;
Mitarbeiterstab; Personal,
Personalwirtschaft

human resources department –
Personalwesen

human resources management –
Personalführung; Personalmanagement

human resources planning –
Personalplanung

humanistic – humanistisch

humanities – Geisteswissenschaften

humo(u)r – Humor

humour – bei Laune halten; Laune

hundred per cent incentive – voller
Leistungsanreiz

husband – Ehegatte

hygiene factor – Hygienefaktor

hygiene engineer (AmE) –
Umweltingenieur
hygienist (AmE) – Umweltbeauftragter
einer Industriefirma
hypothetical – hypothetisch
hysterical person – hysterischer Mensch

I

ID for handicapped persons –
Behindertenausweis
ID for severely handicapped persons –
Schwerbehindertenausweis
idea – Gedankengang; Idee
identification – Identifikation
identification number –
Ausweisnummer
identifier – Identifikator
identity – Identität
identity card (ID) – Personalausweis
idiom – Redewendung
idiomatic expression – Redewendung
idle time – unproduktive Zeit,
Leerlaufzeit
idle time compensation – Ausfalllohn
ill – arbeitsunfähig; krank
illegal employment – illegale
Beschäftigung
illegal strike – gesetzwidriger Streik
illicit work – Schwarzarbeit
illicit worker – Schwarzarbeiter
illiteracy – Analphabetentum
illiterate – analphabetisch
illiterate person – Analphabet
illness – Erkrankung; Krankheit
illustrative material –
Anschauungsmaterial
image – Bild
imaginary date – Datum, fiktives
imagination – Phantasie;
Vorstellungskraft
imaginative – phantasievoll
imagine – sich etwas vorstellen;
visualisieren
imbalanced – unausgeglichen

imbecile – beschränkt
imitation – Nachahmung
immature – unreif
immediate resignation (employee) –
fristlose Kündigung
immediate successor –
Sofort-Nachfolger; unmittelbarer
Nachfolger
immersion – Sachunterricht in der
Fremdsprache
immigrant – Einwanderer; Immigrant(in)
immigrant children – Kinder von
Einwanderern
immigrant worker – Gastarbeiter
immoral – unmoralisch; unsittlich
immunity – Immunität
impair – beeinträchtigen
impaired working ability – verminderte
Erwerbsfähigkeit
impairment – Beeinträchtigung
impartial – unbefangen; unparteiisch
impediment – Erschwernis;
Hinderungsgrund
imperative – unabdingbar
impersonal – unpersönlich
impertinence – Ungezogenheit
impetus – Schwung; Triebkraft
implement – ausführen; durchführen
implementation – Erfüllung,
Durchführung
implicit – implizit
importance – Bedeutung; Wichtigkeit
impossibility – Unmöglichkeit
impractical – praxisfern
improve – verbessern
improvement – Verbesserung
improvement in performance –
Erhöhung (einer Leistung)
improvise – improvisieren
impulse – Impuls
impulsive – impulsiv
imputed income – geldwerter Vorteil
in behalf of (AmE) – im Auftrag von; im
Interesse von
in charge of – Betreuung (z.B. von
Systemen)
in good faith – in gutem Glauben
in equal number – paritätisch besetzt
in excess of – zuviel
in excess of (e.g. luggage in excess of) –
überschüssig

in need of help – hilfsbedürftig
in need of protection – schutzbedürftig
in the bounds of reason – Zumutbarkeit
in-company training – Ausbildung
innerhalb des Unternehmens
**in-company vocational training and
education** – betriebliche Aus- und
Weiterbildung
in-company wage agreement –
Haustarif
in-hospital benefit –
Krankenhausbeihilfe
in-hours – i.D.; innerhalb der Dienstzeit;
während der Dienstzeit
in-hours course – Lehrgang innerhalb
der Dienstzeit (i.D.)
in-hours on-going education –
Fortbildung i.D.; Weiterbildung i.D.
in-house training – Ausbildung im
Unternehmen
in-service training for teachers –
Weiterbildungsmaßnahmen für Lehrer
inability – Unfähigkeit; Unvermögen
inability to adjust adequately –
mangelnde Anpassung
inability to make decisions –
Entscheidungsschwäche
inability to work – Arbeitsunfähigkeit
inaccurate – ungenau
inactive – inaktiv
inactive work relationship –
Arbeitsverhältnis, ruhendes
inadequate – mangelhaft; inadäquat
inadequate health – mangelnde
gesundheitliche Eignung
inadmissible for reasons of age –
zurückgestellt aus Altersgründen
inaugural lecture – Antrittsvortrag
incapable of work – arbeitsunfähig
incapacitation – Entmündigung
incentive – Anreiz; Antrieb;
Leistungsanreiz
incentive bonus – Anreizprämie
incentive payment – Leistungslohn
incentive system – Anreizsystem;
Leistungsentlohnungssystem;
Incentivesystem
incentive wage – Leistungslohn
incentive wage earner – Leistungslöhner
incidental – beiläufig; gelegentlich;
zufällig

incidental costs – Nebenkosten
incidental expenses – Spesen
incidental travel expenses – Reisespesen
inclination – Neigung
include – einbeziehen; einschließen
income – Bezüge; Einkommen;
Einkünfte
income band – Einkommensband
income category – Einkommenstyp
income from employment – Einkünfte
aus nichtselbständiger Arbeit
income guarantee –
Einkommensgarantie
income imputation –
Einkommensanrechnung
income list – Einkommensliste
income policy – Einkommensgestaltung;
Einkommenspolitik
income range – Einkommensrahmen
income scale – Einkommensband
income security – Einkommensgarantie
income statement –
Entgeltbescheinigung
income structure – Einkommensstruktur
income tax – Einkommensteuer;
Lohnsteuer
income tax allowance –
Lohnsteuerfreibetrag
income tax certificate –
Lohnsteuerbescheinigung
income tax class – Lohnsteuerklasse
income tax deduction –
Lohnsteuerabzug
income tax form – Lohnsteuerausweis
income tax law –
Einkommensteuergesetz
income tax notification –
Lohnsteueranmeldung
income tax rate – Einkommensteuertarif
income tax regulations –
Einkommensteuerrichtlinien
income tax statistics –
Lohnsteuerstatistik
income threshold –
Beitragsbemessungsgrenze
income-related expenses –
Werbungskosten
incoming wage type – Zuflusslohnart
incompatible – unvereinbar;
unverträglich

incompetence – Unfähigkeit;
Unvermögen
incompetent – unfähig
inconsiderate – rücksichtslos
incontestable, unchallengable –
unanfechtbar
incorrect – falsch; unrichtig
incorrect behaviour – Fehlverhalten
increase – Erhöhung; Steigerung;
Vermehrung; Zunahme; steigern
increase efficiency – Leistung steigern
increase experience – Erfahrungsbreite
vergrößern
increase factor – Erhöhungsfaktor
increase in basic pay –
Basisbezugserhöhung
increase in growth of population –
Bevölkerungswachstum
increase in performance –
Leistungssteigerung
increase in the cost of living –
Verteuerung der Lebenshaltung
increase in wage scales – Tariferhöhung;
Tariflohnerhöhung
increase of motivation – Steigerung der
Motivation
increase performance – Leistung
steigern
increased insurance –
Höherversicherung
increment – Erhöhung; Steigerung;
Vermehrung; Zunahme
incubator – Geburtshelfer
incumbent – Amtsinhaber;
Stelleninhaber; Lehrstuhlinhaber
incur – zuziehen (sich)
indecisive – unentschlossen
indemnification – Entschädigung
indemnity – Abgeltungsbetrag
indenture – Lehrvertrag
independence – Selbständigkeit;
Eigenständigkeit
independent – selbständig; unabhängig;
eigenständig
independent activity – selbständige
Tätigkeit (nicht versichert in der
gesetzlichen Unfallversicherung)
independent learning phase –
Individualphase
independent plant – selbständiger
Betrieb

independent school – unabhängige
Schule
independent system – privates
Schulsystem
index – Index; Sachverzeichnis
indicate – andeuten
indication – Hinweis
indicator for previous day –
Vortageskennzeichen
indifferent – gleichgültig
indirect – indirekt
indirect labour costs – indirekte
Arbeitskosten
indispensability – Unabdingbarkeit
indispensable – unabdingbar
indistinct (speech) – undeutlich
individual – einzeln; Einzelperson;
Individuum
individual ability – individuelle
Könnerschaft
individual conditions –
Einzelbedingungen
individual contract – Einzelvertrag
individual employment contract –
Einzelarbeitsvertrag
individual incentive wage –
Einzelleistungslohn
individual needs – individuelle
Bedürfnisse
individual performance – Einzelleistung
individual performance and ability –
persönliche Leistungsfähigkeit
individual piecework – Einzelakkord
individual piecework rate –
Einzelakkordlohn
individual skill – individuelle
Könnerschaft
individual statement – Einzelnachweis
individual wage – Individuallohn
individual work – Einzelarbeit
individualization – Individualisierung
indolence – Indolenz
inducement – Mitarbeiterabwerbung
induct – einarbeiten
induction – Einführung eines neuen
Mitarbeiters am Arbeitsplatz
induction course – Einführungskurs
induction training – Einführungstraining
inductive – induktiv
industrial – gewerblich; Industrie...;
industriell

industrial accident – Arbeitsunfall;
Betriebsunfall
industrial action – Arbeitskampf
industrial clerk – Industriekauffrau,
Industriekaufmann (mit
abgeschlossener Lehre)
industrial conflict – Arbeitskonflikt
industrial dispute – Arbeitskampf
industrial electronics technician –
Industrieelektroniker
industrial employee – gewerblicher
Arbeitnehmer
industrial engineering –
Betriebstechnik; Gewerbetechnik
industrial experience –
Industrieerfahrung
industrial law – Arbeitsrecht
industrial mechanic –
Industriemechaniker
industrial medicine – Arbeitsmedizin
industrial peace – Arbeitsfrieden;
Betriebsfrieden
industrial psychologist –
Betriebspsychologe
industrial psychology –
Arbeitspsychologie;
Betriebspsychologie
industrial relations – Verhältnis der
Tarifparteien
industrial revolution – Industrielle
Revolution
industrial safety – Betriebssicherheit
industrial secondment –
betriebsverbundenes Praktikum
(students and teachers on sabbatical
leave can participate in practical
training in an industrial concern for a
limited time)
industrial social work – Sozialberatung
industrial sociology – Betriebssoziologie
industrial student – Werkstudent
industrial technologist –
Industrietechnologe
industrial training – gewerbliche
Ausbildung
industrial tribunal – Arbeitsgericht
industrial union – Industriegewerkschaft
industrial worker – Gewerblicher;
Industriearbeiter
industrialist – Industrieller, Unternehmer

industry – gewerbliche Wirtschaft;
Industrie; Fleiß
industry with high labour costs –
arbeitsintensive Industrie; lohnintensive
Industrie
industry-wide agreement – Tarifvertrag
für den gesamten Industriebereich
industry-wide bargaining –
Tarifverhandlungen für den gesamten
Industriebereich
industry-wide training centre –
überbetriebliche Ausbildungsstätte
ineducable – bildungsunfähig
inefficiency – Ineffizienz;
Wirkungslosigkeit
inefficient – ineffizient; ineffektiv
inequality – Ungleichheit
inequity – Unbilligkeit; Ungerechtigkeit
inexperienced – unerfahren
infancy – frühe Kindheit
infant – Säugling
infant school – Schulkindergarten
infant school child – Abc-Schütze
infantile – kindisch; infantil
infectious disease – Infektionskrankheit
inference – Schlussfolgerung
inferiority – Minderwertigkeit
inferiority complex –
Minderwertigkeitskomplex
inflammable material – brennbares
Material
inflexible – unflexibel
influence – Einfluss
inform on a mutual basis – gegenseitig
informieren
inform on someone – petzen
informal application – formlose
Bewerbung
informal authority – informelle
Weisungsbefugnis
informal communication – informelle
Kommunikation
informal organization – informelle
Organisation
informal policy – informelle Direktive
information – Auskunft; Informieren
(das)
information about pensions –
Rentenauskunft

information and communication technology (ICT) – Informations- und Kommunikationstechnologie
information deficit – Informationsmangel
information flow – Informationsfluss
information power – Macht durch Informationsvorsprung
information procurement – Beschaffung von Informationen
information society – Informationsgesellschaft
information technology – Informationstechnik
information to staff – Information nach unten
information to superiors – Information nach oben
informational role – informatorische Rolle
infotype – Informationstyp
infotype attribute – Infotypeigenschaft
infrastructure – Infrastruktur
ingenuity – Erfindungsgabe
inherent – ureigen
inheritance – Erbe
initial contact – Kontaktaufnahme
initial entry – Ersteintritt
initial job – Ausgangsstelle
initial object – Ausgangsobjekt
initial requirements – Einstiegsvoraussetzungen
initial salary – Anfangsgehalt
initial talk – Eingangsgespräch (An interview between professor and student in the early stages of university studies to clarify interests and main areas of study.)
initial training – erlernter Beruf; Erstausbildung
initiate – beginnen; einführen; ins Leben rufen
initiative – Initiative
injunction – einstweilige Verfügung; gerichtliche Anordnung
injure – beschädigen; verletzen
injurious to health – gesundheitsgefährdend
injury to person – Personenschaden
injustice – Ungerechtigkeit
ink – Tinte

ink cartridge – Tintenpatrone
innate – angeboren; ureigen
innate ability – angeborene Fähigkeit
innate behaviour – angeborenes Verhalten
innovation process – Innovationsprozess
input into gross – Bruttoeinspeisung
inquire – sich erkundigen nach
inquiry – Anfrage
inquiry sheet – Erhebungsbogen
insert – einfügen
insider knowledge – Hintergrundwissen; Marktkenntnisse
insight – Einsicht
insinuate – einflüstern
insistence – Geltendmachung
insolvency – Insolvenz
insolvency insurance – Insolvenzversicherung
inspection – Durchsicht; Einsichtnahme
inspector – Prüfer
inspiration – Inspiration
install – einsetzen; installieren
installment – Abschlagszahlung; Anzahlung; Ratenzahlung
instalment – Teilzahlung
instalment savings – Ratensparen
instant dismissal – Kündigung, fristlose
instinct – Instinkt
instinctive – instinktiv
institute – Institut
Institute of (Science and) Technology – Technische Hochschule, Technische Universität
instruct – anleiten; lehren; unterrichten
instruct subordinates – Mitarbeiter anleiten
instruction – Anleitung; Schulung, Belehrung, Unterweisung
instructions on right to appeal – Rechtsmittelbelehrung
instructive – lehrreich
instruction – Anweisung
instructor – Ausbilder; Ausbildungsmeister; Lehrmeister
instructor aptitude examination – Ausbildereignungsprüfung
instructor's bonus – Ausbilderzulage
insulation – Isolierung
insult – Beleidigung, beleidigen
insular – engstirnig

insurance – Versicherung
insurance against accidents abroad –
Auslandsunfallversicherung
insurance benefit –
Versicherungsleistung
insurance case – Versicherungsfall
insurance company –
Versicherungsgesellschaft,
Versicherungsanstalt
insurance contribution –
Versicherungsbeitrag
insurance coverage –
Versicherungsdeckung;
Versicherungsschutz
insurance coverage of persons –
Personenversicherungsschutz
insurance deduction –
Versicherungsabzug
insurance exempt –
Versicherungsfreiheit
insurance index – Versicherungsindex
insurance lead month –
Versicherungszielmonat
insurance liability limit –
Versicherungspflichtgrenze
insurance period – Versicherungszeit
insurance policy – Versicherungsschein
insurance policy number –
Versicherungsnummer
insurance premium –
Versicherungsprämie
insurance progress – Versicherungslauf
insurance records –
Versicherungsnachweis
insurance tariff – Versicherungstarif
insurance-exempt employment –
versicherungsfreie Beschäftigung
insure – versichern
insured event – Versorgungsfall
insurer – Versicherungsträger
integrated university –
Gesamthochschule
integration – Eingliederung, Aufnahme
integrity – Integrität
intellect – Denkvermögen; Intellekt;
Verstand
intellectual – intellektuell
intellectual property – geistiges
Eigentum
intelligence – Intelligenz

intelligence quotient –
Intelligenzquotient
intelligence test – Intelligenztest
intelligent – aufgeweckt; gescheit;
intelligent
intelligentsia – Intelligenz
intend – beabsichtigen
intensity – Intensität
intensive – intensiv
intensive course – Intensivkurs
intention – Absicht; Zweck
intentional – vorsätzlich
inter-departmental transfer –
bereichsüberschreitende Versetzung
inter-governmental social insurance –
zwischenstaatliches
Sozialversicherungsrecht
interaction – Wechselwirkung
interchangeability – Austauschbarkeit
intercompany – innerbetrieblich
**intercultural relations and
communications** – IRC
interdependency – gegenseitige
Abhängigkeit
interdisciplinary – fachübergreifend;
interdisziplinär
interdiction – Entmündigung
interest – Interesse; Interessenlage; Zins
interest free loan – zinsloses Darlehen
interest group – Interessenvertretung
interest rate – Rechnungszinsfuß
interest test – Interessentest
interested party – Interessent
interface – Schnittstelle
interface competence –
Schnittstellenkompetenz
interfere – stören
interference – Einmischung
interim – provisorisch; vorläufig;
Zwischen...
interim arrangement –
Übergangsregulung
interim reply – Zwischenbescheid
interim reply – Abgabebescheid
intermediary – Vermittler
intermediate school – Mittelstufe;
Unterstufe
internal company matter –
unternehmensintern
internal control – Steuerung,
betriebliche

internal financial statement –
Wirtschaftsrechnung
internal job advertisement –
Stellenausschreibung, interne
internal labour market – firmeninterner
Arbeitsmarkt
internal measures – interne Maßnahmen
internal medical service – Dienst,
werksärztlicher
internal profit center – organisatorisch
selbständige Einheit
internal recruitment – Beschaffung,
interne
international – länderunabhängig
international commuter – Grenzgänger
**international vocational training
competition** – internationaler
Berufswettbewerb
internship – Praktikum
interoffice messenger – Hausbote
interpret – interpretieren; wiedergeben
interpreter – Dolmetscher
interrupt – stören
interrupted work relationship –
Arbeitsverhältnis, unterbrochenes
interruption of work –
Arbeitsunterbrechung
interval – Zeitraum
interval indicator – Intervallkennzeichen
intervention – Vermittlung
interview – Bewerbungsgespräch;
Einstellungsgespräch
intolerance – Intoleranz
intoxication – Trunkenheit
introduce – beginnen; einführen; ins
Leben rufen
introvert – Intravertierte(r)
intuition – Intuition
invalid – Schwerbehinderte(r); ungültig
invalidity – Arbeitsunfähigkeit;
Erwerbsunfähigkeit
inventor – Erfinder
inventor's bonus – Erfindervergütung
inventory – Inventar; Warenbestand;
Warenvorrat
inverse relationship –
Umkehrverknüpfung
investigate – ermitteln; untersuchen
investigation – Untersuchung
investment of capital – Vermögensanlage

investments in education –
Bildungsinvestition
invigilate – eine Prüfung beaufsichtigen;
Examensarbeit beaufsichtigen
invigilation – Prüfungsaufsicht
invitation for an interview – Einladung
zum Vorstellungsgespräch
invigilator – Aufsicht(sperson)
invoicing – Berechnung
involvement – Verwicklung; Verstrickung
ionizing rays – ionisierende Strahlen
irregularity – Ordnungswidrigkeit
irrelevant detail – Nebensächlichkeit
irreplaceable – unersetzbar
isolation – Isolation; Isolierung
issue posting – Entnahmebuchung
itemization – Aufgliederung; Aufteilung
itinerant insurance –
Wanderversicherung
itinerary – Reiseverlauf
Ivy League – Hochschulverband im
Nordosten der U.S.A. (An association
of eight universities and colleges in the
northeast United States, comprising
Brown, Columbia, Cornell, Dartmouth,
Harvard, The University of
Pennsylvania und Yale)

J

janitor (AmE) – Hausmeister
jargon – Jargon
JDS – Arbeitsplatzbefragung
jealousy – Eifersucht; Futterneid
job – Beruf; Stelle; Tätigkeit
job ad(vertisement) – Stellenanzeige;
Stellenausschreibung; interne
Stellenausschreibung; Ausschreibung
job advertisement costs –
Ausschreibungskosten
job analysis – Arbeits- und
Aufgabenanalyse; Arbeitsanalyse
job application – Bewerbung
job category – Berufsgruppe
job centre – Arbeitsamt

job characteristic – Tätigkeitsmerkmal
job connection – Berufsbezug
job content – Arbeitsinhalt
job counselling by the Labour Exchange – Berufsberatung durch das Arbeitsamt
job creation – Arbeitsbeschaffung
job description – Arbeitsplatzbeschreibung; Stellenbeschreibung; Tätigkeitsbeschreibung
job design – Arbeitsgestaltung
job diagnostic survey – Arbeitsplatzbefragung
job enlargement – Aufgabenerweiterung; Job Enlargement (expanding a job)
job enrichment – Job Enrichment
job entry – Berufseinstieg
job evaluation – Arbeitsbewertung; Arbeitsplatzbewertung
job evaluation scale – Arbeitsplatzbewertungsstufe
job evaluation sheet – Arbeitsplatzbewertungsbogen
job evaluation system – Arbeitsplatzbewertungssystem
job familiarization – Einlernen
job feature – Tätigkeitsmerkmal
job file – Stellendatei
job for the future – Zukunftsaufgabe
job freeze – Einstellungsstopp
job guarantee – Beschäftigungsgarantie
job holder – Stelleninhaber
job hopping – häufiger Stellenwechsel
job hunt – Arbeitssuche; Stellensuche
job hunter – Stellensucher
job index – Stellenplan
job instruction – Arbeitsanweisung
job interview – Bewerbungsgespräch; Vorstellungsgespräch
job leasing – Leiharbeit; Zeitarbeit
job loss – Verlust des Arbeitsplatzes
job market – Arbeitsmarkt
job number – Stellennummer
job offers – Stellenangebote
job opportunities – Arbeitsmöglichkeiten; Einstiegschancen
job outline – Berufsbild
job performance – Arbeitsleistung
job placement – Stellenvermittlung

job profile – Tätigkeitsprofil
job prospects – Berufsaussichten
job qualification – Berufsqualifikation
job rating – Arbeitsplatzbewertung
job requirement – Arbeitsplatzanforderung
job requirements – Arbeitsanforderungen
job rotation – Arbeitsplatzwechsel; Job Rotation; Ringtausch
job rotation planning – Stellenwechselplanung
job satisfaction – Arbeitsfreude; Arbeitszufriedenheit
job search – Arbeitssuche; Stellensuche
job security – Arbeitsplatzsicherung; Sicherheit des Arbeitsplatzes
job sharing – Arbeitsplatzteilung; Job-sharing
job splitting – Stellensplitting
job structure – Aufgabenstruktur
job tenure – Dauer der Bertriebszugehörigkeit
job test – Probearbeit
job ticket – Akkordzettel
job title – Funktionsbezeichnung, Tätigkeitsbezeichnung
job work – Akkord
job-holder – Jobholder
job-related instruction – Arbeitsanweisung
job-task analysis – Arbeits- und Aufgabenanalyse; Arbeitsanalyse
job-worker – Akkordarbeiter
jobber – Aushilfe
jobbing – Gelegenheitsarbeit
jobless – arbeitslos
jobless growth – Wirtschaftswachstum ohne neue Arbeitsplätz
joiner – Tischler
joint – gemeinsam; gemeinschaftlich
joint bargaining – gemeinsame Verhandlungen (der Tarifpartner)
joint committee – Betriebskommitee
joint consultation – innerbetriebliche Zusammenarbeit
joint-stock company – Kapitalgesellschaft, Aktiengesellschaft
journal – Journal
journeyman – Geselle
jubilarian – Jubilar

jubilee – Jubilarfeier; Jubiläum, Dienstjubiläum
jubilee payment – Jubiläumszahlung
jubilee present – Jubiläumsgeschenk
judge – beurteilen
judgment – Urteil
jump forward in a career – Karrieresprung
junior college (AmE) – Junior College
junior cycle – Mittelstufe; Unterstufe
junior high school (AmE) – Mittelstufe; Unterstufe, Hauptschule
jurisdiction – Rechtsprechung
just – aufrichtig; fair; gerecht
just-in-time learning – Lernen bei Bedarf
justice – Gerechtigkeit
justified – berechtigt
juvenile – Jugendliche(r)
juvenile aggressiveness – jugendliche Aggressivität
juvenile delinquent – jugendlicher Straftäter
juvenile employees' meeting – Jugendversammlung

K

Kaizen – kontinuierlicher Verbesserungsprozess
keep abreast of – Schritt halten mit; sich auf dem Laufenden halten
keep up to date – Schritt halten mit; sich auf dem Laufenden halten
keep up with – Schritt halten mit; sich auf dem Laufenden halten
key date for retroactive accounting run – Rückrechnungspol
key educational issues – zentrale Bildungsfragen
key experience – Schlüsselerfahrung
key in – Kommt-Buchung
key objective – Gesamtziel
key position – Schlüsselposition

key qualification – Schlüsselqualifikation
key word – Schlüsselwort; Stichwort
keynote – zentrales Thema
keynote speaker – Hauptredner
keynote speech – programmatische Rede
kick – treten
kickback – Schmiergeld
kind – liebenswürdig
kindergarten – Kindergarten
knock down – umrennen
know-all – Alleswisser; Besserwisser
know-how – Know-how
knowledge – Kenntnis(se); Wissen
knowledge base – Wissensbasis
knowledge engineer – Wissensvermittler
knowledge head start – Wissensvorsprung
knowledge in a special field – Fachgebietswissen
knowledge industry – Wissensindustrie
knowledge management – Wissensmanagement
knowledge networks – Wissensnetze
knowledge of human nature – Menschenkenntnis
knowledge of languages – Sprachkenntnisse
knowledge of the facts – Sachkenntnis
knowledge of the market – Marktkenntnisse
knowledge society – Wissensgesellschaft
knowledgeable – kenntnisreich

L

lab(oratory) – Labor(atorium)
label for income tax card – Lohnsteueraufkleber
labo(u)r market problem – Arbeitsmarktproblem
labo(u)r protection meeting – Arbeitsschutztagung
labor protection laws – Arbeitsschutzgesetz

labor utilization rate – Zeitgrad
laboratory worker – Laborant
labour annuity insurance –
 Arbeiterrentenversicherung
labour conflict – Arbeitskonflikt
labour contract – Anstellungsvertrag;
 Arbeitsvertrag
labour costs – Arbeitskosten
labour court – Arbeitsgericht
labour director – Arbeitsdirektor
labour division – Arbeitsteilung
labour division in the industrial society
 – arbeitsteilige Industriegesellschaft
labour force – Arbeitskräfte
labour force reduction – Personalabbau
labour intensive – arbeitsintensiv
labour law – Arbeitsrecht
labour law proceedings –
 Arbeitsrechtsverfahren
labour legislation – Arbeitsgesetzgebung
labour management – betriebliche
 Personalpolitik
labour market – Arbeitsmarkt
labour market research –
 Arbeitsmarktforschung
labour mobility – Arbeitskräftemobilität;
 Mobilität der Arbeitskräfte
labour piracy – Mitarbeiterabwerbung
labour potential – Arbeitskräftepotential
labour relations –
 Arbeitgeber-Arbeitnehmer-Verhältnis
labour relations board – Ausschuss
 Arbeitsrecht
labour relations department –
 Abteilung Arbeitsrecht
labour representative –
 Arbeitnehmervertreter
labour shortage – Arbeitskräftemangel
labour society – Arbeitsgesellschaft
labour supply – verfügbare Arbeitskräfte
labour surplus – Arbeitskräfteüberhang
labour turnover – Arbeitsplatzwechsel;
 Fluktuation
labourer – Handwerker
lack – Mangel
lack of apprenticeships –
 Lehrstellenmangel
lack of interest – Desinteresse
lack of time – Zeitmangel
lack of training positions –
 Lehrstellenmangel

lackadaisical – lustlos; desinterressiert;
 nachlässig
language – Sprache
language ability – Sprachfähigkeit
language development –
 Sprachentwicklung
language identity – sprachliche Identität
language laboratory – Sprachlabor
language sequence – Sprachenfolge
language test – Sprachentest
language training – Sprachausbildung
lapse from one's duty –
 Pflichtverletzung
lapsing of knowledge – Verfall des
 Wissens
large number of applicants –
 Bewerberandrang
large-scale company –
 Großunternehmen
laser beam protection –
 Laserstrahlenschutz
last day of month – Monatsultimo
last day of school – Schulabschlusstag
lasting longer than one calendar day –
 kalendertagübergreifend
late arrival – Verspätung
late developer – Spätentwickler
late entrant – Späteinsteiger
late shift – Nachtschicht; Spätschicht
late starter – Spätzünder
lateral communication –
 Kommunikation innerhalb einer
 Führungsebene
lateral entrant – Seiteneinsteiger
lateral thinker – Querdenker
lateral thinking – unorthodoxe
 Denkmethode
latitude – Handlungsspielraum
lattice organisation – Gitterorganisation
law – Gesetz; Recht; Rechtskunde
lawful – rechtmäßig
law governing employment –
 Beschäftigungsgesetz
law-maker – Gesetzgeber
Law on Co-determination –
 Mitbestimmungsgesetz
lawsuit – Gerichtsverfahren, Rechtsstreit
lay off – freisetzen
layoff benefit – Entschädigung für
 vorübergehende Entlassung
layout regulation – Aufbauvorschrift

layout rule – Aufbauregel
lazy – faul
lead – führen; leiten
leadership – Führung
leadership ability –
Führungseigenschaft; Führungsgabe,
Führungsfähigkeit
leadership by example – Führung durch
Beispiel
leadership competency –
Führungskompetenz
leadership evolution – Entwicklung des
Führungsverhaltens
leadership model – Führungsmodell
leadership position – Führungsposition
leadership responsibility –
Führungsposition
leadership talent – Führungstalent
leading question – Suggestivfrage
leading-edge – Spitzen...
leaflet – Merkblatt; Prospekt;
Werbeschrift
leak – undichte Stelle
lean management – flache
Organisationsform
leap-frogging – Überspringen (Karriere)
learn – lernen
learn by heart – auswendig lernen
learn by rote – auswendig lernen
learners' wage rate – Anlernlohnsatz
learning achievement – Lernerfolg
learning aids – Lernmittel
learning atmosphere – Lernatmosphäre
learning by doing – Lernen durch Tun
learning company – lernendes
Unternehmen (A learning company is
one which establishes a culture of
knowledge by means of systematic
knowledge management and
permanently optimises this culture.)
learning curve – Lernkurve
learning difficulties –
Lernschwierigkeiten
learning experience – Lernerfahrung
learning materials – Lernmaterialien
learning objective – Lernziel
learning organisation – lernende
Organisation
learning pace – Lerngeschwindigkeit
learning phase – Lernphase
learning process – Lernprozess

learning progress – Lernfortschritt
learning resources – Lernressourcen
learning society – Lerngesellschaft
learning system – Lernsystem
learning strategy – Lernstrategie
learning techniques – Lerntechniken
learning theory – Lerntheorie
learning time – Lernzeit
leave – Abgang, Urlaub
leave account – Urlaubskonto
leave accrual – Urlaubsaufbau
leave balance – Urlaubssaldo
leave class – Urlaubsklasse
leave compensation – Urlaubsabfindung;
Urlaubsabgeltung
leave day – Urlaubstag
leave days paid – Urlaubstag,
abgerechneter
leave deduction – Urlaubsabtragung
leave entitlement – Urlaubsanspruch
leave for further training –
Ausbildungsurlaub
leave information – Urlaubsstand
leave of absence – bewilligte
Abwesenheit; entschuldigtes Fehlen
leave period – Urlaubszeit
leave record – Urlaubssatz
leave schedule – Urlaubsplan
leave taken and deducted – Urlaub,
abgerechneter
leave type – Urlaubsart
leave without notice (employee) –
außerordentliche Kündigung
leave without pay – unbezahlter Urlaub
leave year – Urlaubsjahr
Leaving Certificate – Abgangszeugnis;
Abiturzeugnis
lecture – Vorlesung (Uni); Vortrag
lecture hours per week –
Semesterwochenstunden
lecture room – Hörsaal
lecture timetable –
Vorlesungsverzeichnis
lecturer – promovierter
Assistentenzprofessor (a person with an
academic teaching post below the rank
of professor); Dozent
lecturing post – Lehrauftrag
left company – ausgetreten
left-handed person – Linkshänder
legal – gesetzmäßig; rechtmäßig

legal aid – Rechtshilfe
legal action – Rechtsweg
legal basis – Rechtsgrundlage
legal claim – Rechtsanspruch
legal costs insurance –
Rechtsschutzversicherung
legal demand – Rechtsanspruch;
Anspruch
legal department – Rechtsabteilung
legal guardian – Erziehungsberechtigter;
gesetzlicher Vormund
legal holiday – gesetzlicher Feiertag
legal independence – rechtliche
Verselbständigung
legal liability – gesetzliche Haftpflicht
legal provision – Rechtsvorschrift
legal question – Rechtsfrage
legal right – Rechtsanspruch
legal right to future pension payments
– Anwartschaft
legally competent – geschäftsfähig
legally incompetent – geschäftsunfähig
legibility – Lesbarkeit
legislation – Gesetzgebung
legislator – Gesetzgeber
legitimate – rechtmäßig
legitimate power – Macht durch
funktionelle Weisungsbefugnis
leisure activity – Freizeitbeschäftigung
leisure time – Freizeit
lending library – Leihbücherei
length of employment –
Beschäftigungsdauer
length of service – Dienstzeit
length of service group –
Dienstaltersgruppe
length of stay – Aufenthaltsdauer
length of studies – Studiendauer
lenience – Milde; Nachsicht
lesson – Stunde; Unterrichtsstunde;
Unterricht; Unterweisung
letter – Brief; Buchstabe; Schreiben
letter of confinement –
Schwangerschaftsbescheinigung
letter of dismissal – Entlassungspapiere
letter of employment –
Einstellungsschreiben
letter of recommendation –
Empfehlungsbrief;
Empfehlungsschreiben

letter of rejection –
Ablehnungsschreiben
letter of thanks – Dankschreiben
level – Ebene; Schicht; Stand
level of training – Ausbildungsniveau;
Ausbildungsstand
level-headed – ausgeglichen
level of unemployment –
Arbeitslosenquote
leverage – Einfluss(nahme);
Hebelwirkung
lexis – Lexik
liability – Haftpflicht; Haftung
liability to pay contributions –
Beitragspflicht
liability to social security payments –
Abgabepflicht
liable – haftbar
liable for payment – leistungspflichtig
liable to contribution –
beitragspflichtiges Arbeitsentgelt
liable to military service – wehrpflichtig
liable to social security payments –
sozialversicherungspflichtig;
Abgabepflicht
liaison officer – Verbindungsmann;
Vertrauensmann
liberal arts – Geisteswissenschaften
liberal education – Allgemeinbildung
liberal minded – weltoffen
library – Bibliothek
life – Leben; Lebenszeit
life insurance – Lebensversicherung
life insurance policy –
Lebensversicherungsvertrag
life-long learning – lebenslanges Lernen
lighting – Beleuchtung
limit – befristen, einschränken
limit for social insurance –
Sozialversicherungsgrenze
limitation – Einschränkung; Verjährung
limited (liability) company –
Kapitalgesellschaft
limited commercial partnership –
Kommanditgesellschaft (KG)
limited employment – befristetes
Arbeitsverhältnis
limited employment contract –
befristeter Arbeitsvertrag

limited liability company (AmE) – Gesellschaft mit beschränkter Haftung; GmbH

limited part-time work – geringfügige Beschäftigung

limited partner – Kommanditist

limited partnership with limited liability company as general partner – GmbH & Co.KG

limited state of employment – befristetes Arbeitsverhältnis

limited vocabulary – begrenzter Wortschatz

limited work capability – beschränkte Einsatzfähigkeit

line – Zeile

line man – Linienmann

line management – Linien-Führungskräfte

line manager – Fachvorgesetzter; Linienmanager

line of business – Geschäftszweig

line of conduct – Lebenswandel

linear programming – lineare Programmierung

line production – Fließbandarbeit

lingua franca – Umgangssprache; Verkehrssprache

linguistic intelligence – sprachliche Intelligenz

linkage – Verknüpfung

lines – Strafarbeit

liquid funds – liquide Mittel

liquidation – Liquidation

lisp – lispeln

list – Aufstellung

list of employees – Mitarbeiterverzeichnis; Personalverzeichnis

list of members – Mitgliedsverzeichnis

list of names – Anwesenheitsliste; Namensliste

list of staff positions – Stabstellenliste

list of subjects – Fächerliste (Übersicht über Unterrichtsfächer)

listen – hören; zuhören

listening comprehension – Hörverständnis

listlessness – Unlust

literacy – Lese- und Schreibfähigkeit

literal translation – wörtliche Übersetzung

literature – Literatur

litigation – Rechtsstreit

little master (BrE) – Geselle

livelihood – Lebensunterhalt

LLD – Dr.jur.

loan – Darlehen

loan category – Darlehenstyp

loan repayment – Darlehenstilgung

loan repayment instalment – Darlehenstilgungsrate

loan sum – Darlehensbetrag

loan terms – Darlehenskonditionen

loan to an employee – Arbeitnehmerdarlehen

loan type – Darlehensart

lobby – Interessenvertretung

Local Education Authority (LEA) – Schulbehörde

local history – Heimatkunde

local sickness benefit fund – Allgemeine Ortskrankenkasse; AOK

location – Ort; Standort; Wohnort

lock flag – Sperrvermerk

lock-out – Aussperrung

lodging – Beherbergung

log book – Berichtsheft

logical inference – logische Schlussfolgerung

logical thinking – logisches Denken

lone wolf – Eigenbrötler

lonely – einsam

loner – Eigenbrötler

long-term compensation development – langfristige Einkommensentwicklung

long-term memory – Langzeitgedächtnis

long-term personnel planning – langfristige Personalplanung

long-term student – Langzeitstudent

long-term unemployed person – Langzeitarbeitslose(r)

long-term unemployment – Langzeitarbeitslosigkeit

long-winded – weitschweifig

longer holidays – Urlaubsverlängerung; verlängerter Urlaub

look after – betreuen

loose-leaf book – Ringbuch

loser – Verlierer

loss – Ausfall; Verlust; Einbuße

loss of earnings – Verdienstausfall
loss of income – Verlust des Einkommens
loss of job – Arbeitsplatzverlust
loss of power – Machtverlust
loss of privileges – Privilegienverlust
loss of wages – Lohnausfall
loss of working hours – Arbeitsausfall
lost and found – Kramschachtel; Schlamperkiste
lost hour – Ausfallstunde
lost property box – Kramschachtel; Schlamperkiste
lost property office – Fundbüro
lost time – Ausfallprinzip
low level radiation screen – strahlenarmes Gerät
low mental ability – geringe geistige Fähigkeit
low-income earner – Geringverdiener
low-paid employment – Niedriglohn-Beschäftigung
lower class – Unterschicht
lower secondary sector – Sekundarbereich 1
loyal – treu
lump sum – Pauschalzahlung, Pauschale
lump sum charge – Pauschsatz
lump sum for meals – Verpflegungspauschale
lump sum for special expenditures – Sonderausgabenpauschale
lump sum for transportation – Fahrtkostenpauschale
lump sum per night – Übernachtungspauschale
lump-sum accounting – Pauschalabrechnung
lump-sum compensation – Pauschalentschädigung
lump-sum contribution payment – pauschalierte Beitragszahlung
lump-sum deduction – Abzug, pauschaler
lump-sum payment of tax – Pauschalversteuerung
lump-sum rate – Pauschalsatz
lump-sum rule – Pauschalenregelung
lump-sum tax – Pauschalsteuer
lump-sum taxation – Pauschalbesteuerung
lunch break – Mittagspause

lunch coupon – Essensbon
lunchroom (AmE) – Kantine
lure – Lockmittel

M

machine fitter – Maschinenschlosser
macho – Macho
made-to-measure – massgeschneidert
magna cum laude – magna cum laude; mit großem Lob
magnetic board – Magnettafel
main aptitude – Eignungsschwerpunkt
main (business) event – Hauptveranstaltung
main course of studies – Hauptstudium
main focus – Schwerpunkt
main office for applications – Hauptstelle für Bewerbungen
main personnel department – Hauptpersonalabteilung
main personnel office – Hauptpersonalbüro
main schema – Hauptschema
main subject – Hauptfach
main welfare office – Hauptfürsorgestelle
main work area – Hauptarbeitsgebiet
maintain – betreuen; unterhalten (Familie)
maintenance – Wartung; Unterhalt
maintenance allowance – Unterhaltsgeld
maintenance employees – Instandhaltungspersonal
maintenance fitter – Betriebsschlosser
maintenance people – Instandhaltungspersonal
maintenance personnel – Wartungspersonal
major – volljährig
major (subject) (AmE) – Hauptfach
major city allowance – Hauptstadtzuschlag
majority vote – Mehrheitswahl

make a mess of a test – einen Test verpfuschen
make an effort – sich anstrengen; sich bemühen
make known – bekanntmachen
make redundant – freisetzen
make-up-work – Nacharbeit
maladapted behaviour – unangepasstes Verhalten
maladjusted – schwer erziehbar; unausgeglichen
maladjustment – mangelnde Anpassung
malfunction period – Störzeit
malinger – krankfeiern
man of learning – Gelehrter
man of letters – Literat
manage time – Zeit einteilen; Zeit planen
management – Geschäftsleitung; Geschäftsführung
management behaviour margin – Freiraum im Führungsverhalten
management buyout – Management-Buyout
management capability – Führungseigenschaft; Führungsfähigkeit
management concept – Führungskonzept; Managementkonzept
management consultant – Unternehmensberater
management control – Planüberwachung und Steuerung
management culture – Führungskultur
management decisions – Führungsentscheidungen
management experience – Führungserfahrung
management function – Führungsfunktion; Managementfunktion
management group – Führungsebene; Führungskreis; Leitungskreis
management information system – Managementinformationssystem; MIS
management inventory – Führungsinventar (listing of all the managers in an organization with information about their skills and experience)
management job – Leitungsstelle

management know-how – Führungswissen
management level – Führungsebene; Führungskreis; Leitungskreis
management objective – Unternehmensziel
management of one's own concerns – Selbstmanagement
management of visitor groups – Besuchergruppenverwaltung
management organization – Unternehmensordnung
management policy – Firmenrichtlinie
management position – Führungsposition
management potential – Führungspotential
management pressure – von der Führung ausgeübter Druck
management privilege – Vorrecht der Führungskräfte
management procedure – Arbeitsverfahren
management process – Managementprozess
management programme – Managementprogramm (Schulung und Förderung)
management ratios – Kennzahlen
management recruitment from within the company – Gewinnung von Führungskräften aus den eigenen Reihen
management skills – Führungsqualitäten
management structure – Struktur des Führungskreises
management style – Managementverhalten
management succession – Führungskräftenachfolge
management system – Managementsystem
management task – Führungsaufgabe
management team – Führungsteam
management technique – Führungstechnik; Managementtechnik
management tool – Führungsinstrument; Führungswerkzeug
management trainee – Führungsnachwuchs; Führungsnachwuchskraft

management training – Aus- und
Weiterbildung von Führungskräften
manager – Fachreferent; Führungskraft
managerial – leitend
managerial aids – Führungshilfen
managerial authority –
Führungsbefugnis
managerial grid – Bewertungsmatrix für
Führungskräfte
managerial practice – Führungspraxis
managerial resources – Führungshilfen
managerial style – Führungsstil;
Führungsverhalten
managerial training – Ausbildung zur
Führungskraft
Managing Board – Vorstand;
Vorstandsbereich
managing conflict – Konfliktbewältigung
managing director – Geschäftsführer
(GmbH & Co.KG)
mandatory – Gesetzespflicht
manhour – Arbeitsstunde
manipulation – Manipulation
manners – Mores
manning – Besetzen von Stellen
manning tables – Personalkartei
manpower – Arbeitskräfte
manpower assignment –
Arbeitskräfteeinsatz
manpower cutback – Personalabbau
manpower demand –
Arbeitskräftenachfrage
manpower planning –
Personalbedarfsplanung,
Personaleinsatzplanung
manpower planning and development –
Personalplanung und -entwicklung
manpower provision –
Arbeitnehmerüberlassung
manpower provision act –
Arbeitnehmerüberlassungsgesetz
manpower reduction planning –
Personalabbauplanung
manpower requirements –
Arbeitskräftebedarf; Personalbedarf
manpower shortage – Personalmangel
manpower survey –
Arbeitskräftebedarfsplanung
manual – Handbuch; Leitfaden
manual alphabet –
Taubstummenalphabet

manual dexterity – handwerkliches
Können
manual grade – einfacher Dienst
manual labour – körperliche Arbeit
manual work – Handarbeit
manual worker – Arbeiter; Handarbeiter
manufacturing – Fertigung
manufacturing engineering –
Fertigungsaufgaben
manufacturing unit – Fertigungsbetrieb;
Fertigungseinheit
manuscript – Manuskript
map – Landkarte
margin – Spielraum
margin – Freiraum
marginal tax burden – Grenzbelastung
marital partner – Ehegatte, Ehegattin
marital status – Familienstand
mark – benoten; Leistungsnote; Note;
Zensur
market demand – Markt-Nachfrage
market development –
Marktentwicklung
market forecast – Marktprognose
market prospects – Konjunktur;
Wirtschaftslage
market research – Marktforschung
marketing – Marketing
marketing expert – Fachmann für
Marketing
marketing via the media –
Medienmarketing
marking – Korrektur
marking of aims – Zielsetzung
marriage certificate – Heiratsurkunde
married – verheiratet
mass redundancy – Massenentlassung
mass unemployment –
Massenarbeitslosigkeit
master (of a trade) – Meister
master cost center – Stammkostenstelle
master craftsman – Meister
master craftsman's certificate –
Meisterbrief
master craftsmen's college –
Meisterfachschule
master increased requirements –
gestiegene Anforderungen bewältigen
Master of Arts (M.A.) – Magister Artium
Master of Science (M.Sc.) – Magister
Scientiarum

master's degree – Magister
mastery – Beherrschung
matching – Abgleich
mate – Kumpel
material – Werkstoff
material damage – Sachschaden
material technology – Werkstoffkunde;
  Werkstofflehre
materials management –
  Materialwirtschaft
maternity – Mutterschaft
maternity benefit – Mutterschaftsgeld;
  Wochenhilfe
maternity cases – Mutterschaftsfall
maternity grant – Geburtenprämie;
  Geburtenzulage
maternity leave – Mutterschaftsurlaub
maternity pay – Mutterschaftsgeld
maternity protection – Mutterschutz
maternity protection period –
  Mutterschutzfrist
mathematician – Mathematiker
mathematics – Mathematik
mathetics – Mathetik
matriculate – immatrikulieren
matriculation requirements –
  allgemeine Hochschulreife
**matriculation requirements for the
  Fachhochschule** – fachgebundene
  Hochschulreife; Fachhochschulreife
matrix organization –
  Matrix-Organisation
matrix structure – Matrix-Struktur
matter for negotiation –
  Verhandlungsgegenstand
matter of complaint –
  Beschwerde(grund)
matter of form – Formsache
matter-of-factness – Sachlichkeit
mature – reif
maturity – Reife; Reifegrad; Reifeniveau
maximum age – Höchstalter
maximum amount – Höchstbetrag
maximum contribution – Höchstbeitrag
maximum pension – höchste Rente
maximum rate, reduced – Höchstsatz;
  Höchstsatz, ermäßigter
maximum rate for meals –
  Verpflegungshöchstsatz
maximum wage rate – Höchstlohn;
  Spitzenlohn

maximum working hours –
  Höchstarbeitsstunden
meakness – Sanftmut
meal break – Essenspause
meal subsidy – Essenszuschuss
meal voucher – Essensmarke
meals – Verpflegung
meals accounting –
  Verpflegungsabrechnung
meals receipt – Verpflegungsbeleg
mean – geizig; knausrig
means test – Einkommensveranlagung
meaning – Sinngebung, Bedeutung
meaningful learning – wirkliches
  Begreifen
measure – Maßnahme
measure targeted at job creation –
  Arbeitsbeschaffungsmaßnahme
measurement – Messung; Wertung
measurement of times by differences –
  Folgezeitverfahren
measures of development –
  Entwicklungsmaßnahmen
mechanical engineer –
  Maschinenbauingenieur
mechanical engineering –
  Maschinenbau
mechanistic approach – mechanistische
  Vorgehensweise
media – Medien
media didactics – Mediendidaktik
media education – Medienerziehung
media pedagogy – Medienpädagogik
mediate – vermitteln
mediation committee –
  Schlichtungsausschuss
mediator – Schiedsmann; Schiedsrichter;
  Schlichter
medical care – ärztliche Versorgung
medical certificate –
  Arbeitsunfähigkeitsbescheinigung;
  ärztliches Attest
medical examination – ärztliche
  Untersuchung
medical examiner – Vertrauensarzt
medical referee (AmE) – Vertrauensarzt
medical service centre –
  betriebsärztlicher Dienst
mediocre – Mittelmaß
medium – Medium
medium-term – mittelfristig

meeting – Besprechung, Sitzung
meeting of works councils –
  Betriebsräteversammlung
melancholic – Melancholiker(in)
member – Mitglied
member of the exempt staff –
  außertariflicher Mitarbeiter
Member of the Group Executive
  Management – Mitglied des
  Bereichsvorstands
Member of the Managing Board –
  Mitglied des Vorstands
member of the staff – Arbeitnehmer;
  Belegschaftsmitglied
Member of the Supervisory Board –
  Aufsichtsratsmitglied; Mitglied des
  Aufsichtsrats
member of the works council –
  Betriebsratsmitglied
membership – Mitgliedschaft
membership number –
  Mitgliedsnummer
membership period –
  Zugehörigkeitsdauer
memo – Aufzeichnung; Notiz; Vermerk
memorandum – Aktennotiz;
  Aufzeichnung; Notiz; Vermerk
memory – Erinnerung; Gedächtnis
mental ability – geistige Fähigkeit
mental blackout – Bewusstseinsstörung
mental flexibility – geistige Flexibilität
mental illness – Geisteskrankheit
mental withdrawal – innere Kündigung
mentally defective – geistesgestört
mentally handicapped – geistig
  behindert
mention – anschneiden; erwähnen; zur
  Sprache bringen
mentor – Mentor
merit – Leistung; Verdienst
merit bonus – Leistungszulage
merit increase – leistungsbezogene
  Gehaltserhöhung
merit increase according to wage scale
  – tarifliche Leistungszulage
merit-based contract for non-salaried
  employees – Sonderleistungsvertrag
message indicator – Mail-Kennzeichen
metallurgy – Metallurgie
method – Methode; Verfahren
method of work – Arbeitsmethode

method of working – Arbeitsweise
methodical – methodisch
methodical manner – methodische
  Arbeitsweise; methodisches Arbeiten
methodical procedure – methodische
  Arbeitsweise; methodisches Arbeiten
methodicalness – Planmäßigkeit
methodology – Methodik
methods study – Arbeits- und
  Aufgabenanalyse; Arbeitsanalyse
methods time measurement (MTM) –
  Methoden und Zeitstudien (examining
  the way)
meticulous – sorgfältig
Michaelmas term – Herbsttrimester
microsystems electronics technician –
  Mikrosystemelektroniker
microteaching – Microteaching
mid term – Trimesterhälfte
mid-course examination –
  Zwischenprüfung
midday break – Mittagspause
middle class – bürgerlich
middle line – mittleres Management
middle management – mittlerer
  Führungskreis
middle-class – mittelständisch
migrant labour – Gastarbeiter
migrant worker – Saisonarbeiter;
  Wanderarbeitnehmer
mileage money – Kilometergeld
mileage rebate – Kilometerpauschale
miles/kilometer distribution –
  Kilometeraufteilung
miles/kilometer limit – Kilometerlimit
miles/kilometers travelled –
  Fahrleistung
miles/km allowance rule –
  Kilometergeldregelung
miles/km rate – Kilometersatz
miles/kms travelled – Kilometerleistung
military service – Wehrdienst
milling cutter – Fräser
mind – Verstand; Geist
miner's pension scheme –
  knappschaftliche Rentenversicherung
mini-master record – Ministammsatz
minimum – Kapazität, minimale
minimum age – Mindestalter
minimum attendance time –
  Mindestanwesenheitszeit

minimum contribution – Mindestbeitrag
minimum hours – Mindeststunden
minimum living wage –
  Existenzminimum
minimum mark – Mindestnote
minimum time – Mindestzeit
minimum wage – garantierter
  Mindestlohn; Mindestlohn
minimum working hours –
  Mindestarbeitsstunden
Minister of Education – Kultusminister
ministry representative –
  Ministerialbeauftragter
minor – Minderjährige(r)
minor (subject) (AmE) – Nebenfach
minor occupation – Nebenberuf;
  nebenberufliche Tätigkeit
minority – Minderheit; Minorität;
  Unmündigkeit
minority group – Minderheitengruppe
minutes – Protokoll;
  Sitzungsniederschrift
misbehaviour – schlechtes Benehmen;
  Ungezogenheit
miscellaneous – verschieden
mischievous – schelmisch; spitzbübisch;
  verschmitzt
misconduct – schlechtes Benehmen;
  Fehlverhalten
misfit – Aussenseiter(in)
misguided – töricht; unangebracht
misjudgement – Fehlentscheidung
mislead – irreführen
missing consent of the works council –
  fehlende Zustimmung des Betriebsrates
missing work permit – fehlende
  Arbeitserlaubnis
mistake – Fehler; Irrtum
mitch – die Schule schwänzen
mixed school – gemischte Schule
mnemonic – Erinnerungsstütze
mnemonics – Gedächtniskunst;
  Mnemotechnik
mobbing – Psychoterror am Arbeitsplatz;
  Schikane am Arbeitsplatz
mobility – Mobilität
mobility within the firm – interner
  Personalwechsel
mock employment –
  Scheinarbeitsverhältnis
mock examination – Probeprüfung

mock independence –
  Scheinselbständigkeit
mock interview – Probeinterview
model – Modellversuch
model catalog – Musterkatalog
**model for the part-time employment of
  old-age pensioners** –
  Altersteilzeitmodell
model wage type – Musterlohnart
moderate – den Vorsitz führen; mäßig;
  mäßigen; mildern; moderieren
moderator – Diskussionsleiter
moderation – Moderation
modern languages – Neuphilologie
modernize – modernisieren;
  rationalisieren; vereinfachen
modification – Abänderung
modification feature –
  Modifikationsmerkmal
modification indicator –
  Modifikationskennzeichen
modification type – Modifikationstyp
modifier – Modifikator
modular training program – modulares
  Ausbildungsprogramm
modular working time – Arbeitszeit,
  modulare (Employers divide the
  working year, month, week or day into
  working-time components which
  employees can combine as they wish,
  depending on the work-volume
  requirements.)
monitor role – Beobachterrolle
monitoring device –
  Überwachungseinrichtung
monitoring of dates – Terminverfolgung
monotonous – monoton
month already accounted – Todesmonat
monthly balance – Monatssaldo
monthly earnings – Monatsverdienst
monthly factor – Aliquotierungsfaktor
monthly factoring – Aliquotierung
monthly factoring method –
  Aliquotierungsmethode
monthly installment – monatliche Rate
monthly report – Monatsmeldung
monthly salary – Monatsgehalt
monthly wages – Monatslohn
monthly work schedule –
  Monatsarbeitszeitplan

monthly working hours –
Monatsarbeitszeit
moody – launenhaft; launisch
moonlighter – Schwarzarbeiter
morale survey – Untersuchung der
Arbeitsmoral
morning assembly – Morgenandacht
morning shift – Frühschicht
moron – Geistesschwache(r)
mortgage – Hypothek
motel – Gasthof
mother tongue – Muttersprache
**mother-tongue classes for foreign
children** – muttersprachlicher
Unterricht für ausländische Kinder
motivate – Begeisterung schaffen;
motivieren
motivating force – Triebkraft
motivation – Motivation; Leistungswille
motivation technique –
Motivationstechnik
motivation theory – Motivationstheorie
motor mechanic –
Kraftfahrzeugmechaniker
move – Umzug
move-up candidate – Nachrückkandidat
moving power – Triebkraft
multi-purpose wage type –
Kombilohnart
multi-session business event –
Veranstaltung, mehrteilige
multigrade classroom –
Unterrichtsgruppen unterschiedlicher
Altersstufen
multilingual – mehrsprachig
multiple applicant – Mehrfachbewerber
multiple application –
Mehrfachbewerbung
multiple benefits – Mehrfachbezug
multiple birth – Mehrlingsgeburt
multiple choice (m.c.) –
Mehrfachauswahl
multiple choice exercise –
Mehrfachauswahlübung
multiple choice question –
Mehrfachauswahlfrage
multiple employment –
Mehrfachbeschäftigung
multiple payroll – Mehrfachabrechnung
multiple qualification –
Mehrfachqualifikation

multiple regression – multiple
Regression
multiply – multiplizieren
multiply handicapped child –
mehrfachbehindertes Kind
municipal school – städtische Schule
municipal schools inspector –
Stadtschulrat
municipality – Gemeinde
musical intelligence – musikalische
Intelligenz
mute – stumm
mutual – gegenseitig
mutual confidence – gegenseitiges
Vertrauen
mutual indemnity society –
Berufsgenossenschaft
mutual insurance association –
Versicherungsverein auf
Gegenseitigkeit
mutual insurance – Versicherung auf
Gegenseitigkeit

N

nagging – Nörgelei
name – Bezeichnung
name affix – Namenszusatz
name at birth – Geburtsname
name tag – Namensschild
narcissistic person – narzistischer
Mensch
narrow-minded – Tunnelblick;
engstirnig
national economics – Volkswirtschaft
national economy – Volkswirtschaft
national service – Wehrdienst
nationality – Nationalität;
Staatsangehörigkeit
natural – selbstverständlich
natural curiosity – natürliche Neugier
natural endowment – Anlage; natürliche
Begabung
natural leader – natürliche
Führungspersönlichkeit

natural sciences – Naturwissenschaften
natural scientist – Naturwissenschaftler
natural wastage – natürlicher Abgang
von Arbeitskräften
nearest school – nächstgelegene Schule
necessary – notwendig
necessity to take action –
Handlungsbedarf
need for training – Ausbildungsbedarf
needs – Bedarf
needs analysis – Bedarfsanalyse;
Bedarfsermittlung
needs assessment – Bedarfsanalyse
needs orientation – Bedarfsorientierung
negative flexitime balance –
Gleitzeitschulden; Zeitschulden
negative reply – Absage; abschlägiger
Bescheid
negative time recording –
Negativerfassung
neglect – vernachlässigen; versäumen
neglected child – Stiefkind
negligence – Fahrlässigkeit
negotiate – verhandeln
negotiating partner –
Verhandlungspartner
negotiation – Verhandlung
negotiation procedure –
Verhandlungsordnung;
Verhandlungsverfahren
neighbourhood – Nachbarschaft; Nähe
nepotism – Vetternwirtschaft;
Klüngelwirtschaft
nervous – nervös
nervous breakdown –
Nervenzusammenbruch
nervous disorder – nervöse Störung
nervous – Bammel haben
nervousness – Aufregung; Lampenfieber
net – nach allen Abzügen; netto; rein
net amount – Nettobetrag
net calculation – Nettoermittlung
net earnings – Nettobezüge
net income – Nettoeinkommen;
Nettoentgelt
net part – Nettoteil
net pay (payroll) – Auszahlungsbetrag
net payroll – Nettolohnabrechnung
net premium – Nettoprämie
net remuneration – Nettoentgelt
net report payroll – Nettoabrechnung

net retroactive accounting –
Nettorückrechnung
net salary – Nettobezüge
net wage – Nettolohn
network planning technique –
Netzplantechnik
neurotic – neurotisch
new employee – Neueinstellung
new formation – Neubildung
new member – neues Mitglied
new student – Studienanfänger
newcomer – Neuling
next higher ranking manager –
nächsthöherer Vorgesetzter
night school – Abendschule
night shift – Nachtschicht; Spätschicht
night shift bonus – Nachtschichtzulage
night shift heavy work –
Nachtschichtschwerarbeit
night shift regulations –
Nachtschichtgesetz
night-work bonus – Prämie für
Nachtarbeit
night-work – Nachtarbeit
night-work bonus –
Nachtarbeitszuschlag
no marks given – nicht benotet
no school – schulfrei
no-name course booking –
N.N.-Buchung
no-strike clause – Streikverbotsklausel
noise prevention – Lärmschutz
noise rating level –
Lärm-Beurteilungspegel
nominal value – Nominalwert
nominal wages – Nominallohn
nominate – ernennen; nominieren,
aufstellen
nomination – Berufung; Ernennung;
Nominierung, Aufstellung
nomination list – Wahlvorschlag(sliste)
non-academic – Nichtakademiker
non-competition clause –
Wettbewerbsklausel;
Wettbewerbsverbot
non-conformist – unangepasst;
Unangepasstheit
non-contributory – beitragsfrei
non-contributory period – Ausfallzeit
(RV)
non-creditable – anrechnungsfrei

non-criminal offence – Ordnungswidrigkeit

non-enrolled student – Hörer

non-executive director – beratende(r) Direktor(in); Berater

non-exempt personnel – tarifliches Personal

non-exempt staff – Tarifkreis

non-existence – Nichtvorliegen

non-forfeitability (of pension expectancy) – Unverfallbarkeit (bei Pensionsanwartschaft)

non-forfeitability certificate – Unverfallbarkeitsbescheinigung

non-forfeitability limit – Unverfallbarkeitsfrist

non-forfeitable – nicht verfallbar; unverfallbar

non-industrial accident – Nichtbetriebsunfall

non-liability – Haftungsausschluss

non-military service – Ersatzdienst

non-monetary bonus – Naturalzulage

non-monetary compensation – Sachbezüge; sonstige Firmennebenleistungen

non-pay scale – außer Tarif

non-pay scale employee – AT-Angestellter

non-pay-scale – außertariflich

non-productive time – Arbeitsausfall

non-profit educational establishment – gemeinnütziges Bildungswerk

non-profit making – gemeinnützig

non-recurring capital payment – einmalige Auszahlung des Rentenbetrages

non-recurring payments – Einmalbezüge

non-resident participant – Externe(r) eines Internats

non-salaried employee – gewerblich Tätiger; Lohnempfänger

non-salaried employee with a merit-based contract – Mitarbeiter mit einem Sonderleistungsvertrag

non-salaried employee with a special contract – Mitarbeiter mit einem Sonderlohnvertrag

non-standard salary – AT-Gehalt

non-taxable – steuerfrei

non-verbal communication – non-verbale Kommunikation

non-verbal intelligence – non-verbale Intelligenz

non-verbal learning – non-verbales Lernen

non-verbal response – non-verbale Antwort

non-wage labour costs – Lohnnebenkosten

non-waiting list booking – Nicht-Warteliste-Buchung

non-waiting list priority – Nicht-Warteliste-Priorität

non-work-related accident insurance – Nichtbetriebsunfall-Versicherungspflicht

non-working period – Nicht-Arbeit

non-working shift – Freischicht

nonsense word – Nonsenswort; sinnloses Wort

normal period of time for completion of studies – Regelstudienzeit

normal work – Normalarbeit

normal working day – Nicht-Feiertag

normal working time – Normalarbeitszeit

norm – Norm

not under notice to leave – ungekündigt

note – Aufzeichnung; Notiz; Vermerk

notice – bemerken; Kenntnisnahme

notice board – Anschlagtafel; Schwarzes Brett

notice of dismissal – Kündigung (Arbeitsverhältnis)

notice of termination – Kündigung (Arbeitsverhältnis)

notifiable accident at work – anzeigepflichtiger Arbeitsunfall

notification – Bekanntmachung; Mitteilung, Bescheid

notification code – Mitteilungskürzel

notification of absence – Abwesenheitsmitteilung

notification of cancellation of a course – Stornierungsmitteilung

notification of change in terms of employment – Änderungskündigung

notification of change of address or residence – behördliche Abmeldung

notification of illness – Krankmeldung

notification of pregnancy –
Schwangerschaftsmeldung
notification of sickness – Krankmeldung
noxious air – Immissionen
number of aliens – Ausländeranteil
**number of employees in terms of
full-time employment** –
Kapazitätskopfzahl (in Vollzeit
umgerechnete Kopfzahl)
number of graduates –
Akademisierungsgrad
number of pupils – Schülerzahl
number on roll – Schülerzahl
numbers – Stückzahl
numeracy – Rechenkenntnis
numerator – Zähler
nurse – Krankenschwester
nursery – Hort
nursery education – Vorschulbildung
nursery school – Kindergarten
nursery school child – Vorschulkind
nursing insurance – Pflegeversicherung

O

O-Level – mittlere Reife
oath – Eid
obedience – Gehorsam
object – Einwendungen machen;
Gegenstand
object abbreviation – Objektkürzel
object name – Objektbezeichnung
object on loan – Leihgabe
object type – Objekttyp
object-oriented – objektorientiert
objectification – Versachlichung
objection – Reklamation; Widerspruch
objective – Hauptziel; Ziel;
unbeeinflusst; objektiv
objective of course – Kursziel
objectivity – Objektivität; Sachlichkeit
obligation – Pflicht; Verpflichtung
obligation to contribute –
Beitragspflicht

obligation to give information –
Auskunftspflicht
obligation to hire –
Übernahmeverpflichtung
obligation to inform employees –
Informationspflicht
obligation to keep peace –
Friedenspflicht
obligation to make assistance payments
– Unterstützungspflicht
**obligation to make social insurance
payments** – Sozialversicherungspflicht
obligation to register – Meldepflicht
obligatory – obligatorisch; Pflicht...
observation – Beobachtung
observational skill –
Beobachtungsfähigkeit
observe – beobachten
obsession – Besessenheit
obsolescence – Überalterung
obsolete – obsolet
obstinate – eigensinnig
obvious – klar; selbstverständlich
occupancy – Belegung; Besetzung
occupancy time – Belegungszeit
occupation – Beruf; okkupation
occupational – beruflich; Berufs...;
fachlich
occupational accident – Arbeitsunfall;
Betriebsunfall
occupational disability –
Berufsunfähigkeit
occupational disease – Berufskrankheit
occupational future – berufliche
Aussichten
occupational group – Berufsgruppe
occupational hazard – Berufsrisiko
occupational history – Berufsweg
occupational illness – Berufskrankheit
occupational information –
Berufsinformation
occupational medicine – Arbeitsmedizin
occupational prestige –
berufsverbundenes Prestige
occupational requirements –
Berufsanforderungen
occupational safety – Arbeitssicherheit
occupational seniority – Dauer der
Betriebszugehörigkeit; Dienstalter
occupational structure – Berufsstruktur

occupational therapist –
Arbeitstherapeut;
Beschäftigungstherapeut
occupational training –
Berufsausbildung; Berufsbildung
occupied position – Planstelle, besetzte
occupy – beschäftigen
ocean freight – Seefracht
of age – volljährig, mündig
of school age – schulpflichtig
off-setting provision –
Anrechnungsklausel
off-site work – Dienstgang
off-the-job training – Ausbildung, die
nicht am Arbeitsplatz erfolgt
offer – anbieten
offer – Angebot
office – Amt; Büro...; Sekretariat
office administrator – Bürokauffrau,
Bürokaufmann
office boy – Bürogehilfe; Bürogehilfin
office duty – Innendienst
office experience – Büroerfahrung
office hours from...to... – Bürostunden
von ... bis; Parteienverkehr von ... bis ...
Office of Administration – Büro der
Leitung
office of payments (BrE) – Zahlstelle
office practice – Büroerfahrung
office worker – Angestellte(r);
Tarifangestellte(r), kaufmännische(r)
Angestellte(r)
**official absence due to temporary
reassignment** – Freistellung
official certificate – staatliches
Prüfungszeugnis
official holiday – gesetzlicher Feiertag
official matter – dienstliches Anliegen
official regulation – Dienstordnung
officialdom – Beamtentum
officialism – Beamtenmentalität
officially qualified technician – staatlich
geprüfter Techniker
offset benefits – Leistungen anrechnen
OHP – Arbeitsprojektor;
Tageslichtprojektor
old-age and survivor's insurance –
Alters- und Hinterbliebenenversorgung
old-age credit – Altersguthaben
old-age exemption – Altersentlastung

old-age exemption amount –
Altersfreibetrag
old-age pension – Altersrente;
Altersruhegeld
old-age pensioner – Pensionär;
Rentenempfänger; Rentner
omit – auslassen; weglassen
on behalf of (BrE) – im Auftrag von; im
Interesse von
on business – geschäftlich
on call – Rufbereitschaft
on probation – probeweise
on-going education – Fortbildung;
Weiterbildung; nachschulische
Angebote in der Erwachsenenbildung
on-going vocational training –
berufsbegleitender
Fortbildungslehrgang
on-the-job safety – Sicherheit am
Arbeitsplatz
on-the-job safety expert –
Arbeitssicherheitsexperte
on-the-job safety organisation –
Arbeitssicherheitsorganisation
on-the-job training – Ausbildung am
Arbeitsplatz; Lernen am Arbeitsplatz
once-for-all payment – Einmalzahlung
once-only commutation –
Kapitalzahlung (einmalig)
once-only payments/deductions –
Bezüge/Abzüge, einmalige
one-day trip – Reise, eintägige
one-man company –
Ein-Personen-Unternehmen
one-parent family – Alleinerziehende(r)
on-site training – Ausbildung vor Ort
one-time employee compensation –
einmal gezahltes Arbeitsentgelt
one-time premium – Einmalbeitrag (z.B.
zur Versicherung)
one-time remuneration – Einmalentgelt
one-way communication –
Einweg-Kommunikation
one's very own – ureigen
only natural – Selbstverständlichkeit
open – eröffnen
open access – offene Veranstaltung
open book examination –
Bücherbenutzung bei der Prüfung
erlaubt
open day – Tag der offenen Tür

open from... to... – geöffnet von... bis...
open shop – nicht
gewerkschaftspflichtiger Betrieb
(system where an organisation agrees to
employ both union and non-union
members)
open university – offene Universität
open university course – Fernstudium
open-minded – aufgeschlossen
openness – Offenheit
opening time – Öffnungszeit
operating costs – Betriebskosten
operating instructions –
Bedienungsanweisung;
Betriebsanleitungen
operating time – Betriebszeit
operating unit – Betriebsteil
operation – Operation
operational – betriebsfähig; operational
operational accounting –
Betriebsabrechnung
operational learning objective –
operationales Lernziel
operational planning – operative
Planung
operational risk – Betriebsgefahr
opinion poll – Meinungsbefragung
opportunity – Chance; Gelegenheit;
Möglichkeit
opportunity for promotion –
Beförderungschance
opt out – aussteigen; sich dagegen
entscheiden
optimal filling of vacancies – optimale
Stellenbesetzung
optimum – Kapazität, optimale
option – freie Entscheidung
optional question – wahlfreie Frage
option right – Optionsrecht
optional subject – Wahlfach
oral examination – mündliche Prüfung
oral report – mündlicher Bericht
oral skill – Ausdrucksfähigkeit
order – Ordnung; Anweisung; anordnen
order processing centre –
Abwicklungszentrum, Auftragszentrum
order to abstain from alcohol –
Alkoholverbot
order-related workforce planning –
Einsatzplanung, auftragsbezogene
ordinary level – mittlere Reife

ordinary notice of dismissal –
Kündigung, ordentliche
ordinary shareholders' meeting –
ordentliche Hauptversammlung
organigram – Organigramm
organization – Organisation
Organization and Planning –
Organisation und Planung
organization chart – Organisationsplan;
Organisationsschema
organization design –
Organisationsdesign;
Organisationsgestaltung
organization development (OD) –
Organisationsentwicklung (OE)
organization methods –
Organisationstechnik
organizational ability –
Organisationsfähigkeit;
Organisationsgabe
organizational assignment –
Organisationszuordnung
organizational development planning –
Organisationsentwicklungsplanung
organizational form –
Organisationsform
organizational key –
Organisationsschlüssel
organizational level –
Organisationsebene
organizational management –
Organisationsmanagement
organizational needs – organisatorische
Bedürfnisse
organizational personnel structure –
Personalstruktur, organisatorische
organizational plan – Organisationsplan
organizational planning –
Organisationsplanung
organizational purpose –
Geschäftszweck; Unternehmenszweck
organizational structure –
Organisationsstruktur
organizational talent –
Organisationstalent
organizational unit –
Organisationseinheit
organizer – Veranstalter
orientation – Einarbeitung; Orientierung
orientation aid – Orientierungshilfe

orientation and integration of new employees – Orientierung und Integration neuer Mitarbeiter

orientation level – Orientierungsstufe

orientation program – Orientierungsprogramm

origin indicator – Herkunftskennzeichen

original – Originales; ureigen; eigenständig

original income – Anfangseinkommen

original payroll run – Originalabrechnung

orphan's allowance – Waisenrente

orthography – Rechtschreibung

other paid absences – Fehlzeit, sonstige bezahlte

other payments – Bezüge, sonstige

out of order – außer Betrieb; defekt

out one-self – sich bekennen

out-of-hospital expenses – ambulante Krankheitskosten

out-patient treatment – ambulante Behandlung

outcome (of a negotiation) – Ergebnis

outdatedness of knowledge – Wissensverfall

outgoing wage type – Abflusslohnart

outing – Ausflug

outline – Entwurf

outpayment – Auszahlung

outplacement – Outplacement

output – Leistung

outside company – Fremdfirma

outsource – ausgründen; ausgliedern (The procuring of services or products from an outside supplier or manufacturer in order to cut costs.)

outstanding balance – Außenstände; offener Betrag

outward journey – Hinfahrt; Hinreise

outwork – Heimarbeit

outworker – Heimarbeiter

over-achiever – Erfolgsmensch; leistungsorientierter Mensch

over-ambitious person – Streber

over-qualified – überqualifiziert

overall achievement – Gesamtleistung

overall responsibility – Gesamtverantwortung

overall staffing schedule – Gesamtstellenbesetzungsplan

overall wage scale – Lohnrahmentarif

overcoming stress – Stressbewältigung

overestimation – Überschätzung

overhead projector – Arbeitsprojektor; Tageslichtprojektor

overheads – Gemeinkosten

overlap – Überlagerung

overlearning – Überlernen

overload – Überbelastung

overnight stay – Übernachtung

overseas allowance – Überseezulage

overseas bonus – Überseezuschlag

overseas position – Auslandsabordnung; Auslandseinsatz

overstep the boundaries – Kompetenzen überschreiten

overstrain – Überlastung

overstress – Überlastung

overtask – überfordern

overtime – Mehrarbeit; Überstunden

overtime approval – Mehrarbeitsgenehmigung

overtime ban – Überstundenverbot

overtime bonus – Mehrarbeitszuschlag

overtime break – Mehrarbeitspause

overtime compensated by time in lieu of – Mehrarbeitsausgleich

overtime compensation – Mehrarbeitsverrechnung; Mehrzeitausgleich

overtime pay – Überstundenbezahlung; Überstundenlohn

overtime perspective – Mehrarbeitsbetrachtung

overtime premium – Überstundenzuschlag

overview – Übersicht

overview of absences – Abwesenheitsübersicht

overwork – überarbeiten

owner – Inhaber

owner-manager – Geschäftsinhaber

ownership – Eigentum(srecht), Besitzerschaft

Oxbridge – Oxbridge

P

pace – Geschwindigkeit; Tempo
paid absence – Abwesenheit, bezahlt
paid labour – Lohnarbeit
paid public holiday – Feiertag, bezahlter
paid release – bezahlte Freistellung
paid sick leave – Krankheit, bezahlte
paid statutory holidays – bezahlte
  gesetzliche Feiertage
painstaking – sorgfältig
pair formation – Paarbildung
pair type – Paartyp
pal – Kumpel
panel – Expertengruppe
panel discussion – Podiumsgespräch
panel member – Diskussionsteilnehmer
paper – Klausur
paper board – Papiertafel
papers – Unterlagen
paradigm – Musterbeispiel; Paradigma
parameters governing influence –
  Einflussgrößen
paraphrase – interpretieren
parent – Elternteil;
  Erziehungsberechtigter
parent company – Stammhaus
parent counselling – Elternberatung
parent-teacher association –
  Eltern-Lehrer-Ausschuss
parental authority – elterliche Gewalt
parental care – elterliches Sorgerecht
parental choice – elterliche Wahl
parental contacts – Elternkontakte
parental contribution – Elternbeitrag
parental influence – elterlicher Einfluss
parental involvement –
  Elternbeteiligung
parental leave – Erziehungsurlaub
parents – Eltern
Parents' Council – Elternbeirat
parents' night – Elternabend
parity – Gleichheit; Parität
parity (of votes) – Stimmengleichheit
parochial – engstirnig
part – Bestandteil
part of speech – Wortart
part percentage rate – Teilprozentsatz

part timer – Teilzeitarbeiter; Teilzeitkraft
part-time compulsory schooling –
  Teilzeitschulpflicht
part-time course – Teilzeitkurs
part-time employees – Beschäftigte,
  geringfügig
part-time employment – Halbtagsarbeit;
  Teilzeitarbeit
part-time employment for old-age
  pensioners – Altersteilzeitarbeit
part-time job – Halbtagsarbeit;
  Teilzeitarbeit
part-time labour market – Markt für
  Teilzeitkräfte
part-time schooling – Teilzeitunterricht
part-time teacher – Teilzeitlehrer
part-time work – Halbtagsarbeit;
  Teilzeit; Teilzeitarbeit;
part-time work for elder co-workers –
  Teilzeitarbeit für ältere Mitarbeiter
part-time worker – Teilzeitarbeiter;
  Teilzeitkraft
part-time working model –
  Teilzeitmodell
part-time workplace –
  Teilzeitarbeitsplatz
partial interval – Teilintervall
partial remuneration – Teilentgelt
partial retirement – Altersteilzeit
partial salary – Teilgehalt
partially blind – sehbehindert
participant – Kursbesucher; Teilnehmer
participant in a course – Kursteilnehmer
participant profile – Teilnehmerprofil
participate in – besuchen; teilnehmen
  (an)
participation – Beteiligung; Mitwirkung;
  Teilnahme; Partizipation
participation in turnover –
  Umsatzbeteiligung
particulars of employment – Angaben
  zur Tätigkeit
partner – Geschäftsteilhaber; Partner;
  Teilhaber, Gesellschafter
partner to collective bargaining –
  Tarifpartner
partnership – Partnerschaft
pass an exam – eine Prüfung bestehen
pass degree – Universitätsabschluss ohne
  Auszeichnung
pass rate – Bestehensquote

pass with distinction – Universitätsabschluss mit Auszeichnung
pass/no pass – nicht benotet
passed – bestanden
passenger – Mitfahrer
passion – Leidenschaft
passionately – leidenschaftlich
passive – passiv
passive learning – passives Lernen
passive smoking – Passivrauchen
passive vocabulary – passiver Wortschatz
pastime – Zeitvertreib
paternity – Vaterschaft
patriarchal management behaviour – patriarchalisches Führungsverhalten
pattern – Muster
pattern of behaviour – Verhaltensmuster
pattern of learning – Lernmuster
pattern-setting model – Musterabkommen; Mustervereinbarung
pause – Pause
pay – Lohn; Besoldung
pay (for) – bezahlen; entlohnen; vergüten
pay effective – lohnwirksam
pay element – Gehaltsbestandteil
pay increase – Gehaltssteigerung
pay increase programme – Gehaltserhöhungsprogramm
pay pause – Lohnpause
pay rate – Lohnsatz
pay remittance – Gehaltsüberweisung
pay scale – Tarif
pay scale area – Tarifgebiet
pay scale employee – Tarifangestellte(r)
pay scale group – Tarifgruppe
pay scale grouping – Gehaltsgruppenbezeichnung
pay scale indicator – Tarifkennzeichen
pay scale jump due to age – Alterssprung
pay scale jump due to seniority – Dienstaltersprung
pay scale level – Tarifstufe
pay scale reclassification – Umstufung, tarifliche
pay scale structure – Tarifstruktur
pay scale table – Tariftabelle
pay scale type – Tarifart
pay scale valuation – Tarifbewertung

pay slip – Lohnzettel; Verdienstübersicht; Gehaltsabrechnung, Gehaltszettel
pay slip limit – Lohnzettelgrenze
pay statement – Lohnbescheinigung
pay survey – Lohn- und Gehaltsvergleich
pay tribute to someone – Anerkennung zollen.(jdm.)
pay-as-you-earn – Quellenabzugsverfahren
pay-day – Zahltag
PAYE – Quellenabzugsverfahren
payee – Empfänger
paying office – Zahlstelle
payment – Bezahlung, Entgelt
payment amount – Zahlungsbetrag
payment date – Zahlungszeitpunkt
payment during illness – Fortzahlung im Krankheitsfall
payment in kind – Deputat; Naturalleistung(en); Sachleistung
payment in lieu of holidays – Barabgeltung
payment method – Bezugsmethode
payment of bonuses – Vergütung von Zuschlägen
payment of contributions – Beitragsentrichtung
payment of overtime – Vergütung von Überstunden
payment of remuneration – Entgeltzahlung
payment of retrospective contributions – Nachversicherung
payment of salaries – Gehaltszahlung
payment of wages – Lohnauszahlung; Löhnung; Lohnzahlung
payment on account – Abschlagszahlung; Anzahlung; Ratenzahlung
payment period – Auszahlungsperiode
payment principle – Bezugsprinzip
payment provision – Vergütungsregelung
payment scheme – Entgeltsystem
payments – Bezüge
payments/deductions – Bezüge/Abzüge
payments from retroactive accounting – Lohnnachgenuss
payments spanning more than one year – Bezüge, mehrjährige
payments system – Entlohnungsmethode

payout – Auszahlung
payroll – Gehaltsliste; Lohn- und
  Gehaltsliste
payroll account – Lohnkonto
payroll account form –
  Lohnkontoformular
payroll accounting – Lohn- und
  Gehaltsabrechnung;
payroll accounting area –
  Personalabrechnungskreis
**payroll accounting for reduced
  working** – Kurzarbeitsabrechnung
**payroll accounting for salaried
  employees** – Gehaltsabrechnung
payroll accounting for the hourly paid
  – Lohnabrechnung
payroll accounting parameter –
  Personalabrechnungsparameter
payroll accounting system –
  Abrechnungssystem
payroll administrator – Sachbearbeiter
  für Abrechnung
payroll basis – Abrechnungsbasis
payroll calendar – Abrechnungskalender
payroll category – Abrechnungstyp
payroll clerk – Abrechner
payroll cluster – Abrechnungscluster
payroll control record –
  Abrechnungsverwaltungssatz
payroll customizing –
  Abrechnungsanpassung
payroll cycle – Abrechnungszyklus
payroll data – Daten,
  abrechnungsrelevante
payroll day – Abrechnungstag
payroll deductions – Lohnabzüge
payroll department – Lohnbüro;
  Lohnverwaltung
payroll driver – Abrechnungstreiber
payroll form – Abrechnungsformular
payroll fringe costs – Lohnnebenkosten
payroll hours – Abrechnungsstunden
payroll journal – Lohnjournal
payroll month – Abrechnungsmonat
payroll office – Lohnbüro
payroll officer – Abrechner
payroll past –
  Abrechnungsvergangenheit
payroll period –
  Lohnabrechnungszeitraum
payroll period – Abrechnungsperiode

payroll program –
  Abrechnungsprogramm
payroll record(s) (AmE) –
  Lohnunterlagen
payroll relevancy –
  Abrechnungsrelevanz
payroll result – Abrechnungsergebnis
payroll rule – Abrechnungsregel
payroll schema – Abrechnungsschema
payroll sheet – Gehaltsliste; Lohn- und
  Gehaltsliste
payroll subunit – Abrechnungskreis
payroll summary (AmE) –
  Lohnübersicht
payroll system (AmE) –
  Personalabrechnung
payroll tax (AmE) –
  Arbeitgeberlohnanteil
payroll type – Abrechnungsart
payroll unit – Abrechnungseinheit
payroll variant – Abrechnungsvariante
peak load – Spitzenbelastung
peak performance – Spitzenleistung;
  Höchstleistung
peak time – Arbeitsspitze
pedagogue – Pädagoge
pedagogy – Erziehungslehre;
  Erziehungswissenschaft
pedant – Pedant(in)
pedantic – pedantisch
pedantry – Pedanterie
peer – Ebenbürtige(r); Gleiche(r)
peer assessment – Bewertung durch
  Ebenbürtige
peer group – Peer-Gruppe
peer pressure – Druck unter
  Ebenbürtigen
pegged wages – Indexlohn
penalty – Bußgeld (bei Verstößen); Strafe
penalty for breach of contract –
  Vertragsstrafe
pencil – Bleistift; Stift
pencil case – Federmäppchen
pencil sharpener – Spitzer
penetration – Durchdringung
**pension, disability and surviving
  dependents' provision** – Alters- und
  Hinterlassenenversicherung
**pension adjustment to the national
  average wage** – Rentenanpassung;
  Ruhegehaltsanpassung

pension advisory department –
Rentenauskunft
pension applicant – Rentenantragssteller
pension application – Rentenantrag
pension as of.... – Rentenbeginn
pension benefit account –
Pensionsleistungskonto
pension computation –
Rentenberechnung
pension entitlement – Pensionsanspruch
pension expectancy –
Pensionsanwartschaft;
Rentenanwartschaft
pension for occupational invalidity –
Berufsunfähigkeitsrente
pension from the social insurance –
Versichertenrente
pension fund – Pensionskasse
**pension fund of a professional
association** –
berufsgenossenschaftliche
Pensionskasse
pension increment –
Rentensteigerungsbetrag
pension insurance fund –
Rentenversicherung
pension insurance institution –
Rentenversicherungsträger
pension insurance number –
Rentenversicherungsnummer
pension liability –
Ruhegeldverpflichtung
pension notice – Rentenbescheid
pension paid abroad – Auslandsrente
pension payment – Pensionszahlung
pension reduction – Rentenkürzung
pension reserves – Pensionsrückstellung
pension right – Rentenanspruch
pension rights for marital partners –
Ehegattenversorgungsanspruch
pension scale – Pensionsband;
Pensionsstaffel
pension scheme – Ruhegeldordnung,
Rentenversicherung
pension year – Rentenjahr
pension-paying institution –
Rententräger; Versorgungsträger
pensionable age – Pensionsalter;
Rentenalter
pensionable income – ruhegeldfähiges
Einkommen

pensioner – Pensionär;
Rentenempfänger; Rentner
pensioning warrant – Direktzusage;
Pensionszusage
pensions advisory department –
Rentenberatungsstelle
pensions and related benefits –
Versorgungsbezüge
people-centred leadership –
kooperativer Führungsstil
peptalk – Anfeuerung
per capita expenditure –
Pro-Kopf-Ausgaben
per diem sickness indemnity –
Krankentagegeld
per-diem (travel) – Pauschale
per-diem allowance – Tagegeld;
Tagessatz
per-diem allowance insurance –
Krankentagegeldversicherung
percentage of foreigners –
Ausländeranteil
percentage rate of increase –
Erhöhungsprozentsatz
perception – Wahrnehmung
perform – abschneiden; leisten
performance – Abschneiden; Leistung
performance ability – Befähigung;
Fähigkeit; Leistungsfähigkeit
performance appraisal –
Mitarbeiterbeurteilung
performance appraisal system –
Leistungsbeurteilungssystem
performance comparison –
Leistungsvergleich
performance criterion –
Leistungskriterium
performance deficit –
Leistungsrückstand
perfomance evaluation –
Leistungsbeurteilung
performance expense report –
Leistungsabrechnung
performance guarantee –
Leistungsgarantie
performance linked – leistungsorientiert
performance linked bonus –
Leistungszulage
performance measures –
Leistungsmaßstäbe; Maßstab für
Leistung

performance principle –
Leistungsprinzip
performance readiness –
Leistungsbereitschaft
performance related compensation –
leistungsbezogene
Einkommensgestaltung
performance related income –
leistungsgerechtes Einkommen
performance related salary promotion
– leistungsbedingte Gehaltserhöhung
performance results – Leistungsergebnis
performance system – Leistungssystem
performance test – Leistungstest
performance under stress –
Belastbarkeit
performance-related –
leistungsabhängig
performer – Leistungsträger
period – Stunde; Unterrichtsstunde
period (BrE) – Unterrichtseinheit
period abroad – Auslandsaufenthalt
period of accrual – Zuflussperiode
period of continued pay –
Lohnfortzahlungsfrist
period of employment –
Beschäftigungszeit;
Betriebszugehörigkeit
period of notice – Kündigungsfrist
period of vocational adjustment –
Einarbeitungszeit
period parameter – Periodenparameter
period split – Zeitraumsplitt
period time sheet –
Multimomentaufnahme
period work schedule –
Periodenarbeitszeitplan
period-based population –
Bewegungsmasse
periodically recurring special payment
– Sonderzahlung, periodisch
wiederkehrende
periods of employment –
Beschäftigungszeiten
peripatetic – umherlaufend
peripatetic teacher – Lehrer, der an
mehreren Schulen unterrichtet
peripheral condition – Randbedingung
**Permanent Committee of the Ministers
of Education** –
Kultusministerkonferenz

permanent employment –
Dauerarbeitsvertrag; feste Anstellung;
Dauerbeschäftigung
permanent incapacity to work –
Erwerbsunfähigkeit
permanent job contract –
Dauerarbeitsvertrag; feste Anstellung
**permanent readiness (willingness) to
learn** – ständige Lernbereitschaft
permanent residence – Wohnsitz,
ständiger
permanent residence card –
Daueraufenthaltsgenehmigung
permanent staff – Stammpersonal
permanent travel costs –
Dauerreisekosten
permanent work contract –
unbefristeter Arbeitsvertrag
permanent work relationship –
Arbeitsverhältnis, unbefristetes
permissible – zulässig
permission – Erlaubnis; Genehmigung,
Bewilligung
permissive atmosphere – tolerante
Atmosphäre
permit – gestatten
perplexity – Ratlosigkeit
perseverance – Ausdauer;
Durchhaltefähigkeit
persistance – Beharrlichkeit
person being retrained – Umschüler
person employed – Beschäftigter
person entitled to child allowance –
Kindergeldberechtigter
person group – Personengruppe
person in charge – Sachbearbeiter;
Verantwortliche(r)
**person interested solely in his own
subject** – Fachidiot
**person subject to compulsory
insurance** – Versicherungspflichtige(r)
person responsible – Verantwortliche(r)
person responsible for time recording –
Zeiterfassungsbeauftragter
**person who has commercial power of
attorney** – Handlungsträger
person who looks after someone –
Betreuer(in)
person with security clearance –
Geheimnisträger

person writing post-doctoral thesis –
Habilitand (This thesis is necessary in
Germany to qualify as a university
lecturer.)

personal absence calendar –
Fehlzeitenkalender, persönlicher

personal aptitude – persönliche Eignung

personal assessment basis – persönliche
Bemessungsgrundlage

personal authority – personelle
Weisungsbefugnis

personal data – Angaben zur Person;
Daten zur Person

personal discussion –
Vier-Augen-Gespräch

personal effects – persönliche Habe

personal environment – Lebensumfeld

personal history – Lebenslauf

personal leave – Pflegeurlaub

personal mastery – persönliche
Meisterschaft

personal matter – persönliches Anliegen

personal predisposition – persönliche
Voraussetzung

personal responsibility – persönliche
Verantwortung;
Eigenverantwortlichkeit

personal shift plan – Dienstplan,
persönlicher

personal shift schedule – Schichtplan,
persönlicher

personal tax allowance – Freibetrag,
persönlicher

personal tax exemption for retirement
– Versorgungsfreibetrag

personality – Persönlichkeit

personality characteristic –
Persönlichkeitsmerkmal

personality formation –
Persönlichkeitsbildung

personality test – Persönlichkeitstest

personality type – Persönlichkeitstyp

personally responsible –
eigenverantwortlich;
Eigenverantwortung

personnel – Belegschaft; Mitarbeiterstab;
Personal

personnel administration –
Personaladministration;
Personalverwaltung

Personnel Advisory Office – Referat
Personal (RefPers)

personnel affairs of top management –
Personalangelegenheiten der
Firmenleitung

personnel allocation –
Personaldisposition

personnel and social expenses –
Personal- und Sozialaufwand

personnel appraisal sheet –
Beurteilungsbogen

personnel area – Personalbereich

**personnel assessment and trend
procedures** – Personalbemessung und
Trendverfahren

personnel assistant for a special unit –
Bereichsbetreuer

personnel calculation rule –
Abrechnungsregel

personnel calculation schema –
Personalrechenschema

personnel change notifications –
Personalveränderungsmitteilung

personnel control record –
Personalverwaltungssatz

personnel controlling –
Personal-Controlling

personnel cost assignment –
Personalkostenzuordnung

personnel cost planning –
Personalkostenplanung

personnel costs – Personalkosten

personnel country grouping –
Ländergruppierung für Personal

personnel cutback –
Personalstandreduzierung

personnel data – Personaldaten

personnel decisions –
Personalentscheidungen

personnel department –
Personalabteilung

personnel development –
Personalentwicklung

personnel development plan –
Personalentwicklungsplan

personnel development planning –
Personalentwicklungsplanung

personnel disposition –
Personaldisposition

personnel division (AmE) –
Hauptpersonalabteilung

personnel evaluation –
Personalbewertung
personnel event – Personalmaßnahme
personnel expense report –
Personalabrechnung
personnel expenses –
Personalaufwendungen
personnel file – Personalakte
personnel group – Personalgruppe
personnel guardianship –
Personalbetreuung
personnel information system –
Personalinformationssystem
personnel intensive – personalintensiv
personnel inventory –
Personalausstattung; Personalbestand
personnel layoff – Personalfreisetzung
personnel management –
Führungskultur; Personalführung
personnel manager –
Personalführungskraft
personnel number – Personalnummer
personnel office – Personalbüro
personnel officer – Personalreferent;
Personalsachbearbeiter
personnel order – Personalauftrag
personnel organization –
Personalorganisation
personnel placement – Personaleinsatz
personnel plan – Personalplan
personnel planning – Personalplanung
personnel policy – Personalpolitik
personnel questionnaire –
Personalfragebogen
personnel re-deployment – Umsetzung
(beim Personaleinsatz)
personnel record – Personalakte;
Personalunterlagen
personnel recruitment – Anwerbung;
Personalbeschaffung;
Personalakquisition
personnel report – Personalbericht
personnel reporting system –
Personalberichterstattungssystem
personnel responsibility –
Personalverantwortung
personnel review – Beurteilung
personnel review procedure –
Beurteilungsverfahren,
Beurteilungssystem
personnel situation – Personaldecke

personnel strategy – Personalstrategie
personnel structure – Personalstruktur
personnel survey – Personalausstattung
personnel tasks – personalfachliche
Aufgaben
personnel work – Personalarbeit
persons entitled to imputation credit –
Person, anrechnungsfähige
persons leaving – Austritte, Abgang
perspective – Perspektive
persuade – überreden; überzeugen
persuasiveness – Überzeugungsfähigkeit,
Uberzeugungskraft
pertinent – einschlägig
PhD – Dr. phil.
phlegmatic person – Phlegmatiker(in)
phobia-ridden person – phobischer
Mensch
phoney – Blender
physical ability – körperliche Fähigkeit
physical and mental development –
körperliche und geistige Entwicklung
physical defect – körperlicher Schaden
physical education – Sport;
Leibesübungen
physical examination – Musterung
physical science – Naturwissenschaften
physically handicapped –
körperbehindert
physicist – Physiker
physics – Physik
physiological needs – physiologische
Bedürfnisse
picket – Streikposten; Streikwache halten
"pick-up" method – Methode der
selbständigen Einarbeitung
picture – Bild
piece-worker – Akkordarbeiter
piecerate – Akkordlohn,
Akkordgrundlohn, Akkordrichtsatz,
Stücklohnsatz
piecerate bonus – Akkordprämie
piecerate formula – Akkordsatz
piecerate work – Geldakkord
piecework – Akkord
piecework area – Akkordbereich
piecework pay – Akkordbezahlung;
Gedingelohn
piecework system –
Akkord(lohn)system;
Stück(lohn)system

piecework wage – Akkordlohn, Akkordgrundlohn, Akkordrichtsatz
pilot course – Pilotkurs
pilot program – Pilotprogramm
pilot project – Pilotprojekt
place – vermitteln
place (on a course) – Platz (Kurs)
place of birth – Geburtsort
place of employment – Arbeitsstätte
place of training – Ausbildungsstelle
place of work – Arbeitsort; Arbeitsstätte; Betriebsstätte
placed on waiting list – zurückgesetzt
placement and promotion of the employee – Einsatz und Förderung des Mitarbeiters
placement of a child – Unterbringung eines Kindes
placement officer – Vermittler
placement test – Einstufungstest
placing an order – Auftragsvergabe
placing service – Arbeitsvermittlung
plagiarism – Ideenklau
plain – klar
plaintiff – Antragsteller; Kläger
plan – Plan
plan for the promotion of women – Frauenförderplan
plan of instruction – Unterweisungsplan
plan scenario – Planungsversion
plan status – Planstatus
plan to fill vacancies – Besetzungsplan; Stellenbesetzungsplan
plan version – Planvariante
planned – geplant
planned hour – Sollstunde
planned labor costs – Sollohnkosten
planned pay – Sollbezüge
planned remuneration – Sollbezahlung
planned time – Planzeit
planned time pair – Sollpaar
planned work – Sollarbeit
planned working hour – Sollarbeitsstunde
planned working time – Sollarbeitszeit
planning aids – Planungshilfe
planning group – Planungsgruppe
planning of capacity – Kapazitätsplanung

planning of promotion – Entwicklungsplanung; Förderungsplanung
planning premise – Planungsprämisse; Planungsvorgabe
planning technique – Planungstechnik
plant – Betrieb; Werk
plant category – Werktyp
plant closure – Betriebsstillegung; Werkschließung
plant data collection – Betriebsdatenerfassung
plant fire department – Werksfeuerwehr
play group – Spielgruppe
plant holidays – Betriebsurlaub
plant ID card – Werksausweis
plant magazine – Werkzeitschrift
plant mechanic – Anlagenmechaniker
plant section – Betriebsteil
plant security – Werkschutz
plant shut-down – Betriebsferien
plant site – Betriebsgelände
plant subject to approval – genehmigungspflichtige Anlage
play time – große Pause
plant tour – Betriebsbesichtigung
play school – Kindergarten
play truant – die Schule schwänzen
playground – Pausenhof; Schulhof
plenary session – Plenarsitzung
poach employees – Mitarbeiter abwerben
poaching – Mitarbeiterabwerbung
poaching campaign – Abwerbekampagne
pocket calculator – Taschenrechner
pocket money – Taschengeld
point in time – Zeitpunkt
point of friction – Reibungspunkt
point rating system – Punktebewertungssystem
point-in-time population – Bestandsmasse
pointer – Zeigestock
poisoning – Vergiftung
policy month – Versicherungsmonat
policy year – Versicherungsjahr
polite – höflich
politeness – Höflichkeit
pollution control technician – Umwelttechniker
polyglot – vielsprachig, mehrsprachig

polytechnic – Polytechnikum;
  polytechnische Hochschule
Polytechnic College – Fachhochschule
  (BRD)
pool – Vorrat
pool of junior executives –
  Nachwuchspotential
poor concentration – konzentrationsarm
poor learner – lernschwach
population growth –
  Bevölkerungswachstum
portal-to-portal pay – Fahrtkosten
porter – Pförtner
portfolio planning – Portfolio-Planung
portion of wages – Lohnanteil
position – Position; Stelle
position authorized (in the budget) –
  Planstelle
position becoming vacant – Freiwerden
  einer Stelle
position description –
  Aufgabenbeschreibung;
  Planstellenbeschreibung
position hierarchy –
  Planstellenhierarchie
position of trust – Vertrauensstellung
position title – Funktionsbezeichnung;
  Positionsbezeichnung
positive flexitime balance –
  Gleitzeitguthaben; Zeitguthaben;
  Gleitsaldoüberschuss
positive reply – Zusage
positive time recording –
  Positivdatenerfassung; Positiverfassung
possibility – Chance; Gelegenheit;
  Möglichkeit
post – Position; Stelle
post-doctoral lecturing qualification –
  Habilitation
post-doctoral thesis –
  Habilitationsschrift (required for
  qualification as a university lecturer)
post-graduate course – Studium zur
  Erlangung eines höheren akademischen
  Grades
post-graduate studies – weiterführende
  Studien
post-secondary course – Hochschulkurs
post-secondary Technical College –
  Fachhochschule (BRD)

postal giro account number –
  Postgironummer
poster advertising – Plakatwerbung
postpayment of contributions –
  Beitragsnachentrichtung
postpone – aufschieben
postponement – Aufschub
potential – Potential
potential assessment –
  Potentialbeurteilung
potential earnings –
  Verdienstmöglichkeiten
potential for success – Erfolgpotential
power – Autorität; Macht; Vollmacht;
  Machtbefugnis
power electrician – Starkstromelektriker
**power engineering electronics
  technician** – Energieelektroniker
power of attorney – Vollmacht; Prokura
power of concentration –
  Konzentrationsfähigkeit
power of convincing –
  Durchsetzungsvermögen
power of expression –
  Ausdrucksfähigkeit;
  Ausdrucksvermögen
power of hearing – Hörvermögen
power of perception –
  Wahrnehmungsfähigkeit
power of procuration – Prokura
powerful (e.g. computer) –
  leistungsfähig
power of imagination –
  Vorstellungsvermögen
PR – öffentliche Beziehungen; Pflege der
  guten Beziehungen
practical – Praktikum; versiert
practical activity – praktische Tätigkeit
practical experience – fachpraktische
  Kenntnisse; praktische Erfahrung;
  Praxiserfahrung
practice – einüben
practice period – Übungsperiode
practical person – Praktiker
practical studies in a company –
  Firmenpraktikum
practical subject matters – praktische
  Studieninhalte
practical training – Praktikum
practical training abroad –
  Auslandspraktikum

practical tuition – fachpraktischer Unterricht
practically orientated education – praxisnahe Ausbildung
practice – Praxis
practice book – Arbeitsbuch
practice period – Einarbeitung
practician – Praktiker
practise – üben
praise – loben; preisen; rühmen
pre-book – vormerken
pre-booking – Vormerkung
pre-booking list – Vormerkungsliste
pre-school age – Vorschulalter
pre-school child – Kind im Vorschulalter
pre-school education – Vorschulerziehung
pre-school level – Vorschulbildung
pre-schooling – Vorschulerziehung
pre-vocational course – Berufsvorbereitungslehrgang
pre-vocational education year – Berufsgrundschuljahr
pre-vocational training – berufsvorbereitende Maßnahme
pre-vocational training programme – Berufsvorbereitungsprogramm
pre-vocational year – Berufsvorbereitungsjahr
precautionary checkup – Vorsorgeuntersuchung
precautionary measure – Schutzmaßnahme
precise manner of expression – präzise Ausdrucksweise
precision engineer – Feinmechaniker
precision engineering – Feinmechanik
precision mechanism technology – Feinwerktechnik
precocious – frühreif
precondition – Voraussetzung
predecessor – Vorgänger
predict – vorhersagen; voraussagen
predictable – berechenbar
prediliciton – Vorliebe
prefect – Präfekt; Vertrauensschüler
preference – Vorliebe
preference profile – Wunschprofil
preferential price – Vorzugspreis
preferred seizure – Pfändung, bevorrechtigte

pregnancy – Schwangerschaft
pregnant – in anderen Umständen, schwanger
prejudice – Vorurteil
preliminary entry of travel expense data – Reisekostenvorerfassung
preliminary examination – Vorprüfung
preliminary work – Vorleistung
premature admittance – vorzeitige Aufnahme
premature disability – vorzeitige Invalidität
premature return to work – Arbeitsversuch (missglückter)
premium – Prämie
premium (e.g. insurance premium) – Prämie (z.B. Versicherungsprämie/ Leistungsprämie)
premium function – Prämienfunktion
premium modifier – Prämienmodifikator
premium pay – Prämienlohn
premium rate – Prämiensatz
premium wage system – Prämienlohnsystem
preparation for a professional activity – Vorbereitung auf eine Berufstätigkeit
preparation for managerial tasks – Vorbereitung auf Führungsaufgaben
preparation for retirement – Vorbereitung auf den Ruhestand
preparation of decisions – Entscheidungsvorbereitung
preparatory – propädeutisch
preparatory course – Vorbereitungskurs
preparatory school – private Vorbereitungsschule für eine Public School (BrE); private Vorbereitungsschule für die Hochschule (AmE)
preparatory work – Vorarbeit
prepare – vorbereiten
prerequisite – Voraussetzung; Eingangsvorausetzung
preretirement reduced working hours – reduzierte Arbeitszeit vor der Pensionierung
preschool – Vorschul...
prescription – Verjährung
preselection – Vorauswahl
presence – Anwesenheit

present value – Barwert;
Berechnungsgrundlage bei
Pensionsrückstellungen
present value factor – Barwertfaktor
presentation – Darstellung; Präsentation
presentation ability –
Darstellungsfähigkeit
president – Präsident;
Vorstandsvorsitzender eines
Unternehmens; Vorsitzende(r)
presumable – voraussichtlich
prevailing wage – gültiger Lohnsatz
prevalence – Vorherrschen; weite
Verbreitung
prevent – verhindern
prevention – Prävention
preventive medical care –
Vorsorgeleistung
preventive medical checkup –
Vorsorgeuntersuchung
preventive medical programme(s) –
Gesundheitsvorsorge
previous conviction – Vorstrafe
previous day – Vortag
previous employer – Vorarbeitgeber
previous month's claim – Forderung aus
Vormonat
previously convicted – vorbestraft
price advantage – Kursvorteil
pride – Stolz; Hochmut
primary career – Hauptlaufbahn
primary education – Besuch der
Grundschule; Grundschulunterricht
primary level – Primärstufe
primary school (BrE) – Grundschule;
Volksschule
primary school age – Grundschulalter
primary school teacher –
Grundschullehrer
principal – Oberstudiendirektor(in);
Schulleiter(in); Rektor(in)
principal (Fachschule) –
Studiendirektor(in); Rektor(in)
principal occupation – Haupttätigkeit
principle – Grundgedanke; Prinzip
principle of averages –
Durchschnittsprinzip
principle of equal treatment –
Gleichbehandlungsgrundsatz
principle of law – Rechtsgrundsatz

principle of management –
Managementgrundsatz
principles of management –
Führungsgrundsätze; Führungsprinzip
print-out (EDP) – Ausdruck
printed material – gedrucktes Material
prior planning – Vorausplanung
privacy – Privatleben
private accident insurance – private
Unfallversicherung
private college – privates Lehrinstitut
private education – Ausbildung in einer
Privatschule
private life – Privatleben
private health insurance – private
Krankenversicherung
private health insurance fund –
Ersatzkasse
private insurance premium –
Privatversicherungsbeitrag
private lessons during the holidays –
Nachhilfeunterricht während der Ferien
private life insurance – private
Lebensversicherung
private school – Privatschule
private study – Selbststudium
private teacher – Pauker; Privatlehrer
private tuition – Nachhilfeunterricht
private tuition group –
Nachhilfeunterrichtsgruppe
privately insured – Privatversicherter
privilege – Privileg; Sonderrecht
prize day – Tag der Preisverleihung
prize distribution – Preisverteilung
pro-active – vorsorglich Maßnahmen
ergreifen
probability – Wahrscheinlichkeit
probation – Bewährung
probationary period – Probezeit
probationer – Praktikant
problem area – Problembereich
problem child – Sorgenkind, schwer
erziehbar
problem-oriented – problemorientiert
procedural rules – Geschäftsordnung
procedure – Methode; Verfahren;
Vorgehen, Vorgehensweise
process – Prozess; Verfahren; verarbeiten
process engineering – Technologie;
Verfahrenstechnik

process management –
Prozessmanagement
process of selection – Auswahlverfahren
process optimization –
Prozessoptimierung
processing class – Verarbeitungsklasse
processing of applications –
Bewerbungsbearbeitung
processing of averages –
Durchschnittsverarbeitung
processing type – Verarbeitungstyp
procuration – Prokura
procurement – Beschaffen; Beschaffung
product optimization –
Produktoptimierung
product training – Produktschulung
production – Fertigung
production control –
Fertigungssteuerung
production cycle – Arbeitstakt
production director – Betriebsdirektor
production planning –
Fertigungsplanung
production time ticket –
Leistungslohnschein
productive – ergiebig; produktiv
productive hours – Produktivstunden
productivity – Produktivität
productivity wage increase –
Produktivitätsanteil;
profession – Beruf
profession group – Berufskategorie
professional – beruflich; Berufs...;
fachlich
professional association –
Berufsgenossenschaft
professional attitude – Berufsauffassung
professional ban – Berufsverbot
professional body – berufsständische
Körperschaft; Berufsverband
professional career – beruflicher
Werdegang; Berufslaufbahn
professional classes – höhere
Berufsstände
professional competence – berufliche
Kompetenz
professional discretion –
Schweigepflicht
professional environment – berufliches
Umfeld
professional ethics – Berufsethos

professional expenses – Berufsunkosten
professional experience –
Berufserfahrung; Berufspraxis
professional future – berufliche Zukunft
professional grouser – Berufsnörgler
professional management –
professionelles Management
professional manager – in
Managementaufgaben ausgebildete
Führungskraft
professional orientation –
Berufsorientierung
professional prerequisite – berufliche
Voraussetzung
professional prospects –
Berufsaussichten
professional qualification consultation
– Eignungsberatung
professional requirements –
Berufserfordernisse
professional situation – Berufssituation
professional success – beruflicher Erfolg
professionally experienced –
berufserfahren
professions – freie Berufe
professor – Professor; Ordinarius
profile comparison – Profilvergleich
profit – erlangen; Gewinn; gewinnen;
profitieren
profit center – Profit Center
profit-related income share –
erfolgsabhängiger Einkommensanteil
profit-sharing – Erfolgsbeteiligung;
Gewinnbeteiligung
prognosis – Prognose; Vorhersage
program – Programm
program change – Programmänderung
programme – Programm
programme budget – Programmbudget
**programme for improvement of
on-the-job safety** –
Arbeitssicherheitsprogramm
programmed instruction –
programmierte Unterweisung
programmer – Programmierer
programming – Programmierung
programming course –
Programmierkurs
programming language –
Programmiersprache

progress to next level – aufrücken; versetzen; Versetzung
progressive bonus – Stufenbonus
progressive educational policy – zukunftsorientierte Bildungspolitik
progressive training – Stufenausbildung
progressive wage rate – Progressivlohn
prohibit – verbieten
prohibition of employment – Beschäftigungsverbot
prohibition of smoking – Rauchverbot
prohibition of use – Verwendungsverbot
project manager – Projektmanager
project staff plan – Projektmitarbeiterplan
project work – Projektarbeit
projection – Hochrechnung
projector – Projektor
prolix – weitschweifig
promise – Versprechen; Zusage
promise of a pension – Ruhegeldzusage
promise of reintegration – Wiedereingliederungszusage
promote – fördern
promote systematically – systematisch fördern
promotion – Beförderung; Förderung
promotion chart – Beförderungsplan
promotion level – F-Stufe; Förderungsstufen
promotion of employment – Beschäftigungsförderung
promotion of vocational training – Ausbildungsförderung
promotion of young people – Nachwuchsförderung
promotion prospects – Aufstiegschancen
promotion quota – Förderungsleistung (promotion for those who best promote others)
promotion strategy – Förderstrategie
promotion through new assignments – Förderung durch neue Aufgaben
promotional appraisal system – Förderungsbeurteilungssystem
promotional development of employees – Förderung der Mitarbeiter
promotional guidelines – Förderungsleitlinien
promotional measures – Förderungsmaßnahmen

promotional opportunities – Aufstiegschancen
promotional transfer – Förderungsversetzung
promotions file – Förderdatei; Förderkartei
prompt – Stichwort geben
prone to illness – krankheitsanfällig
pronunciation – Aussprache
proof – Beweis; Beweismaterial; Beweismittel
propensity to disturb – Störpotential
propensity to violence – Gewaltbereitschaft
proper – sachgerecht, vorschriftsmäßig
proposal catalog – Vorschlagskatalog
propose – vorschlagen
propose change – Plan vorschlagen
proprietor – Geschäftsinhaber, Inhaber
prospective – voraussichtlich
prospective beneficiary – Anwartschaftsberechtigter
prospects for the future – Zukunftsaussichten
prospectus – Merkblatt; Prospekt; Werbeschrift
protect – schützen
protecting mask – Schutzmaske
protection against body damage – Körperschutz
protection against radiation – Strahlenschutz
protection against unlawful dismissal – Kündigungsschutz
protection against unwarranted notice – Kündigungsschutz
protection in case of explosion – Explosionsschutz
protection of labour – Arbeitsschutz
protection of social data – Schutz der Sozialdaten
protection of vested rights – Besitzstandwahrung
protective clothing – Schutzkleidung
protective coat – Schutzmantel
protective glasses – Schutzbrille
protective gloves – Schutzhandschuhe
protective hood – Schutzhaube
protective shoes – Sicherheitsschuhe
protective suit – Schutzanzug

protest – Reklamation; Widerspruch; aufmucken

proverb – Sprichwort

provide – versorgen

provided that – vorbehaltlich

provincial – provinziell

provision – (gesetzliche) Regelung; Vorsorge; Versorgung

provision for years of service – Dienstjahresregelung

provisional – provisorisch; vorläufig; Zwischen...

provisional lump sum – Vorsorgepauschale

provisional notification – Zwischenbescheid

provisory care – Vorsorge

provocative – provokant

provost – Dekan

pseudo beginner – falsche(r) Anfänger(in)

psychic strain – psychische Belastung

psychological projections – psychologische Prognoseverfahren

psychological working conditions – psychologische Arbeitsbedingungen

psychologist – Psychologe

psychology – Psychologie

puberty – Pubertät

public administration – öffentliche Verwaltung

public body – Körperschaft des öffentlichen Rechts

public education sector – öffentlicher Bildungsbereich

public funds – öffentliche Mittel

public holiday – Feiertag; gesetzlicher Feiertag

public holiday calendar – Feiertagskalender

public holiday class – Feiertagsklasse

Public Holiday Loss-of-Income Act – Feiertagsausfallgesetz

public holiday type – Feiertagsart

public pension expectancy – Ruhegeldanwartschaft

public relations – öffentliche Beziehungen

public review board – Revisionsbehörde

public servant – Staatsbeamte(r)

public school (AmE) – öffentliche Schule; staatliche Schule

Public School – höhere Privatschule mit Internat

public school examination – erfolgreicher Hauptschulabschluss (In the English School system, most pupils at the end of their fifth years take some form of public examination in around seven subjects. These exams give those who pass them a qualification that is recognized by employers and others.)

public schools (s. Grafik) – Schulwesen, privates

public service – öffentliche Versorgungsleistung; öffentlicher Dienst

public service (AmE) – Staatsdienst

publication – Veröffentlichung

published financial statements – Handelsbilanz

puerility – Infantilität

punch-up – Prügelei

punctuality – Pünktlichkeit; Termintreue

punishment – Strafe; Bestrafung

punishment exercise – Strafarbeit

pupil – Schüler; Zögling

pupil at a senior high school – Oberstufenschüler

pupil in senior cycle – Oberstufenschüler

pupil-teacher ratio – Schüler-Lehrer-Zahlenverhältnis

purchasing department – Einkauf

purchasing power – Kaufkraft

purchasing-power allowance – Kaufkraftausgleich

purchasing-power parity – Kaufkraftparität

pure research – Grundlagenforschung

purpose – Absicht; Zweck

purpose of distinction – Unterscheidungszweck

purpose of training – Bildungszweck

purview – Aufgabengebiet

push – stoßen

pusher – Streber

put on hold (eg. applicant) – zurückstellen

put on the sick list – krankschreiben

put up – aufrücken; versetzen; errichten

Q

qualification – Befähigung(szeugnis); Qualifikation
qualification deficit – Qualifikationsdefizit
qualification measure – Qualifizierungsmaßnahme
qualification to teach – Lehrbefähigung
qualifications – berufliche Bildung
qualifications and requirements – Qualifikationen und Anforderungen
qualifications catalog – Qualifikationskatalog
qualifications loophole – Qualifikationslücke
qualifications network – Qualifikationsnetz
qualifications pool – Qualifikationspool
qualifications profile – Qualifikationsprofil
qualifications structure – Qualifikationsstruktur
qualified – qualifiziert
qualified management responsibilities – anspruchsvolle Führungsaufgaben
qualified teacher – Lehrer mit anerkanntem Studienabschluss
qualified teacher status in the EU – EU-weit anerkannte Lehrbefähigung
qualify someone for/as – befähigen; qualifizieren
qualifying period – Anwartschaftszeit; Wartezeit
qualifying vocational training – qualifizierende Berufsausbildung
qualitative personnel planning – qualitative Personalplanung
quality – Qualität
quality assurance – Qualitätssicherung
quality bonus – Qualitätsprämie
quality circle – Qualitätsgruppe
quality control – Qualitätskontrolle
quality feature – Qualitätsmerkmal
quality improvement – Qualitätsverbesserung
quality of work life – Qualität des Arbeitslebens

quantitative output – Mengenleistung
quantity – Quantität
quantity of research – Forschungsleistung (The quantity of research is frequently measured by the number of publications, either by an individual, the department or the faculty.)
quantity-based piecerate work – Stückakkord
quarell – Zank; hadern, zanken
quarterly salary – Vierteljahresgehalt
quasi full-time employment situation – arbeitnehmerähnliches Verhältnis
querulous person – Nörgler; Querulant
question – Frage
question of detail – Detailfrage
questionnaire – Fragebogen
quick on the uptake – schnell von Begriff
quorum – beschlussfähige Anzahl; Quorum
quota deduction – Kontingentabtragung
quota of hours – Stundenkontingent
quota processing – Kontingentverarbeitung
quota type – Kontingenttyp
quote – zitieren

R

racial – Rassen...
racial discrimination – Rassendiskriminierung
raiding – Mitgliederabwerbung
raise – erhöhen; Erhöhung
random – zufällig
random sample – Stichprobe
range group – Bereichsgruppe
range of competence – Handlungsspielraum
range of courses – Kursangebot
range of managerial styles – Variationsbreite der Führungsstile
rank – Rangstufe, Rang

rank comparison – Rangvergleich
rank concept – Rangbegriff
rank designation – Rangbezeichnung
rank equivalency – rangliche
  Gleichstellung
rank for specialists – fachlicher Rang
rank order – Hierarchie; Rangordnung
rank order system –
  Rangordnungssystem
rank for managerial staff – dispositiver
  Rang
rank structure – Rangstruktur
ranking – Hierarchie; Rangordnung
ranking description – Rangbezeichnung
ranking system – Rangsystem
rapid reading – Schnelllesen
rare – rar; selten
rate – Betrag-pro-Einheit
rate of increase – Steigerungssatz
ratee (AmE) – Anwärter; Kandidat;
  Prüfling
rating – Einstufung
rating of employees – Einstufung nach
  Leistung
rationality – Rationalität
rationalization – Rationalisierung
rationalize – modernisieren;
  rationalisieren; vereinfachen
rationalize away – wegrationalisieren
re-entrant – Wiedereinsteiger
re-entry – Wiedereintritt
re-hiring – Wiedereinstellung;
  Wiedereintritt
reaction – Reaktion
read – lesen
reader (BrE) – Fibel; Lesebuch;
  Dozent(in) (Now, in some British
  universities, (the title of) a lecturer of
  the highest grade below professor.)
readiness – Bereitschaft
reading ability – Lesefähigkeit
reading aloud – laut vorlesen
reading difficulty – Leseschwierigkeit
reading habit – Lesegewohnheit
reading list – Leseliste
reading material – Lesestoff
reading skills – Lesefertigkeiten
reading speed – Lesegeschwindigkeit
reading understanding –
  Leseverständnis
reading vocabulary – Lesewortschatz

readjust – neu ordnen
readjustment allowance –
  Überbrückungszulage
ready money – Bargeld
real income – Realeinkommen
real wage – Reallohn
Realschule – Realschule
rear – aufziehen; großziehen
reason – Beweggrund; Veranlassung;
  Begründung
reason for dismissal – Entlassungsgrund
reason for giving notice –
  Kündigungsgrund
reasonable demand – zumutbar
reasonable – einsichtig
reasoning power – Denkfähigkeit;
  Kombinationsgabe
reassignment to wage group –
  Umgruppierung
reassignment to wage level – Umstufung
rebook – umbuchen
rebooking – Umbuchung
rebooking confirmation –
  Umbuchungsbestätigung
rebooking notification –
  Umbuchungsmitteilung
rebuke – abmahnen; Abmahnung; Tadel;
  tadeln
recalculation – Aufrollung
recalculation difference –
  Aufrollungsdifferenz
recalculation of contributions –
  Beitragsnachberechnung
recall – sich erinnern an; sich ins
  Gedächtnis rufen
recapitulate – kurz wiederholen;
  rekapitulieren
receipt – Quittung, Beleg
receipt accounting – Belegabrechnung
receipt for entertainment expenses –
  Bewirtungsbeleg
receipt for overnight stay –
  Unterkunftsbeleg
received pronunciation –
  hochsprachliche Aussprache
receptive – aufgeschlossen
recession – Rezession
recessive – rückläufig
recipient of benefits –
  Leistungsempfänger
recipient – Empfänger

reciprocal – gegenseitig
reciprocity indicator –
Gegenseitigkeitskennzeichen
recite – rezitieren
reclamation – Beschwerde
reclassification – Neubewertung;
Neueinstufung; Umgruppierung
reclassification type –
Umgruppierungsart
recognition – (Wieder)erkennen
recognize – erkennen
recognized profession – anerkannter
Beruf
recognized trade – anerkannter Beruf
reconciliation of interests –
Interessenausgleich
reconsider – überdenken
record layout field – Satzbettfeld
record of achievement –
Gesamtbeurteilung
record sheet – Aufzeichnungsbogen
recording – Aufzeichnung
recreation – Entspannung; Erholung
recreation centre – Erholungsheim
recreation places – Erholungsplätze
recruit – anwerben
recruitment – Anwerbung;
Personalbeschaffung
recruitment problems –
Nachwuchssorgen
recruitment instrument –
Beschaffungsinstrument
recruitment planning –
Personalbeschaffungsplanung
rector – Rektor(in)
recurring deductions – Abzüge,
wiederkehrende
recurring payments/ deductions –
Bezüge/Abzüge, wiederkehrende
red tape – Bürokratismus
red-brick university – neuere Universität
redemption – Tilgungsrate
redemption money – Abgeltungsbetrag
redemption value – Rückkaufswert
redeploy – umgruppieren;
umstrukturieren
redirect – umbestellen
redirected for reasons of competence –
Abgabe zuständigkeitshalber
redo – nacharbeiten

reduced contribution rate – Beitragssatz
(ermäßigt)
reduced earnings –
Einkommensverringerung
reduced incentive – degressive Prämie
reduced output – Leistungsminderung
reduced qualifying period –
Wartezeitverkürzung
reduced working hours – Kurzarbeit
reduction – Kürzung
reduction in earning capacity –
Erwerbsminderung
reduction in working hours –
Arbeitszeitverkürzung
reduction method – Kürzungsmethode
reduction numerator – Kürzungszähler
reduction of staff – Personalabbau
**regulation of reduction in working
hours** – Arbeitszeitverkürzungsregelung
reduction regulation –
Kürzungsvorschrift
reduction rule – Kürzungsregel
redundancy – Verlust des Arbeitsplatzes
redundancy payment – Abfindung;
Entlassungsentschädigung
redundancy payments scheme –
Sozialplan
redundancy scheme – Abfindungsplan
redundant – überflüssig; weitschweifig
redundant labour –
Arbeitskräfteüberschuss
reemploy – wiederanstellen;
wiedereinstellen
reemployment – Wiederbeschäftigung;
Wiedereinstellung
reengage – wiederanstellen;
wiedereinstellen
reenrolment – Neueinschreibung
refectory – Mensa; Speiseraum;
Speisesaal
referee – Schiedsmann; Schiedsrichter;
Schlichter
reference (from employer) – Referenz;
Zeugnis (für Angestellte);
Arbeitszeugnis
reference number – Ordnungszahl
reference person – Bezugsperson
reference planning – Referenzplanung
reference principle –
Vergangenheitsprinzip

reference value – Bezugswert;
Referenzwert
referent power – Macht durch
Anerkennung
referral program – direktes Recruiting
referring to a further department –
Abgabebescheid
reflect – widerspiegeln; überdenken
reflux – Rückfluss
refractory – störrisch
refresh – auffrischen
refresher course – Auffrischungskurs;
Wiederholungskurs
refund – rückerstatten; zurückzahlen
refund amount – Erstattungsbetrag
refund class – Erstattungsklasse
refund group – Erstattungsgruppe
refund of costs – Kostenerstattung
refund rate – Erstattungssatz
refusal – Verweigerung
refusal to obey –
Gehorsamsverweigerung
refusal to work – Arbeitsverweigerung
region – Gebiet
region key – Regionenschlüssel
region near international border –
Grenzregion
regional employment office –
Landesarbeitsamt
regional health insurance fund –
Gebietskrankenkasse
regional labour court of appeal –
Landesarbeitsgericht
regional social insurance office –
Landesversicherungsanstalt
register – registrieren
registrar – höchste(r)
Verwaltungsbeamte(r)
registration – Anmeldung
registration certificate –
Aufenthaltserlaubnis
registration of applicants –
Bewerbererfassung
regression analysis – Regressionsanalyse
regular – normal; ordnungsgemäß;
regelmäßig; regulär
regular customer – Stammkunde
regular overtime – regelmäßige
Überstunden
regular pay – Bezüge, laufende

regular payments/deductions – Bezüge/
Abzüge, laufende
regular support payments – Unterhalt,
laufender
regularization – (gesetzliche) Regelung
regulate a just salary –
Einkommensfindung
regulation – (gesetzliche) Regelung
regulations as laid down in collective
agreement – tarifvertragliche
Regelungen
regulations governing industrial
relations – Betriebsverfassung
regulations on travel expenses –
Reisekostenrichtlinien
rehabilitation – Rehabilitation
rehabilitation measure –
Rehabilitationsmaßnahme
reimburse – rückerstatten; zurückzahlen
reimbursement – Erstattung;
Rückerstattung; Rückzahlung
reimbursement of contributions –
Beitragserstattung
reimbursement of costs –
Kostenerstattung
reimbursement of fares to/from school
– Erstattung der Schulwegkosten
reimbursement of travel expenses –
Reisekostenvergütung
reinforce – verstärken
reinforcement – Verstärkung
reinstate – wiedereinsetzen
reinstatement value –
Versicherungsnennwert
reintegration – Wiedereingliederung
reintegration into working life –
berufliche Wiedereingliederung
reinterpretation – Umdeutung
reject – ablehnen; zurückweisen
rejected – abgelehnt
rejection – Ablehnung; Zurückweisung
related person – Familienangehörige
related to payroll – abrechnungsbezogen
relating (to) – in Zusammenhang mit
relation – Beziehung; Verhältnis
relational level – Beziehungsebene
relationship – Verknüpfung
relationship to child –
Kindschaftsverhältnis
relaxation exercise – Lockerungsübung

relaxation through self-hypnosis –
autogenes Training
release from work – Arbeitsbefreiung
relevancy test – Relevanztest
relevancy to increase –
Erhöhungsrelevanz
relevant to payment –
bezahlungsrelevant
relevant to payroll –
abrechnungsrelevant
relevant to the job – berufsbezogen
reliability – Zuverlässigkeit
reliable – zuverlässig
relief – Abhilfe
relief fund – Unterstützungskasse
relief fund regulations –
Unterstützungskassenrichtlinien
relief man – Springer
relief person – Springer
relief shift – Springerschicht
religion – Religion
religious denomination – Konfession
relocate – umziehen
reluctance – Unlust
remaining balance of payment –
Auszahlungsbetragsrest
remaining leave – Resturlaub
remaining leave entitlement –
Resturlaubsanspruch
remaining time to maturity of a loan –
Restlaufzeit
remedial classes – Ergänzungsunterricht
remedial teaching – Förderunterricht
remedy – Abhilfe
remember – erinnern; sich erinnern
reminder date – Mahndatum
reminder of appointment –
Terminerinnerung
remittance of social insurance –
Beitragsabführung
removal – Umzug
removal allowance (BrE) –
Umzugskostenbeihilfe
removal costs – Umzugskosten
removal costs allowance –
Umzugsbeihilfe
remove – umziehen
remunerate – bezahlen; entlohnen;
vergüten
remuneration – Arbeitsentgelt;
Bezahlung; Entgelt

remuneration for employee inventions
– Erfindervergütung
remuneration for meals –
Essensentschädigung
remuneration for temporary personnel
– Aushilfsbezüge
remuneration in kind – Sachbezüge;
sonstige Firmennebenleistungen
remuneration management –
Vergütungsverwaltung
remuneration matrix –
Vergütungsmatrix
remuneration of work – Vergütung der
Arbeit
remuneration statistics – Lohn- und
Salärstatistik
remuneration system –
Vergütungssystem
remuneration – Vergütung
rendition of services (AmE) –
Dienstleistung
renew – erneuern
rent – Miete
rent allowance – Mietzuschuss
renunciation – Verzicht
renunciation of pay – Lohnverzicht
reopen – wieder in Betrieb nehmen;
wiedereröffnen
reorganisational measure –
Reorganisationsmaßnahme
reorganize – umorganisieren
reorganization – Neuordnung
repayment – Rückerstattung;
Rückzahlung
repayment installment – Tilgungsrate
repayment pattern – Tilgungsrhythmus
repayment rate – Tilgungssatz
repayment requirement –
Rückzahlungsverpflichtung
repeat – wiederholen, wiedergeben
repeat a year – sitzenbleiben
repeat exam(ination) –
Wiederholungsprüfung
repeated sickness – Mehrfacherkrankung
repetition – Wiederholung
repetitive work – Routinearbeit
replace – ersetzen
replacement – Ersatz
replacement needs – Ersatzbedarf

replacement ratio – Verhältnis der Einkommen vor und nach der Pensionierung

replacement teacher – Lehrervertretung

reply – Antwort; antworten; Bescheid

report – Jahreszeugnis; Report; Zeugnis

report book – Berichtsheft

report card – Schülerbogen

reporting – Berichtswesen

reporting path – Berichtsweg

reporting procedure – Meldeverfahren

reporting procedure for social insurance – Meldeverfahren in der Sozialversicherung

represent – vertreten

representation of juvenile employees – Jugendvertretung

representative – Vertreter; Repräsentant

representative role – Verhalten gegenüber Außenstehenden

repressive – repressiv

reprimand – Tadel; tadeln; verwarnen; Verweis erteilen

reproduce – wiedergeben

repudiate – nicht anerkennen

repudiation – Nichtanerkennung

reputation – Leumund; Reputation; Ruf

request – Anforderung, Aufforderung

request for extension – Verlängerungsantrag

request for (additional or new) personnel – Personalanforderung (Anforderung aus einer Abteilung nach neuem oder zusätzlichem Personal)

request for reapplication – Aufforderung zur Wiederbewerbung

request for refund – Erstattungsantrag

require too little of someone – unterfordern

required course – Pflichtfach

required working hours – Sollarbeitszeit

requirement – Bedarf

requirement category – Anforderungstyp

requirement type – Anforderungsart

requirements and qualifications profile – Anforderungs- und Qualifikationsprofil

requirements of a compensation system – Erfordernisse eines Einkommenssystems

requirements of personnel policies – Anforderungen an die Personalpolitik

requirements profile – Anforderungsprofil

requirements type – Bedarfsart

requisition – Anforderung; Ersuchen

rescue plan – Rettungsplan

rescue route – Rettungsweg

research – Forschung

Research and Development (R&D) – Forschung und Entwicklung (F&E)

research centre – Forschungszentrum

research cooperation – Forschungskooperation

research fellow – Forschungsstipendiat(in)

research institute – Forschungsstätte, Versuchsanstalt

research results – Forschungsergebnisse

research student – Doktorant(in)

research supervisor – Doktorvater

research team – Forschergruppe

research worker – Forscher

resentment – Ressentiment

reservation – Belegung; Buchung; Vormerkung; Vorbehalt; Reserviertheit

reservations list – Vormerkungsliste

reserve – buchen, reservieren, vormerken; Reserviertheit

reserve duty training exercise – Wehrübung

reserve pool employee – Springer

reserved responsibility – Verantwortung für nicht delegierte Aufgaben

reserves for vacation bonus – Urlaubsgeldrückstellung

reservoir of ideas – Ideenpotential

reset candidate – Rücksetzkandidat

resettlement measure – Wiedereingliederungsmaßnahme

residence – Aufenthalt(sort); Wohnort; Wohnsitz

residence abroad – Auslandsaufenthalt

residence permit – Aufenthaltserlaubnis; Aufenthaltsgenehmigung

residence state tax – Wohnsitzsteuer (USA)

resident participant – Teilnehmer von außerhalb, der am Studienort wohnt

residual – Rest...

resignation – eigene Kündigung; Kündigung durch den Arbeitnehmer
resignation (letter of) – Entlassungsgesuch
resigned – resigniert
resilient – belastbar
resistance – Widerstand
resistance to change – Widerstand gegen Neuerungen
resit an exam – Prüfung wiederholen
resolute – entschlossen
resolution – Beschluss; Resolution; Vorsatz
resource allocation – Geldmittelzuteilung
resource reservation – Ressourcenbelegung
resource type – Ressourcentyp
resources – Hilfsmittel; Hilfsquellen; Ressourcen
resources planning – Ressourcenplanung
respond – antworten
response – Antwort
responsibility – Fachverantwortung; Verantwortlichkeit
responsible – verantwortlich
responsible authorities – zuständige Stelle
responsible for – Betreuung (z.B. von Systemen)
rest cure – Erholungskur
rest period – Ruhezeit
rest room – Pausenraum
restraint clause – Konkurrenzklausel
restraint of trade – Konkurrenzverbot
restrict – beschränken; einschränken
restricted entry – Numerus Clausus
restricted vocabulary – begrenzter Wortschatz
restriction – Einschränkung
restriction on admission – Zulassungsbeschränkung
restrictive – hemmend
restructure – umgruppieren; umstrukturieren
result – Ergebnis; Folge; Konsequenz; Resultat
result wage type – Ergebnislohnart
results assessment – Ergebnisbewertung
resume – Lebenslauf; wiederaufnehmen
résumé – Zusammenfassung

resumption – Wiederaufnahme
retarded – zurückgeblieben
retention – Einbehaltung
retention period – Aufbewahrungsfrist
retire – in den Ruhestand treten; jemanden in den Ruhestand versetzen
retired employee – Pensionär; Rentenempfänger; Rentner
retirement – Pensionierung; Ruhestand; Versetzung in den Ruhestand
retirement age – Pensionsalter; Rentenalter
retirement annuity – Versicherungsrente
retirement benefit – Altersrente; Altersruhegeld
retirement benefits agreement – Ruhegehaltsabkommen
retirement pay – Pensionszahlung
retirement pension – Altersruhegeld; Altersversorgung
retirement pension expectancy – Anwartschaft auf Ruhegeld
retirement plan – Pensionsplan
retraction – Zurücknahme
retrain – umschulen; umlernen
retraining – Umschulung
retraining measures – Umschulungsmaßnahmen
retroactive – rückwirkend
retroactive accounting – Rückrechnung
retroactive accounting category – Rückrechnungstyp
retroactive accounting date – Rückrechnungsdatum
retroactive accounting difference – Rückrechnungsdifferenz
retroactive accounting indicator – Rückrechnungserkennung
retroactive accounting limit – Rückrechnungsgrenze
retroactive accounting month – Rückrechnungsmonat
retroactive accounting period – Rückrechnungsperiode
retroactive accounting relevance – Rückrechnungsrelevanz
retroactive accounting result – Rückrechnungsergebnis
retroactive accounting trigger – Rückrechnungsanstoß

retroactive accounting type –
Rückrechnungsart
retroactive wage increase –
rückwirkende Lohnerhöhung
retrospective – rückwirkend
return journey – Rückfahrt
return trip – Rückreise
returning board (AmE) – Wahlvorstand
revenue – Einkünfte; Einnahmen
revenue expenditure – Betriebsausgaben
reversal – Zurücknahme
review – Kontrolle; kontrollieren;
Überprüfung
review lesson – Wiederholungslektion
revise – revidieren; überarbeiten
revised edition – Neufassung
revision office for travel expenses –
Reisekostenprüfstelle
revocation – Widerruf
revoke – widerrufen
reward – Belohnung
reward system – Belohnungssystem
rework – überarbeiten
ride-sharing group – Fahrgemeinschaft
right – Anspruch
right of benefits – Leistungsanspruch
right of complaint – Beschwerderecht
right of nomination – Vorschlagsrecht
right of return – Rückgaberecht
right to be heard – Anhörungsrecht
right to educate – Erziehungsrecht
right to free collective bargaining –
Tarifautonomie
right to give notice – Kündigungsrecht
right to holidays – Urlaubsanspruch
right to impose tax – Besteuerungsrecht
right to information – Informationsrecht
right to issue instructions –
Weisungsbefugnis; Weisungsrecht
right to negotiate – Verhandlungsrecht
right to pension – Anspruch auf Rente;
Ruhegeldanspruch
right to strike – Streikrecht
right to vote – Stimmrecht;
Wahlberechtigung
rightful claims of relatives –
Anspruchsberechtigung von
Angehörigen
rigid – starr; steif
rigidity – Starrheit
rise – steigen

rise in wages – Ansteigen der Löhne
(allgemeine Entwicklung)
risk – wagen
risk factor – Risikofaktor
rival – Mitbewerber
rivalry – Rivalität
road safety – Verkehrssicherheit
role – Rolle
role game – Rollenspiel
role understanding – Rollenverständnis
roll – Anwesenheitsliste; Namensliste
room data – Raumdaten
room equipment – Raumausstattung
room for maneuver –
Handlungsspielraum
room reservation – Raumbelegung
room reservation type –
Raumbelegungsart
room reservations planning –
Raumbelegungsplanung
roster – Dienstplan; Stundentafel
rotate – rotieren
rotate an employee – einen Mitarbeiter
nacheinander in verschiedenen
Firmenbereichen einsetzen
rotated part-time work – Turnusteilzeit
rotating shift – Wechselschicht
rotating shift premium –
Wechselschichtprämie
rotating shift schedule – Schichtplan,
rollierender
rotation – Turnus; turnusmäßiger
Wechsel
rough-house – Rauferei
round down – abrunden
round of inspection – Betriebsbegehung
round up – aufrunden
**rounding-up and rounding-down
accounting** – Restgroschenbehandlung
routine – Routine
routine work – Routinearbeit
row – Zeile; Zank
royalties – Lizenzgebühren
rubber – Radiergummi
rude – flegelhaft
rudeness – Ungezogenheit
rule – Regel
rule of law – Rechtsvorschrift
rule of thumb – Faustregel
rule type – Regeltyp

rules on working hours –
Arbeitszeitregelungen
rumour – Gerücht; Spekulation
run a course – einen Kurs abhalten
run-off (AmE) – Stichwahl
runner – Bote

S

sabbatical – Langzeiturlaub (An
employee who relinquishes one twelfth
of his salary but continues to work
full-time acquires a right to four weeks
additional holidays. If these are saved
over a period of three years he has a
right to three months paid holidays.)
sabbatical year – akademischer Urlaub;
Ferienjahr eines Professors
sack – entlassen; feuern; hinauswerfen
sacrifice a career – Karriereverzicht
safeguarding existing jobs – Erhaltung
von Arbeitsplätzen
safety – Sicherheit
safety code – Sicherheitsvorschriften
safety control – Sicherheitskontrolle
safety engineer – Sicherheitsingenieur
safety equipment – Schutzausrüstung
safety expert – Sicherheitsfachkraft
safety facility – Sicherheitseinrichtung
safety guard – Schutzgitter
safety needs – Sicherheitsbedürfnisse
safety officer – Sicherheitsbeauftragter
safety precautions –
Sicherheitsvorkehrungen
safety regulations –
Sicherheitsbestimmungen
salaried employee – Angestellte(r)
Salaried Employees' Insurance Act –
Angestelltenversicherungsgesetz
salaried graduate – erwerbstätiger
Hochschulabsolvent
salary – Gehalt; Tarifgehalt; Besoldung
**salary according to marital status and
number of children** – Gehalt nach
Familienstand und Kinderzahl

salary adjustment – Gehaltsanpassung;
Gehaltskorrektur
salary advance – Gehaltsvorschuss
salary bracket – Gehaltsgruppe
salary credit account – Gehaltskonto
salary demands – Gehaltsforderung;
Gehaltswünsche
salary demotion – Rückstufung im
Gehalt
salary differential – Gehaltsgefälle
salary expectations – Gehaltserwartung
salary finding – Gehaltsfindung
salary frame – Gehaltsrahmen
salary group – Gehaltsgruppe;
Tarifgruppe
salary increase – Gehaltserhöhung
salary jump – Gehaltssprung
salary level – Gehaltsniveau;
Gehaltsstufe; Rang
salary level change – Bezüge, Änderung
der
salary limit – Einkommensgrenze
salary matrix – Gehaltsmatrix
salary payment – Gehaltszahlung
salary policy – Gehaltspolitik
salary position – Rang
salary range – Gehaltsspanne
salary review list – Gehaltslesungsliste
salary revision – Gehaltsüberprüfung
salary slip – Gehaltszettel
salary statement – Gehaltsabrechnung
salary structure – Gehaltsstruktur
salary type – Gehaltsart
sales – Umsatz
sales and marketing functions –
kaufmännische Vertriebsaufgaben
sales forecast – Verkaufsprognose
sales success – Vertriebserfolg
sample – Auswahl; Muster
sample file – Mustermappe
sanction – Billigung; Zustimmung
sandwich course – Sandwich-Lehrgang
sanguine person – Sanguiniker(in)
satchel – Schulranzen
satisfaction – Befriedigung; Genugtuung;
Zufriedenheit
satisfactory – befriedigend
Saturday work – Samstagsarbeit
saving through equity investments –
Aktiensparen
savings amount – Sparbetrag

savings bonus – Sparzulage
savings plan – Sparplan
savings premium – Sparprämie
savings supplement –
    Arbeitnehmersparzulage
scale – Skale; Tabelle
scapegoat – Sündenbock
scarce – knapp
scared – Bammel haben
schedule – Arbeitsplan; Terminplan
schedule model – Ablaufmuster
schedule planning – Terminplanung
schedule-dependent access control –
    Zutrittskontrolle, zeitliche
scheme – Schema
scholar – Gelehrter; Studierende(r)
scholarship – Stipendium
Scholastic Aptitude Test (SAT) (AmE) –
    Eignungstest für amerikanische
    Universitäten
school – Schule
school abroad – Auslandsschule
school adviser – Schulberater
school atmosphere – Schulatmosphäre
school badge – Schulemblem
school bag – Schulranzen
school beginner – Schulanfänger
school bell – Schulglocke
school board – Schulamt
school boycott – Unterrichtsboykott
school bus – Schulbus
school child – Schulkind
school choir – Schulchor
school committee – Schulausschuss
school counsellor – Schulberater
school department – Schulreferat
school enrolment – Einschulung
school fees – Schulgeld
school for the (physically) handicapped
    – Schule für Behinderte und Kranke
school friend – Schulkamerad
school garden – Schulgarten
school holidays – Schulferien
school inspector – Schulrat
school leaver – Schulabgänger
school leaving age – Schulabgangsalter;
    Schulentlassungsalter
school leaving certificate –
    Schulabschluss
school meals – Schulspeisung
school medical officer – Schularzt

school milk – Pausenmilch
school motto – Wahlspruch der Schule
school outing – Schulausflug
school prayer – Schulgebet
school registration – Einschreibung;
    Immatrikulation
school regulations – Schulordnung
school report – Jahreszeugnis; Zeugnis
school report for trainees –
    Praktikantenzeugnis
school spokesman – Schulsprecher
school sports – Schulsport
school sports grounds –
    Schulsportanlage
school uniform – Schüleruniform;
    Schuluniform
school year – Schuljahr
schoolbag – Schulmappe
schoolbook – Schulbuch
schoolboy – Schulbube, Schüler
schooling – Anleitung; Ausbildung;
    Schulung
schoolmaster – Schulmaster
schoolmate – Schulkamerad
schools' inspectorate –
    Schulaufsichtsbehörde
schoolyard – Pausenhof; Schulhof
science and arts – Wissenschaft
scientific – wissenschaftlich
scientific management –
    wissenschaftliche
    Unternehmensführung
scientific method – Wissenschaftlichkeit
scientific society – wissenschaftliche
    Gesellschaft
scientist – Wissenschaftler
scold – schelten; schimpfen
scope – Freiraum
scope of agreement – Geltungsbereich
    eines (Tarif)vertrages
score – Punkte erzielen
scrap – raufen; Rauferei
screen – Bewerber auswählen; Bewerber
    sieben
screen applicants – Bewerber
    auswählen; Bewerber sieben
search for employment –
    Arbeitsplatzsuche; Suche nach Arbeit
season – Saison
season ticket – Zeitfahrkarte

season ticket for schoolchildren – Schülermonatskarte
seasonal fluctuation – Schwankung, saisonale
seasonal worker – Saisonarbeiter; Wanderarbeitnehmer
seat of learning – Lehrstätte
second – abordnen
second degree – Zweitstudium
second job – Nebenbeschäftigung; Nebentätigkeit, Nebenberuf, nebenberufliche Tätigkeit
second language – Zweitsprache
secondary career model – Nebenlaufbahn
secondary education – Gymnasialwesen
Secondary Examinations Council (SEC) – externe Prüfungskommission
secondary grammar school – höhere staatliche Schule
secondary illness – Zweiterkrankung
secondary income – Zusatzverdienst
secondary modern pupil (BrE) – Realschüler
secondary modern school – Realschule, Hauptschule
secondary modern school teacher – Realschullehrer
secondary residence – Zweitwohnsitz
secondary school – Sekundarschule; weiterführende Schule
secondary school emphasizing business – Wirtschaftswissenschaftliches Gymnasium (WWG)
secondary school emphasizing social sciences – sozialwissenschaftliches Gymnasium (SWG)
secondary school for adults – Kolleg (Adults with vocational training can attend this type of secondary school and have the possibility of qualifying for university.)
secondary school leaver – Absolvent einer weiterführenden Schule
secondary school pupil – Schüler einer weiterführenden Schule
secondary school teacher – Studienassessor; Studienrat(rätin)
secondary school teaching – höheres Lehramt

secondment – zeitweilige Versetzung, Abordnung
secondment payment – Abordnungsgeld; Abordnungskosten
secrecy – Geheimhaltung
secret ballot – geheime Abstimmung
secretarial experience – Sekretariatspraxis
secretary – Sekretär(in)
sector – Fachgebiet
security – Sicherheit; Bürgschaft
security control – Kontrolle (an der Pforte)
seize – pfänden
seizure number – Pfändungsnummer
seizure of personal property – Sachpfändung
seizure of support payments – Unterhaltspfändung
seizure type – Pfändungsart
select – aussuchen; Auswahl treffen; auswählen
selection – Auswahl; Wahl
selection criterion – Auswahlkriterium
select from – exzerpieren
selection – Auslese
selection guideline – Auswahlrichtlinie
selection of personnel – Personalauswahl
selection pressure – Auslesedruck
selection procedure – Auswahlverfahren
selection test – Auswahltest
selective employment tax (BrE) – Arbeitgeberlohnanteil
self instruction – Selbstunterweisung
self-access centre – Selbstlernzentrum
self-appraisal – Selbsteinschätzung
self-assessment – Selbsteinschätzung
self-assessment test – Selbsteinstufungstest
self-assurance – Selbstbewusstsein
self-awareness – Selbsterkenntnis
self-centred behaviour – egozentrisches Verhalten
self-confidence – Selbstbewusstsein; Selbstvertrauen; Selbstsicherheit
self-confident – selbstsicher
self-conscious – befangen
self-control – Selbstkontrolle
self-controlled – beherrscht
self-depreciation – bescheiden
self-determination – Selbstbestimmung

self-directed learning – selbstgesteuertes Lernen

self-discipline – Selbstdisziplin

self-employed – selbständig erwerbstätig; Selbständige(r), freiberuflich

self-employer – selbständig Tätiger, Freiberufler

self-esteem – Selbstachtung; Selbstwertgefühl

self-evident – selbstverständlich

self-expression – Ausdruck der eigenen Persönlichkeit

self-fulfilment – Selbsterfüllung

self-governing school – Schule mit Selbstverwaltung

self-government board of the social insurance institutions – Selbstverwaltung der Sozialversicherungsträger

self-help – Selbsthilfe

self-help group – Selbsthilfegruppe

self-initiative – Eigeninitiative

self-interest – Eigennutz

self-knowledge – Selbsterkenntnis

self-organisation – Selbstorganisation

self-pay patient – Selbstzahler

self-protection – Selbstschutz

self-reliant – selbständig

self-respect – Selbstachtung

self-responsibility – Eigenverantwortung

self-tutoring station – Selbstlernplatz

seller – Verkäufer(in)

semi-skilled occupation – Anlernberuf

semi-skilled work – Anlerntätigkeit

semi-skilled worker – angelernter Arbeiter

seminar – Seminar

seminar and convention management – Veranstaltungsmanagement

seminar paper – Referat

send down – relegieren

sending country – Entsendeland

senior consultant – Chefberater

Senior Director – Berater

Senior Engineer – Oberingenieur

senior grammar school level – gymnasiale Oberstufe

senior high school (AmE) – Oberstufe

senior management – obere Betriebsführung; obere Führungsebene

senior manager – leitender Angestellter

senior teacher – Oberstudienrat(rätin)

Senior Vice President – stellvertretendes Vorstandsmitglied

senior year – letztes Studienjahr

seniority – Dienstalter

seniority roster – Dienstaltertabelle

sense of duty – Pflichtbewusstsein

sense of guilt – Schuldgefühl

sense of hearing – Hörsinn

sense of smell – Geruchssinn

sense of superiority – Eigendünkel

sense of touch – Tastsinn

sensitiveness – Sensibilität

sensitivity – Sensibilität; Empfindsamkeit

sensitivity training – Sensitivitätstraining

sensitization – Sensibilisierung

sensory – sensorisch; Sinnes-

sentence – Satz

separate(d) – gesondert; trennen

separation – Abgang

separation allowance – Trennungsentschädigung; Trennungsgeld

separation compensation – Trennungsgeld

sequence – Reihenfolge

sequence of operation – Arbeitsablaufplan

serendipity – Spürsinn

serious – ernst

serve an apprenticeship – eine Lehre absolvieren

servant – Bedienstete(r)

service awards – Dienstalterzulage

service contract – Dienstvertrag

service industry – Dienstleistungsindustrie

service engineer – Dienstleistungsingenieur

service level – Dienstgrad

service rating – Einstufung nach Dienstalter

service sector – Dienstleistungsbranche; Dienstleistungssektor

services – Dienstleistung

servile – kriecherisch

session – Teilveranstaltung

set a time limit – befristen

set objectives – Arbeitsziele festsetzen

set of persons – Personenkreis,
Personengruppe
settle – erledigen; fixieren; regeln
settlement – Einigung; Vereinbarung
settlement of hardship cases –
Härtefallregelung; Härteregelung
settling-in period – Einarbeitungszeit
setup time – Rüstzeit
several-day trip – Reise, mehrtägige
severance pay reserves –
Abfindungsrücklagen
severance payment – Abfindung
severely disabled – schwerbehindert
severely disabled person –
Schwerbehinderte(r)
severely handicapped – schwerbehindert
Severely Handicapped Persons Act –
Schwerbehindertengesetz
severely handicapped person –
Schwerbehinderte(r)
sex – Geschlecht
sexual harassment – sexuelle
Belästigung
shake-up – Personalumbau
shake-out – Aussondern ineffizienter
Mitarbeiter
shame – Scham
share – Aktie
share in decisions – sich an
Entscheidungen beteiligen
share in wages – Lohnanteil
share liable to income tax – Anteil,
lohnsteuerpflichtiger
shared vision – gemeinsam getragene
Vision
shareholder – Aktionär; Anteilseigner;
Gesellschafter
sharp-witted – scharfsinnig
sheet metal worker – Blechschlosser
shift – Dienst
shift bonus – Schichtzulage
shift change compensation –
Nachteilsausgleich
shift differential – Schichtzulage
shift foreman – Schichtführer
shift indicator – Schichtkennzeichen
shift planning – Dienstplanung
shift rotation – Schichtwechsel
shift schedule – Schichtplan
shift substitution – Schichtvertretung
shift worker – Schichtarbeiter

shift(work) – Schicht(arbeit)
shipping – Fracht; Transport; Verladung
shop assistant – Verkäufer(in)
shop rule – Betriebsordnung
shop steward – Gewerkschaftsvertreter
im Unternehmen
short leave – Kurzurlaub
short list – engere Auswahlliste
short time – verkürzte Arbeitszeit
short-term – kurzfristig
short-term document – Kurzzeitbeleg
short-term employment – kurzzeitiges
Beschäftigungsverhältnis
short-term memory –
Kurzzeitgedächtnis
short-time allowance – Kurzarbeitergeld
short-time vacation – Kurzzeiturlaub
short-time work – Kurzarbeit
short-time worker – Kurzarbeiter(in),
Teilbeschäftigte(r)
short-winded – kurzatmig
shortage – Mangel
shortage of work – Arbeitsmangel
shortfall – Minderzeit
shorthand – Kurzschrift
shorthand typist – Stenotypist(in)
shutdown – Betriebsschließung
shy – scheu; schüchtern
sick – arbeitsunfähig; krank
sick day – Krankheitstag
sick leave – Dienstbefreiung wegen
Krankheit
sick leave grace period – Karenzurlaub
sick pay – Krankengeld
sick pay supplement –
Krankengeldzuschuss
sick pay supplement period –
Krankengeldzuschussfrist
sick room – Sanitätsraum
sickness – Erkrankung
sickness benefit – Krankengeld
sickness certificate – Krankenschein
sickness certificate printout –
Krankenscheinausdruck
sickness rate – Krankheitsquote
sickness tracking –
Krankheitsverfolgung
side job – Nebentätigkeit
sideline – Nebenberuf; nebenberufliche
Tätigkeit

Siemens engineering student's program
– Ingenieurkreis
silence – Stillschweigen
simple maintenance – einfache Pflege
simple organization structure –
Organisationsgrundstruktur
simplify – vereinfachen
simulate – verstellen
simulation – Simulation; Verstellung
sincerity – Aufrichtigkeit
sine qua non – unerlässlich
sinecure – Sinekure
single – ledig
single booking – Einzelbuchung
Single European Market – Europäischer
Binnenmarkt
single plant bargaining agreement –
Betriebsvereinbarung
single premium – Einmalbeitrag (z.B.
zur Versicherung)
single receipt – Einzelbeleg
single record entry – Einzelerfassung
single task – Einzelaufgabe
single work – Einzelarbeit
single-minded – zielstrebig
sit a law exam before a stipulated date –
Freischuss
sit an exam – eine Prüfung machen
sit-down strike – Sitzstreik
sit-in – Betriebsbesetzung
situation on the labour market –
Arbeitsmarktlage
sixth form – Kollegstufe
sixth form college – Oberstufe
sixth form student – Kollegstufenschüler
skeleton agreement –
Rahmenabkommen
skeleton staff – auf eine Mindestzahl
reduziertes Personal
skeleton time – Rahmenzeit
skeleton training schedule –
Ausbildungsrahmenplan
skill – Können; Geschick
skilled labour – Fachkraft; Spezialist
skilled staff – Fachpersonal
skilled worker – Facharbeiter
skills – Berufsqualifikation; Fähigkeiten;
Können
skim (over) – überfliegen (beim Lesen)
skin disease – Hautkrankheit

skip work – blaumachen;
Kurzzeiterkrankung
slap – schlagen
sleeping partner – stiller Teilhaber
slide – Dia; Folie
sliding wage scale – gleitende Lohnskala
slip – Panne
slow – schwerfällig (von Begriff)
slow on the uptake – begriffsstutzig
slow on the uptake – eine lange Leitung
haben; schwer von Begriff; Spätzünder
slums – Elendsviertel; Slums
smart-aleck – Besserwisser
smart-ass (AmE) – Besserwisser
smooth – reibungslos
sneak – petzen
sociable – kontaktfreudig
social advisory office – Sozialberatung;
Sozialberater
social affairs – Sozialaufgaben
social benefits – Sozialleistungen
social competence – Sozialkompetenz
social environment – soziales Umfeld
social expenditure –
Sozialaufwendungen
social expenses – Sozialkosten
social facilities – Sozialeinrichtungen
social growth of a child –
Gesamtentwicklung eines Kindes
social insurance – Sozialversicherung
social insurance agency –
Sozialversicherungsträger;
Sozialleistungsträger
social insurance agreement –
Sozialversicherungsabkommen
social insurance booklet –
Sozialversicherungsheft
social insurance contribution –
Sozialversicherungsbeitrag;
Sozialabgaben
social insurance law –
Sozialversicherungsgesetz
social insurance legislation –
Sozialversicherungsrecht
social insurance number –
Sozialversicherungsnummer
social insurance statement –
Sozialversicherungsnachweis
social intelligence – soziale Intelligenz
social isolation – soziale Isolation

social legislation – soziale
    Gesetzgebung; Sozialrecht
social legislation code –
    Sozialgesetzbuch
social loafing – soziales Faulenzen
social plan – Sozialplan
social policy – Gesellschaftspolitik;
    Sozialpolitik
social science – Sozialkunde
social sciences – Sozialwissenschaften
social security – soziale Sicherheit
social security (insurance) –
    Angestelltenversicherung
social security contribution (AmE) –
    Sozialversicherungsbeitrag
social security data – Daten,
    beitragsrechtliche
social security scheme –
    Sozialversicherungssystem
social security statement – Sozial-
    versicherungsentgeltbescheinigung
social services office – Sozialamt
social statistics – Sozialstatistik
social welfare supplement –
    Sozialzulage
social withdrawal – sozialer Rückzug
social (work) measures – soziale
    Maßnahmen
socialization – Sozialisierung
socially atrophied – sozial erfolgsarm
    sein
society – Verein
socio-economics – Sozioökonomie
socio-political – gesellschaftspolitisch
soft factor (social competence) –
    weicher Faktor (nicht fachliche
    Kompetenz)
software company – Software-Haus
sole proprietor – Alleininhaber (GmbH
    & Co. KG); Einzelkaufmann
sole wage earner – Alleinverdiener
solely responsible – alleinverantwortlich
solidarity – Solidarität
solidarity contribution –
    Solidaritätsbeitrag
solution – Lösung
solvent – zahlungsfähig
sophomore (AmE) – Student im zweiten
    Studienjahr
sorority – Studentinnenvereinigung
source of income – Erwerbsquelle

source of information –
    Informationsquelle
source of interference – Störquelle
space of time – Zeitraum
span of attention –
    Konzentrationsspanne
span of control – Kontrollspanne
span of management – Führungsspanne
span of supervision – Kontrollspanne
spare time – Freizeit
spatial – räumlich
speaker – Referent
special agreement – Sondervereinbarung
special allowance – Sonderzuwendung
special area – Fachgebiet; Sachgebiet
special assignments – Sonderaufgaben
special bonus – Sonderzulage;
    Sonderzuwendung
special class – Sondergruppe
special company leave – Freistellung,
    betriebliche
special course – Sonderlehrgang
special department for.... –
    Fachabteilung
special education – Sondererziehung
special field – Fachrichtung
special knowledge – Fachkenntnisse
special needs students – Sonderschüler
special payment calculations –
    Sonderzahlungslauf
special payroll run – Sonderabrechnung
special permission –
    Sondergenehmigung
special project – Sonderprojekt
special regulation – Sonderregelung
special remuneration – Sondervergütung
special rule – Sonderregelung
special training – Fachausbildung;
    Sonderausbildung
special wage contract –
    Sonderlohnvertrag
special work – Sonderarbeit
specialised secondary school –
    Fachgymnasium
specialist – Fachkraft; Spezialist
specialist area – Fachgebiet
specialist literature – Fachliteratur
specialist support – fachliche Betreuung
specialist teacher – Fachlehrer
specialist training –
    Spezialistenausbildung

specialists 248

specialists – Fachleute
specialized competence – fachliche
Zuständigkeit
specialized on-going training –
fachliche Weiterbildung
specialized technical knowledge –
technisches Fachwissen
specialized textbook – Fachbuch
specially gifted children – besonders
begabte Kinder
specific objective – operatives Ziel
specific requirements – fachliche
Anforderungen
specification – Spezifikation
specify – näher angeben; spezifizieren
speech – Sprache; Ansprache, Rede
speech impediment –
Sprachbehinderung
speed – Geschwindigkeit; Tempo
speed reading – Schnellesen
speech therapy – Logopädie
speed of learning – Lerntempo
speedup – Leistungssteigerung
spell – buchstabieren
spelling test – Buchstabiertest
sphere of action – Wirkungsbereich
sphere of activity – Arbeitsgebiet;
Aufgabenbereich
spin-off – Nebenprodukt; Spinoff
spirit of the times – Zeitgeist
split accuracy – Splitt-Genauigkeit
split indicator – Splittkennzeichen
split period – Splittzeitraum
spokesman on educational affairs –
bildungspolitischer Sprecher
spokesman role – Sprecherrolle
sponge – Schwamm
sponsor – Kostenträger; Förderer;
Sponsor (an individual or group with
the power to sanction or legitimize a
project)
spontaneous – Spontanität; spontan
sports – Sport
sports facilities – Sportanlagen
sports grounds – Sportanlagen
sports teacher – Sportlehrer
sports-day – Sportfest
spouse – Ehegatte; Ehegattin
spouse (husband/wife) – Ehepartner(in)
spread – Verteilung auf
spread (over) – verteilen

spring term – Frühlingstrimester
Spoonerism – lustiger Versprecher (The
name of William Archibald Spooner,
British clergyman and educationalist is
forever associated with a nervous
tendency to transpose initial letters or
half-syllables in speech, e.g. "a
half-warmed fish" for "a half-formed
wish".)
stable – stabil
stable wage – fester Lohn
staff – Arbeitsstab; Belegschaft;
Mitarbeiterstab, Bediensteter,
Mitarbeiterkreis
staff bond – Mitarbeiterbindung
staff cut – Personalabbau
staff cutback – Personaleinsparung
staff dialogue – Mitarbeitergespräch
staff function – Stabsfunktion
staff increase – Personalaufbau
staff indicator – Stabskennzeichen
staff list – Personalverzeichnis
staff man – Stabsmann
staff manager – Personalsachbearbeiter
staff meeting –
Belegschaftsversammlung
staff membership in a company or firm
– Betriebszugehörigkeit
staff outing – Betriebsausflug
staff participation –
Mitarbeiterbeteiligung
staff position – Stabstelle
staff promotion – Personalentwicklung
staff relationship –
Belegschaftsatmosphäre
staff report – Belegschaftsbericht
staff representation – Personalvertretung
staff savings – Personaleinsparung
staff turnover – Personalwechsel
staff unit – Stabsabteilung
staff-student ratio –
Lehrer-Schüler-Verhältnis
staff-students council –
Schülermitverantwortung;
Schülermitverwaltung
staffing adjustment –
Personalanpassung,
Personalbestandsanpassung
staffing percentage –
Besetzungsprozentsatz

staffing schedule – Planstellenbesetzungsplan

staffing schedule for management – Leitungsstellenbesetzungsplan

staffroom – Lehrerzimmer

stage – Arbeitsstufe

stage-fright – Aufregung; Lampenfieber

staggered hours – gestaffelte Arbeitszeit

stamina – Ausdauer; Durchhaltefähigkeit; Stehvermögen

stammer – stottern

stammerer – Stotterer

stand-by duty – Bereitschaftsdienst

stand-in – Vertretung

standard – Leistungsstandard; Niveau; Norm; Standard

standard agreement provisions – Tarifbedingungen

standard annual leave – Tarifurlaub

standard bonus – Tarifzulage

standard company number – Betriebsnummer, einheitliche

standard day type – Standard-Tagestyp

standard file – Mustermappe

standard file maintenance – Pflege der Mustermappe

standard hourly pay – Tarifstundenlohn

standard hourly pay rate – Stammstundenlohnsatz

standard monthly pay – Tarifmonatslohn

standard of attainment – Leistungsgrad; Leistungsniveau; Leistungsnorm

standard of living – Lebensstandard

standard of performance – Leistungsgrad; Leistungsniveau; Leistungsnorm

standard output – Soll-Leistung

standard performance – Vorgabeleistung; Leistungskriterium

standard time – Vorgabezeit

standard wage group – Lohngruppe, tarifliche

standard wage level – Lohnstufe, tarifliche

standard wage maintenance – Verdienstsicherung

standard week – Standardwoche

standardization – Normierung

standardized test – normierter Test

standing – Ansehen

standing order – Dauerauftrag

standing orders – Geschäftsordnung

star pupil – bester Schüler

start in a job – Berufseinstieg

start in working life – Eintritt ins Erwerbsleben

start of the school year – Schuljahresbeginn

start of working hours – Arbeitszeitbeginn

Start-up – Geschäfts-Neugründung

starting of work – Dienstantritt; Dienstaufnahme

starting rate – Anfangslohn; Einstell-Lohn

starting salary – Anfangsgehalt; Eingangsgehalt

starting time – Arbeitsanfang; Arbeitsbeginn

starting wage – Anfangslohn

state – Land

state benefits office – Versorgungsamt

state of employment – Anstellung; Arbeitsverhältnis; Beschäftigung

state of the economy – Konjunktur; Wirtschaftslage

state school – öffentliche Schule; staatliche Schule

state sponsored retraining – staatlich finanzierte Umschulung

state-governed secondary grammar school – staatliches Gymnasium

state-of-the-art – Stand der Technik

state-registered teacher – staatlich geprüfter Lehrer

statement – Bescheinigung

statement for wage withholding tax – Steuernachweis

statement of account – Abrechnung

statement of contributions paid – Beitragsnachweis

statement of earnings – Verdienstbescheinigung; Verdienstnachweis

statement of gross wages earned – Bruttolohnnachweis

statement of housing subsidy – Wohngeldbescheinigung

statement wage type – Ausweislohnart

statistical projection – statistisches Prognoseverfahren

statistics – Statistik
status – Status
status feature – Statusmerkmal
status of residence permit – Aufenthaltsstatus
status-oriented – statusorientiert
statute – Satzung
statutory – satzungsgemäß
statutory health insurance – gesetzliche Krankenversicherung
statutory membership – Pflichtmitgliedschaft
statutory net amount – Netto, gesetzliches
statutory pension insurance – Rentenversicherungspflicht
statutory pension insurance fund – gesetzliche Rentenversicherung
statutory salary scale – Bundesangestelltentarif
statutory source – Rechtsgrundlage
stay – Aufenthalt
stay down – sitzenbleiben
staying power – Stehvermögen
steady – stetig
steepening incentive – progressive Produktionsprämie
stenography – Kurzschrift
step – Maßnahme
stepchild – Stiefkind
stiff – steif
stimulating – anregend
stimulus – Anregung; Ansporn; Stimulus
stock market rate – Börsenkurs
stock option – Aktienbezugsrecht, Optionsrecht
stock ownership – Aktienbesitz
stockholder – Aktionär; Anteilseigner; Gesellschafter
stopgap – Aushilfskraft; Gelegenheitsarbeiter
stopover – Zwischenziel
stopover night – Zwischenzielübernachtung
stoppage – Betriebsausfall
stoppage of work – Arbeitsunterbrechung
stopping time – Arbeitsschluss
storage – Lagerung
store – ablegen, lagern
store-room clerk – Lagerverwalter

storehouse – Lager
storeman – Lagerarbeiter
stores – Lagerwesen
straight – Gradlinigkeit
straight from the horse's mouth – Information aus erster Hand
strain – Belastung
strained – angespannt
strategic business unit (SBU) – strategische Geschäftseinheit (SGE)
strategic mission – Marktaufgabe; Mission
strategic planning – Geschäftspolitik; strategische Planung; Unternehmenspolitik
strategy – Strategie
streaming – Einteilung in Leistungsgruppen
streamline – modernisieren; rationalisieren; vereinfachen
strength – Stärke
strength of character – Charakterfestigkeit
strength of vision – Sehvermögen
stress – nervliche Belastung
strict – streng
strike – Streik
strike benefit – Streikgeld
strike benefits – Streikvergütung
strike control – Streikregelung
strike fund – Streikfonds; Streikkasse
strike insurance – Streikversicherung
strike vote – Urabstimmung
strike-breaker – Streikbrecher
strong point – hervorragende Eigenschaft; Stärke
structural change – Strukturwandel
structural engineering – Hochbau
structure – Struktur
structure depth – Strukturtiefe
structure of the tasks – Aufgabenstruktur
structured teaching – strukturierter Unterricht
structuring of operations – Ablauforganisation
stubborn – eigensinnig; starrsinnig
student – Hörer; Schüler
student council – Studentenrat
student adviser – Studienberater
student grant – Studienbeihilfe

student limitation – Numerus Clausus
student taking his diploma examination – Diplomand
student teacher – Studienreferendar
student who is taking or has taken the "Abitur" examination – Abiturient
students hostel – Studentenheim; Studentenwohnheim
students' representation – Studentenvertretung
studies at a Fachhochschule – Fachhochschulstudium
study – Arbeitszimmer; studieren
study a subject – Fachgebiet studieren
study aids – Studienhilfen
study group – Arbeitsgruppe; Arbeitskreis
study habits – Studiengewohnheiten
study leave – Bildungsurlaub
study period – angesetzte Arbeitsstunde
stumbling block – Hindernis, Hemmschuh
stutter – stottern
stutterer – Stotterer
style of learning – Lernstil
style of teaching – Unterrichtsstil
subject – Fach; Lehrgegenstand; Sachgebiet; Studienfach
subject area – Fachrichtung
subject matter – Unterrichtsinhalt; Unterrichtsstoff
subject to – vorbehalten
subject to approval – genehmigungspflichtig
subject to compulsory insurance – versicherungspflichtig
subject to directives – weisungsabhängig; weisungsgebunden
subject to notice – kündbar
subject to social insurance contribution – sozialversicherungspflichtig
subject-specific – fachspezifisch
submit – einreichen
submitted – beantragt
subordinate – Untergebene(r)
subordination – Unterordnung
subschema – Teilschema
subscription right – Bezugsrecht
subsequent improvement – Nachbesserung
subsequent payment – Nachzahlung

subsequent sickness – Folgekrankheit
subsequent statement – Folgebescheinigung
subsequent time ticket – Folgelohnschein
subsequent wage type – Folgelohnart
subsidiary – Tochtergesellschaft; verbundenes Unternehmen
subsidiary agreement – Nebenabrede
subsidiary subject – Nebenfach
subsidize – bezuschussen
subsidy – Zuschuss
substitute – Stellvertreter; Vertreter
substitute attendee – Ersatzteilnehmer
substitute teacher – Ersatzlehrer
substitution – Vertretung
substitution hour – Vertretungsstunde
subtask – Teilaufgabe
subtle – subtil
subtract – abziehen
subtype – Informationssubtyp
succeed – Erfolg haben
succession candidate – Nachfolgekandidat
succession planning – Nachfolgeplanung; Nachwuchsplanung
successor – Nachfolger
sue a company – eine Firma verklagen
sufficient – ausreichend
suggestion – Ratschlag
suggestion box – Vorschlagskasten
suggestion for improvement – Verbesserungsvorschlag
suggestion scheme – Verbesserungsvorschlagswesen; Vorschlagswesen
suggestopaedia – Suggestopädie
suitability – Eignung
suitability percentage – Eignungsprozentsatz
suitability profile – Eignungsprofil
suitable – geeignet, passend
suitable to the situation – situationsgerecht
sum – Betrag
summa cum laude – mit höchstem Lob; summa cum laude
summarized form – Kompaktformular
summary – Kurzfassung; Zusammenfassung

summary dismissal (employer) –
außerordentliche Kündigung; fristlose
Kündigung; – fristlose Entlassung
summer school – Sommerschule
summer term – Sommersemester
summer vacation – Sommerurlaub
Sunday work – Sonntagsarbeit
superannuation – Ruhegehalt
superficial – flüchtig; oberflächlich
superfluous – überflüssig
superintendant (of exams) –
Aufsicht(sperson)
superior – Vorgesetzter; souverän
superior objectives – übergeordnete
Ziele
superior's statement – Stellungnahme
des Vorgesetzten
supervise an exam – eine Prüfung
beaufsichtigen; Examensarbeit
beaufsichtigen
supervision – Beaufsichtigung;
Prüfungsaufsicht
supervisor – Aufsicht
supervisory authority –
Aufsichtsbehörde
supervisory behaviour –
Vorgesetztenverhalten
Supervisory Board – Aufsichtsrat
supervisory job – Führungsstelle
supervisory school authority –
Schulamt
supervisory training –
Vorgesetztenschulung
supplementary benefits –
Zusatzversorgung
supplementary compensation –
Nachzahlung
supplementary course – Zusatzkurs
supplementary disability pension –
Invalidenzusatzrente
supplementary holiday – Zusatzurlaub
supplementary income –
Nebeneinkommen; Nebeneinnahmen
supplementary insurance –
Zusatzversicherung
supplementary pension fund –
Zusatzversorgungskasse
supplementary pension institute –
Zusatzversorgungseinrichtung
supplementary vacation (AmE) –
Zusatzurlaub

supply – versorgen; Versorgung; Angebot
supply and demand – Angebot und
Nachfrage
supply facilities –
Versorgungseinrichtung
supply of temporary workers –
Arbeitnehmerüberlassung
supply teacher – Aushilfslehrer
support – unterstützen
support payment arrears –
Unterhaltsrückstand
suppose – annehmen
supranational social insurance law –
überstaatliches
Sozialversicherungsrecht
supreme Labour Court –
Bundesarbeitsgericht
surety – Bürgschaft
surname – Familienname; Zuname
surname prefix – Vorsatz
surpass – hervorragen; sich auszeichnen;
übertreffen
surplus – Überschuss
surplus of graduates – Überangebot an
Akademikern
survey – Bewertung, alternative;
Erhebung; Übersicht
survey of earnings – Verdiensterhebung
survey of earnings – Lohn- und
Gehaltserhebung
survey of established posts –
Planstellenübersicht
surviving dependent – Hinterbliebener
surviving dependents' insurance –
Hinterbliebenenversorgung
survivors – Verbleibende
survivor's pension –
Hinterbliebenenrente
suspend – aussetzen; eine Tätigkeit
vorübergehend einstellen
suspensive condition – aufschiebende
Bedingung
sweated labour – ausgebeutete Arbeiter
swing the lead – krankfeiern
swot – büffeln, pauken
syllable – Silbe
syllabus – Lehrplan; Studienplan;
Unterrichtsplan
syllabus design – Aufstellen eines
Lehrplans
symbolic account – Konto, symbolisches

symbolic account assignment – Kontierung, symbolische
sympathetic action – Solidaritätsmaßnahme
sympathy strike – Solidaritätsstreik; Sympathiestreik
symposium – Symposion
synergy – Synergie
syntactic construction – Syntagma
synonym – Synonym
synthesis – Synthese
synthetic – synthetisch
system of thought – Denksystem
system of values – Wertesystem
systematic promotion – systematische Förderung
systems integration engineer – Fachinformatiker
systems technique – Anlagentechnik
systems thinking – systemisches Denken

T

table – Tabelle
table value – Tabellenwert
taboo – Tabu
tactful – taktvoll
tactile – taktil; Tast-
tactless – taktlos
take a day in lieu of – abfeiern
take a degree – akademischen Grad erwerben
take a doctor's degree – promovieren
take a subject – Fachgebiet studieren
take an exam – eine Prüfung machen
take effect – in Kraft treten
take instructions from – Anweisungen nehmen
take into account – berücksichtigen
take into consideration – berücksichtigen
take off the university register – exmatrikulieren
take on – einstellen

take orders from – Anweisungen nehmen
take over – übernehmen
take over new tasks – neue Aufgaben übernehmen
take sides – Partei ergreifen
take up work – Dienstantritt; Dienstaufnahme
take-home pay – Effektivlohn; Nettolohn
taken for granted – Selbstverständlichkeit
talent – Begabung
talent for languages – sprachbegabt
talented – begabt
tape recorder – Tonbandgerät
tardiness – Langsamkeit; Unpünktlichkeit
target – Hauptziel; Ziel; Zielvorgabe; Zielvorstellung
target area – Zielbereich
target group – Zielgruppe
target object – Zielobjekt
target oriented – zielbezogen; zielorientiert
target plan – Sollplan
target setting – Zielfestlegung
target specification – Sollvorgabe
target time – Sollzeit
targeted employment policy – gezielte Einstellungspolitik
tariff agreement – Tarifvertrag
tariff staff – Tarifkreis
task – Arbeitszuweisung am Arbeitsplatz; Aufgabe
task analysis – Aufgabenanalyse
task behaviour – aufgabenorientierte Führung
task catalog – Aufgabenkatalog
task complex – Aufgabenkomplex
task fulfillment – Aufgabenerfüllung
task group – Aufgabengruppe
task planning – Arbeitsplanung; Planung
task profile – Aufgabenprofil
task structuring – Aufgabengliederung
tautology – Tautologie
tax – Steuer
tax allowance – Freibetrag
tax authorities – Steuerbehörden
tax basis – Besteuerungsgrundlage
tax bracket – Steuerklasse

tax calculation procedure – Steuerberechnungsverfahren

tax declaration – Steuererklärung

tax deduction – Abzug, steuerlicher

tax equalization payment – Steuerausgleichszahlung

tax exemption – Steuererlass; Steuerbefreiung

tax free – steuerfrei

tax group – Steuerklasse

tax office – Steueramt

tax payer's account number – Steuernummer

tax period – Steuerzeitraum

tax refund – Steuererstattung

tax return – Steuererklärung

tax scale – Steuertarif

tax statement – Steuerbescheid

tax withholding – Steuereinbehaltung

tax-deductible – steuerabzugsfähig

tax-deductible amount for car – Kfz-Pauschale

tax-exempt – steuerfrei

tax-free amount – Steuerfreibetrag

tax-free Christmas allowance – Weihnachtsfreibetrag

tax-free income – lohnsteuerfrei

tax-privileged – steuerbegünstigt

taxable income – Einkommen, versteuerbares

taxonomy – Taxonomie

teach – anleiten; lehren; unterrichten

teacher – Lehrer(in); Lehrkraft, Schulmeister(in), Erzieher(in)

teacher at a school for handicapped – Sonderschullehrer(in)

teacher candidate – Lehramtsbewerber

teacher centred lesson – Frontalunterricht

teacher on probation – Referendarzeit

teacher training – pädagogische Ausbildung

Teacher Training College (T.T.C.) – Pädagogische Hochschule (PH)

teacher's book – Lehrerhandbuch; Lehrerhandreichungen

teacher's diploma – Lehrbefähigung

teachers' organization – Lehrerverband

teaching aids – Lehrmittel

teaching assistant – Assistenzlehrer

teaching job – Lehrberuf

teaching load – Unterrichtsstunden

teaching load per week – Unterrichtsstunden je Woche

teaching material – Unterrichtsmaterial

teaching media – Lehrmedien

teaching method – Lehrmethode; Unterrichtsmethode

teaching practice – Unterrichtserfahrung

teaching profession – Lehrberuf

teaching skills – Fertigkeiten des Unterrichtens

teaching staff – Lehrerkollegium

teaching the highly talented – Begabtenförderung (Teachers are frequently insufficiently trained to be able to teach the highly talented pupils satisfactorily.)

teaching unit – Unterrichtseinheit

teachware – Lehrmaterial

team ability – Teamfähigkeit

team spirit – Gruppengeist; Teamgeist

team teaching – Unterricht im Team

teamwork – Gruppenarbeit; Teamarbeit

teamwork performance – Verhalten in der Zusammenarbeit

tease – necken; hänseln

technical – technisch; fachspezifisch

Technical College – Fachhochschule (BRD); Berufsschule

technical consultant – Fachberater

technical courses – technisches Kolleg

technical drawing – technisches Zeichnen

technical employees – technisch Tätige

Technical Engineering College – Technische Hochschule; Technische Universität

technical knowledge – Fachwissen

technical questions – Fachfragen

technical school – berufsbildende Schule

technical school for the handicapped – Berufsschule für Behinderte

technical skill – fachliches Können

technical supervisor – technischer Aufsichtsbeamter

technical term – Fachausdruck

technical training – technische Ausbildung

Technical University – Technische Hochschule; Technische Universität

technical wage type – Lohnart,
  technische
technician – Techniker
technique – Methode; Verfahren
technocrat – Technokrat
technology – Technik
technostructure – Technostruktur
tedious – weitschweifig
teens – Teenageralter
telecommunication engineering –
  Fernmeldetechnik
telecommunications electronics –
  Fernmeldeelektronik
teleprocessing – Datenfernverarbeitung
tele outworker – Teleheimarbeiter
tele-learning – Telelernen
teleworker – Telearbeiter
teleworking – Telearbeit
tell (someone) off (for something) –
  jemanden ausschimpfen (für etwas)
tell tales – petzen
telltale – Petzer
temporary employment – Leiharbeit;
  Zeitarbeit
temporary employment agency –
  Arbeitnehmerüberlassungsfirma
temporary employment on loan basis –
  Leiharbeitsverhältnis
temporary freezing of payments –
  Zahlungsverbot, vorläufiges
temporary help (AmE) –
  Leiharbeitskraft
temporary incapacity – vorübergehende
  Arbeitsunfähigkeit
temporary lay-off – vorübergehende
  Entlassung; zeitweilige Freisetzung
temporary personnel – Aushilfe
temporary solution – Notlösung
temporary student employee –
  Werkstudent
temporary work – Aushilfstätigkeit
temporary work relationship –
  Arbeitsverhältnis, befristetes
temporary worker – Aushilfskraft;
  Gelegenheitsarbeiter
tense – angespannt
tension – Spannung
tentatively – versuchsweise
tenure – Amtszeit
tenure-track-professor – befristete
  Professur

tenured employment – unkündbare
  Stellung
term – Bedingung; Semester,
  Gültigkeitsdauer, Gültigkeitszeitraum;
  Laufzeit (z.B. einer Versicherung,
  Rente)
term of limitation – Verjährungsfrist
term of office – Amtszeit
term of prescription – Verjährungsfrist
term paper – Semesterarbeit
term report – Zwischenzeugnis
term spent in industry – praktisches
  Studiensemester
terminableness – Kündbarkeit
termination – Beendigung; Kündigung
termination of employment – Austritt;
  Dienstenthebung; Entlassung
termination of pregnancy –
  Schwangerschaftsabbruch
termination payment –
  Abgangsentschädigung
termination with notice – ordentliche
  Kündigung
terms and conditions of insurance –
  Versicherungskonditionen
terms of admission –
  Zulassungsbestimmung
tertiary education – tertiäre Bildung
test – ausprobieren; Erfolgskontrolle;
  Kurzprüfung
test aid – Prüfhilfe
test department – Prüffeld
test equipment – Experimentiergerät
test evaluation – Auswertung von Tests
test material – Prüfungsmaterial;
  Testunterlagen
test paper – Klassenarbeit
test procedure – Testabwicklung
test run (AmE) – Prüflauf
test score – Testergebnis
testator – Erblasser
testimonial – Empfehlungsbrief;
  Empfehlungsschreiben;
  Führungszeugnis
textbook – Lehrbuch
13th monthly pay – 13. Monatsgehalt
the same age – Gleichaltrige(r)
the second way of gaining university –
  Zweiter Bildungsweg
theme – Thema
theoretical – theoretisch

theoretical basis – theoretische Grundlagen
theoretical knowledge – theoretische Kenntnisse
theoretical tuition – fachtheoretischer Unterricht
theorist – Theoretiker
therapeutic – therapeutisch
therapeutic treatment – Heilbehandlung
therapy – Therapie
thesis – Diplomarbeit
thesis defense – Rigorosum
thesis supervisor – Doktorvater
think-tank – Denkfabrik
thinking – Denken
thinking in business categories – unternehmerisches Denken
thinking in narrow categories – Schubladendenken
thinking process – Denkprozess
think over – überdenken
third-level education – tertiäre Bildung
third-level sector – Tertiärbereich
third-party action against execution – Widerspruchsklage
third-party liability insurance – Haftpflichtversicherung
third-party payroll accounting – Lohn- und Gehaltsabrechnung (fremde)
thirst for action – Tatendrang
thorough – eingehend; gründlich
thoroughness – Gründlichkeit; Sorgfalt
thought – Gedanke; Gedankengang
thoughtful – nachdenklich
threat – Bedrohung
360°-feedback – periphäre Beurteilung
three R's (reading (w)riting, (a)rithmetic) – Lesen, Schreiben, Rechnen
three-period average – Dreiperiodendurchschnitt
threshold – Schwelle
threshold performance – Leistungsgrenze; Prämiengrenze
through the ranks – Ochsentour
tie – Stimmengleichheit
time accounting – Zeitabrechnung
time administrator – Zeitbeauftragter
time agreement – Zeitvereinbarung
time and motion study – Arbeitsstudie; Zeit- und Bewegungsstudie

time authorization group – Zeitberechtigungsgruppe
time balance – Zeitsaldo
time balance correction – Zeitsaldokorrektur
time basis – Zeitunterlage
time bonus – Zeitzuschlag
time card – Lohnkarte; Stechkarte; Stempelkarte
time category – Zeittyp
time clock – Stechuhr; Stempeluhr
time constraint – Zeitbindung
time credit – Zeitguthaben; Zurechnungszeit
time data – Zeitdaten
time data entry – Zeitdatenerfassung
time data management – Zeitdatenverwaltung
time data processing – Zeitdatenverarbeitung
time deduction – Zeitabschlag
time evaluation – Zeitauswertung
time evaluation run – Zeitauswertungslauf
time evaluation schema – Zeitauswertungsschema
time field – Uhrzeitfeld
time group – Zeitgruppe
time in lieu – Freizeitausgleich
time leveling – Zeitabgleich
time line – Zeitplan; Zyklogramm
time limit – Frist
time management – Zeitwirtschaft
time measurement – Zeitmessung
time model – Zeitmodell
time off – Arbeitsbefreiung; Ausfallzeit
time office – Arbeitsstudienabteilung
time offset – Zeitausgleich
time periods – Zeitraster
time posting – Zeitbuchung
time pressure – Zeitdruck
time problem – Zeitproblem
time processing – Zeitverarbeitung
time quota – Zeitkontingent
time rate – Zeitlohn
time recording – Zeiterfassung
time recording device – Zeiterfassungsgerät
time recording system – Zeiterfassungssystem

time recording terminal – Zeiterfassungsterminal
time report – Stundenzettel
time result – Zeitergebnis
time rule – Uhrzeit-Regelung
time savings – Zeitersparnis
time segment – Zeitabschnitt
time series analysis – Zeitreihenanalyse
time span – Zeitspanne
time stamp – Zeitstempel
time standard – Zeitvorgabe
time statement form – Zeitnachweisformular
time study – Zeitstudie
time study engineer – Arbeitsstudien-Ingenieur
time study observation sheet – Zeitstudien-Beobachtungsbogen
time study sheet – Zeitaufnahmebogen
time substitution – Uhrzeitvertretung
time table – Stundenplan
time target – Zeitvorgabe
time ticket – Lohnschein
time travelled – Wegezeit
time under articles – Referendarzeit
time unit – Zeiteinheit
time wage – Zeitlohn
time wage type – Zeitlohnart
time-based delimitation – Abgrenzung, zeitliche
time-based piecerate work – Zeitakkord
time-table planning – Stundenplangestaltung
time-unit code – Dauercode
timekeeping – Zeitkontrolle
timetable – Stundenplan
timid – ängstlich; furchtsam
timing – richtiger Zeitpunkt; Timing
tip – Trinkgeld
to be logged – belegrelevant
to bear the cost – Kosten tragen
to issue (someone with) a certificate – bescheinigen
to leave – austreten
to perform follow-up course work – nachbereiten (eines Kurses)
to show symptoms of fatigue – Ermüdung zeigen
to withdraw – ausscheiden
token strike – Warnstreik
tolerance – Toleranz

tolerance interval – Toleranzspanne
tolerate – dulden
tolerated – geduldet
tool maker – Werkzeugmacher
tool mechanic – Werkzeugmechaniker
toolsetter – Einrichter
top earner – Spitzenverdiener
top efficiency – Spitzenleistung
top manager – Direktionsmitglied; Spitzenführungskraft
top of the class – Klassenbester
top position – Spitzenposition
top wage rate – Höchstlohn; Spitzenlohn
top wages – Höchstlohn; Spitzenlohn
topic – Thema
total – Gesamt...
total costs for education and training – Bildungsaufwand
total exposure course – Intensivkurs
total gross amount – Bruttosammelergebnis; Gesamtbrutto
total income – Gesamteinkommen
total of all contributory and non-contributory periods to social insurance – Versicherungsverlauf
total output – Gesamtleistung
total permanent incapacity – vollständige und dauernde Arbeitsunfähigkeit
total salary – Gesamtgehalt
total social security contribution – Gesamtsozialversicherungsbeitrag
total staff costs – Personalaufwand
total wages – Gesamtlohn
total work force – Gesamtbelegschaft
total working hours – Gesamtarbeitszeit
touch – berühren
tough negotiator – schwieriger Verhandlungspartner
tour a company – einen Betrieb erkunden
track record – bisherige Karriere
trade – Beruf; Gewerbe; handwerklicher Beruf
trade dispute – Arbeitskampf; Tarifkonflikt
trade fair – Messe
trade inspection board – Gewerbeaufsichtsamt
Trade Regulation Act – Gewerbeordnung

trade school – berufsbildende Schule;
Berufsschule; Gewerbeschule
trade secret – Betriebsgeheimnis;
Geschäftsgeheimnis
trade supervisory authority –
Gewerbeaufsichtsbehörde
trade supervisory officer –
Gewerbeaufsichtsbeamter
trade tax – Gewerbesteuer
trade union – Gewerkschaft
trade union movement –
Gewerkschaftsbewegung
tradesman – Handwerker
train – schulen
train on the job – einarbeiten
trained apprenticeship – abgeschlossene
Lehre
trained staff – Fachpersonal
trainee – Auszubildende;
Auszubildende(r); Azubi; Volontär
trainee (AmE) – Praktikant
trainee for personnel management –
Personalführungsnachwuchs(kraft)
trainee dropout – Ausbildungsabbrecher
trainee place – Lehrstelle
traineeship – Referendarzeit
trainer – Ausbilder; Ausbildungsmeister;
Lehrmeister
training – Ausbildung; berufliche
Bildung; Schulung
training allowance –
Ausbildungsbeihilfe;
Ausbildungsvergütung
training and continuing education –
Aus- und Weiterbildung
training and development measures –
Maßnahmen zur Aus- und
Weiterbildung
training and development of personnel
– Aus- und Weiterbildung der
Mitarbeiter
training and education system –
Bildungssystem
training and information – Bildung und
Information
training as a skilled worker –
Facharbeiterausbildung
training at a technical college –
Fachschulausbildung
training bonus – Ausbildungszulage
training by stages – Stufenausbildung

training centre – Ausbildungsstätte;
Schulungszentrum
training company – Ausbildungsbetrieb
training concept – Ausbildungskonzept
training concluded – ausgelernt
training contract – Ausbildungsvertrag;
Berufsausbildungsvertrag
training costs – Ausbildungsaufwand
training course – Kurs
training department –
Ausbildungsabteilung
training directive – Ausbildungsordnung
training firm – Ausbildungsbetrieb
training goal – Ausbildungszeit
training manager – Bildungsmanager
training material – Schulungsunterlagen
training measure –
Ausbildungsmaßnahme
training objective – Ausbildungszeit
training occupation – Ausbildungsberuf
training off-the-job – Fortbildung
außerhalb des Arbeitsplatzes
training on-the-job – Fortbildung am
Arbeitsplatz
training package – Schulungspaket
training period – Anlernzeit
training position – Ausbildungsplatz;
Ausbildungsstelle
training possibilities –
Bildungsmöglichkeiten
training possibilities on offer –
Ausbildungsmarkt
training profile – Ausbildungsprofil
training program – Ausbildungsplan;
Ausbildungsinhalt
training programme –
Ausbildungsprogramm;
Fortbildungsprogramm
training regulations –
Ausbildungsordnung
training schedule – Ausbildungsplan
training scheme –
Ausbildungsprogramm;
Fortbildungsprogramm
training school for master craftsmen –
Meisterschule
training sequence – Ausbildungsablauf
training status – Ausbildungsverhältnis
training supervisor – Lehrgangsleiter
training system – Ausbildungssystem
training workshop – Lehrwerkstatt

trait – (Charakter)Zug; Merkmal
traitor – Verräter
transfer – versetzen; Versetzung
transfer month – Überweisungsmonat
transfer request – Versetzungsgesuch
transfer to a foreign country –
  Auslandsentsendung
transfer to the salary payroll –
  Übernahme in das
  Angestelltenverhältnis
transferee abroad – ein ins Ausland
  versetzter Mitarbeiter
transitional allowance –
  Überbrückungsbeihilfe;
  Übergangsbeihilfe; Überbrückungsgeld
transitional payment –
  Übergangszahlung (Pensionäre)
translate – übersetzen
transparency – Folie
transport allowance – Fahrgeldzuschuss
transportation allowance –
  Fahrtkostenzuschuss
transportation cost accounting –
  Fahrtkostenabrechnung
transportation costs – Fahrtkosten
transportation expenses – Fahrgeld
transportation receipt –
  Fahrtkostenbeleg
travel – Reise
travel accident – Wegeunfall
travel accounting area –
  Reise-Abrechnungskreis
travel expense management –
  Reisekostenmanagement
travel expense posting –
  Reisekostenbuchung
travel expense regulations –
  Reisekostenbestimmungen
travel expense report –
  Reisekostenabrechnung
travel expense statement –
  Reisekostennachweis
travel expenses – Reisekosten
travel privileges – Reiseprivilegien
travelling allowance – Reisevergütung
traversee – Quereinsteiger
treat – behandeln
trend extrapolation – Trendberechnung;
  Trendextrapolation
trend towards university graduates –
  Akademikertrend

trend towards university studies –
  Akademisierungstrend
trimester – Trimester
trial and error – empirische Lösung
trial and error method – empirische
  Methode
Trinity term – Sommertrimester
trip abroad – Auslandsreise
trip application – Reiseantrag
trip area – Reisebereich
trip data – Reisedaten
trip destination – Reiseziel
trip duration – Reisezeit
trip home – Grenzübertritt, Rückreise
trip lasting less than one day – Reise,
  untertägige
trip number – Reisenummer
trip out – Grenzübertritt, Hinreise
trip schema – Reiseschema
trip status – Reisestatus
trip type – Reiseart
trouble-maker – Nörgler; Querulant;
  Störenfried; Störer
truancy officer (BrE) – Beamter einer
  Schulbehörde, der Fälle von
  Schulschwänzen untersucht
truant – Schwänzer, Schulschwänzer
trust – Vertrauen; Zutrauen
trust (AmE) – Konzern
trust deed – Treuhandvertrag (Urkunde)
trust fund – Treuhandfonds;
  Treuhandvermögen
trustee – Treuhänder
trusting – gutgläubig
trustworthy – vertrauenswürdig
tuckshop – Bonbonladen
tuition – Unterricht
tunnel vision – Betriebsblindheit
turner – Dreher
turnover – Umsatz
tutor – Studienbetreuer; Tutor
tutorial – Unterrichtsstunde mit einem
  Tutor
tutoring – Lernbetreuung
two tier – zweistufig
two-fold burden – Doppelbelastung
two-way communication – doppelt
  gerichtete Kommunikation
type of birth – Geburtsart
type of communication –
  Kommunikationsart

type of deadline date – Terminart
type of increase – Erhöhungstyp
type of school – Schulart
typist – Schreibkraft

U

ultimatum – Ultimatum
umbrella agreement –
Manteltarifvertrag
unable to work – arbeitsunfähig;
erwerbsunfähig
unapproved overtime – Überzeit
unbiased – unparteiisch;
unvoreingenommen
uncertainty – Unsicherheit;
Verunsicherung
uncommitted – nicht engagiert
uncompromising – kompromisslos
unconditional – bedingungslos
unconscious – unbewusst
uncontrolled – unbeherrscht,
unkontrolliert
undeclared employment –
Schwarzarbeit
under-achiever – einer, der hinter den
Erwartungen zurückbleibt
undergraduate – Student
underperforming – leistungsschwach
understaffed profession – Mangelberuf
understandable – verständlich
understanding – Auffassungsgabe;
Verständnis; Verständigung; einsichtig
understanding of one's role –
Rollenverständnis
undetermined attendance – Teilnahme,
offene
uneducated – ungebildet
unemployed – arbeitslos
unemployed person – Arbeitslose(r)
unemployment – Arbeitslosigkeit
unemployment assistance –
Arbeitslosenhilfe
unemployment benefit –
Arbeitslosengeld

unemployment compensation –
Arbeitslosenunterstützung
unemployment insurance –
Arbeitslosenversicherung
unemployment insurance –
Arbeitslosenversicherung
unemployment insurance fund –
Arbeitslosenkasse
unemployment payment –
Arbeitslosenunterstützung
unemployment rate – Arbeitslosenrate
unemployment relief – Arbeitslosenhilfe
unequivocal – eindeutig
unfavourable – ungünstig
uniform – Schuluniform
unfounded anxiety – unbegründete
Angst
ungifted – unbegabt
uninhibited – hemmungslos
union dues – Gewerkschaftsbeiträge
union policy – Gewerkschaftspolitik
union rates – Tarifgehalt
union representative –
Gewerkschaftsvertreter
unit – Einheit, Kapitel
unit labour costs – Lohnstückkosten
unit wage costs – Lohnstückkosten
university – Hochschule; Universität
university board of trustees –
Hochschulrat (The Board of Trustees is
composed of representatives from
business and the sciences.)
university calendar –
Vorlesungsverzeichnis
university catalogue –
Vorlesungsverzeichnis
university course – Hochschulkurs
university degree – Hochschulabschluss;
Universitätsabschluss
university education –
Hochschul(aus)bildung
university fees – Studiengebühren
university for applied science –
Fachhochschule
university graduate – Akademiker;
Hochschulabsolvent
university grounds – Universitätsgelände
university lecturer – Hochschullehrer
university training –
Universitätsausbildung

unlimited employment – unbefristetes Dienstverhältnis
unmannerly – ruppig; flegelhaft
unmarried – ledig
unnecessary – überflüssig
unobtrusive – unauffällig
unoccupied – unbesetzt
unoccupied position – Planstelle, unbesetzte
unpaid – Pause, unbezahlte; unbezahlt
unpaid absence – Abwesenheit, unbezahlt
unpaid day of sick leave – Karenztag
unpopular – unbeliebt
unpredictable – unberechenbar
unprejudiced – vorurteilsfrei
unqualifed work – Mcjob
unread – unbelesen
unreasonable demand – unzumutbar
unreliable – unzuverlässig
unruly – aufsässig; ungebärdig
unsatisfactory – ungenügend
unseizable – unpfändbar
unskilled work – Arbeit, für die eine Lehre nicht nötig ist; ungelernte Tätigkeit
unskilled worker – Hilfsarbeiter; ungelernte Arbeitskraft
unsolicited applicant – Spontanbewerber
unsolicited applicant group – Spontanbewerbergruppe
untalented – unbegabt
up-the-ladder qualification – Stufenqualifikation
upbringing – Erziehung (zu Hause); Kinderstube
update – auf den neuesten Stand bringen
upgrade – höher einstufen
upgrading – Aufgruppierung; Höherstufung
upkeep – Wartung
upper limit – Obergrenze
upper secondary sector – Sekundarbereich 2
urgent case – dringender Fall
usage – Praxis
use – benutzen; Gebrauch
use of media – Medieneinsatz
use of vehicle – Fahrzeugnutzung
user group – Benutzergruppe
user master data – Benutzerstamm

user wage type – Benutzerlohnart
utility table – Hilfstabelle

V

vacancies – Stellenangebot
vacancy – freie Stelle; offene Stelle; Vakanz
vacant – vakant, unbesetzt
vacant position – unbesetzte Planstelle
vacation – Erholungsurlaub; Ferien; Semesterferien
vacation address – Ferienadresse
vacation allowance – Urlaubsgeld
vacation allowance (AmE) – Urlaubsabgeltung
vacation bonus – Urlaubsgeld
vacation pay regulations – Urlaubsgeldordnung
vacation replacement (AmE) – Urlaubsvertretung
vacation schedule (AmE) – Urlaubsplan
vacation school – Sommerschule
vacation with pay – bezahlter Urlaub
valid – gültig
validation check – Plausibilitätsprüfung
validity – Gültigkeit
validity of ratings – Ausprägung der Merkmale
validity period – Gültigkeitszeitraum, Gültigkeitsdauer
valuation – Bewertung, indirekte
valuation assignment – Bewertungszuordnung
valuation basis – Bewertungsgrundlage
valuation of averages – Durchschnittsbewertung
valuation principle – Bewertungsprinzip
valuation rate – Bewertungssatz
valuation wage type – Bewertungslohnart
value – Wert
vandal – Vandale
vanity – Eitelkeit
variable budget – variables Budget

variable income elements – variable
Entgeltbestandteile
variable payments/deductions –
Bezüge/Abzüge, variable
variance – Abweichung; Varianz
vendor – Verkäufer; Verkäuferin
ventilation – Belüftung
venture – wagen
venture capital – Wagniskapital
verbal ability – sprachlicher Ausdruck
verbal violence – verbale Gewalt
verdict – Urteil
verification – Nachprüfung
vernacular – Alltagssprache;
Landessprache
versatile – vielseitig
versatility – Vielseitigkeit
version number – Versionsnummer
vertical decentralization – vertikale
Dezentralisierung
vested pension right – unverfallbare
Pensionsanwartschaft
vested right – Unverfallbarkeit (z.B. von
Ruhegeld)
vested rights – sozialer Besitzstand
vestibule school (AmE) – Lehrwerkstatt
vesting – Leistungsübertragung;
Unverfallbarkeit (z.B. von Ruhegeld)
vesting rule – Übertragungsregel
vesting schedule – Übertragungsplan
vexation – Ärgernis
vice-chancellor – Rektor
Vice President – Mitglied des
Bereichsvorstands (SAG)
vicinity – Nachbarschaft; Nähe
victimise – schikanieren
video (tape) recorder –
Fernseh-Aufzeichnungsgerät
virtual company – virtuelles
Unternehmen
vision – Sehkraft; Vision
visionary – visionär
visiting card – Karte; Visitenkarte
visiting lecturer – Gastdozent
visiting professor – Gastprofessor
visiting teacher – Gastlehrer
visual aids – Anschauungsmaterial
visual memory – bildhaftes Gedächtnis
visualize – sich etwas vorstellen;
visualisieren
visualizing ability – Visualisierung

vocabulary – Wortschatz
vocabulary drill – Wortschatzübung
vocabulary test – Vokabeltest
vocationally oriented – berufsbezogen
vocational category –
Ausbildungsgruppe
vocational competence –
Berufsqualifikation
vocational counsellor – Berufsberater
vocational course – Berufsschullehrgang
vocational education –
Berufsausbildung; Berufsbildung
vocational field – Berufsfeld
vocational guidance – Berufsberatung
vocational retraining – berufliche
Umschulung
vocational school – berufsbildende
Schule; Berufsschule
vocational school certificate –
Berufsschulzeugnis
vocational school teacher –
Berufsschullehrer
vocational training – berufliche Bildung;
Berufsausbildung
Vocational Training Act –
Berufsbildungsgesetz
vocational training centre –
Berufsbildungszentrum
void contract – nichtige Vereinbarung
volatile – sprunghaft
Volkshochschule – Volkshochschule
(VHS)
voluntary – freiwillig
voluntary bonus – Zulage, freiwillige
voluntary membership – freiwillige
Mitgliedschaft
voluntary repetition – freiwillige
Wiederholung
voluntary social contributions –
Gehaltsnebenleistungen
voluntary termination – freiwillige
Beendigung
volunteer – Volontär
vote – abstimmen; Stimme abgeben
voters' list – Wählerliste
voting – Abstimmung
voting by secret ballots – geheime Wahl
voting by separate ballots – getrennte
Wahl
voting paper – Stimmzettel; Wahlzettel

voting right – Stimmrecht;
  Wahlberechtigung
vowel – Selbstlaut; Vokal

W

wage/pay agreement – Tarifvertrag
wage – Lohn
wage adjustment – Lohnausgleich;
  Lohnangleichung
wage advance – Lohnvorschuss
wage agreement – Lohnabkommen
wage assignment – Lohnabtretung
wage calculation – Lohnfindung
wage ceiling – Höchstlohn
wage comparison – Lohnvergleich
wage costs – Lohnkosten
wage curve – Lohnkurve
wage cut – Lohnkürzung
wage demands – Lohnforderungen
wage differential – Lohngefälle
wage dispute – Lohnkonflikt
wage drift – Lohnauftrieb; Lohndrift
wage element – Lohnbestandteil
wage finding – Lohnfestsetzung
wage fluctuation – Lohnbewegungen;
  Lohnschwankungen
wage freeze – Lohnstopp
wage graph – Lohnkurve
wage group – Lohngruppe
wage group analysis –
  Tarifgruppenstrukturanalyse
wage hours – Lohnstunden
wage increase – Lohnerhöhung;
  Lohnsteigerung
wage index – Lohnindex
wage labor – Lohnarbeit
wage level – Lohnniveau
wage negotiations – Lohnverhandlungen
wage packet – Lohntüte
wage payments – Lohnbezüge
wage policy – Lohnpolitik
wage pressure – Lohndruck
wage rate – Lohnsatz; Lohntarif

wage reduction – Lohnkürzung;
  Lohnsenkung
wage requirements – Lohnforderungen
wage scale – Lohnabstufung; Lohnskala;
  Lohnstufe; Tarif
wage scale claim – Tarifforderung
wage scale statistics –
  Lohnstufenstatistik
wage scale system – Tarifsystem
wage settlement – Tarifabschluss
wage slip – Gehaltszettel; Lohnstreifen
wage spread – Lohnspanne
wage statement – Lohnausweis
wage structure – Lohngefüge;
  Lohnstruktur
wage table – Lohntabelle
wage tax card – Lohnsteuerkarte
wage tax notification –
  Lohnsteuervoranmeldung
wage trend – Lohnentwicklung
wage type – Lohnart
wage type category – Lohnartentyp
wage type coding –
  Lohnartenschlüsselung
wage type generation –
  Lohnartengenerierung
wage type key – Lohnartenschlüssel
wage type model – Lohnartenmuster
wage type modifier –
  Lohnartmodifikator
wage type statement –
  Lohnartennachweis
wage type structure – Lohnartenstruktur
wage type table – Lohnartentabelle
wage type text – Lohnartentext
wage type valuation –
  Lohnartenbewertung
wage(-scale) policy – Tarifpolitik
wage-dependant – lohnabhängig
wage-earner – gewerblich Tätiger;
  Lohnempfänger
wage-price relationship –
  Lohn-Preis-Verhältnis
wage-price spiral – Lohn-Preis-Spirale
wage-related document –
  Lohnunterlage(n)
wages – Besoldung
wages and salary history – Lohn- und
  Gehaltsentwicklung
wages policy – Lohngestaltung
wages slip – Lohnabrechnung

wages system – Lohnform
waiting period – Wartezeit
waiting-list booking – Wartelistebuchung
waiting-list priority – Wartelistepriorität
waiver – Verzicht
walkout – Arbeitsniederlegung
war disablement – Kriegsbeschädigung
war disablement pension – Kriegsbeschädigtenrente
warehouse – Lager
warming-up exercise – Aufwärmübung
warn – mahnen; verwarnen
warning – Verwarnung
warrantee – Sicherheitsempfänger
wastage rate – Abgangsrate
wave of graduates – Akademisierungswelle
way of living – Lebenswandel
way of working – Arbeitsweise
way to school – Schulweg
weak incentive – schwacher Anreiz
weak intellectual capacity – Denkschwäche
weak learner – lernschwach
weak point – Defizit (persönliche); Schwäche
weakening – Schwächung
weakness – Schwäche
wealth of imagination – Einfallsreichtum
weekday number – Werktagnummer
weekly wage – Wochenlohn
weekly working hours – wöchentliche Arbeitszeit
weekly working hours according to – tarifliche Wochenarbeitszeit
welding shop – Schweißerei
welfare – Sozialfürsorge
welfare allowance – Sozialzulage
welfare benefits – freiwillige Sozialleistungen
welfare facilities – soziale Einrichtungen
welfare fund – Unterstützungsfonds
welfare office – Fürsorgestelle
welfare responsibility – Fürsorgepflicht
welfare responsibility of the employer – Fürsorgeverantwortung des Arbeitgebers
well-balanced – ausgeglichen
well-founded – fundiert

well-paid jobs – erster Arbeitsmarkt
whistle-blower – Verräter (A whistle-blower is one who draws public attention to misdeeds within his own scientific community.)
white collar crime – Wirtschaftsverbrechen
white collar worker – Angestellte(r); Tarifangestellte(r)
whole day classes – Ganztagsunterricht
whole day school – Ganztagsschule
wider ranging profession – Aufbauberuf
widow's pension – Witwenrente
widow's pension for divorced partner – Geschiedenen-Witwenrente
widowed – verwitwet
widower's pension – Witwerrente
wife – Ehefrau; Ehegattin; Ehepartner(in)
wildcat strike – wilder Streik
willing – bereit
willing to be trained – ausbildungswillig
willing to learn – lernbereit
willing to work – arbeitswillig
willingness – Bereitschaft
willingness to cooperate – Kooperationsbereitschaft
willingness to learn – Lernbereitschaft
willingness to take a risk – Risikobereitschaft
willpower – Willensstärke
win – gewinnen
winner – Gewinner
wise – weise
withdraw – zurückziehen
withdrawal – Zurücknahme
withdrawal of termination – Kündigungsrücknahme
withhold – einbehalten (Steuern)
withhold from wages – vom Lohn einbehalten
withholding statement – Lohnsteuerbescheinigung
withholding tax – Quellensteuer
without a perspective – Perspektivlosigkeit
without a trade – berufslos
without pay – unbezahlt
without time limit – fristlos
woman returner – Berufsrückkehrerin
word – Wort
wording – Formulierung

work – Arbeit; Tätigkeit
work accident – Arbeitsunfall;
Betriebsunfall
work addiction – Arbeitssucht
work area – Arbeitsbereich
work as a sales representative – im
Außendienst arbeiten
work assessment –
Arbeitsplatzbewertung (judging the
value)
work assignment – Arbeitszuweisung am
Arbeitsplatz; Aufgabe
work attitude – Arbeitsverhalten
work behaviour – Arbeitsverhalten
work book – Arbeitsbuch
work break – Pause
work break schedule –
Arbeitspausenplan
work center – Arbeitsplatz
work center and job administration –
Arbeitsplatz- und Stellenverwaltung
work center and job description –
Arbeitsplatz- und Stellenbeschreibung
work center and job grading –
Arbeitsplatz- und Stellenbewertung
work center bonus – Arbeitsplatzzulage;
Zulage, arbeitsplatzbezogene
work center concentration –
Arbeitsplatzkonzentration, maximale
work center data – Arbeitsplatzdaten
work center substitution –
Arbeitsplatzvertretung
work clothes – Berufskleidung
work conflict – Arbeitskonflikt
work contract – Anstellungsverhältnis
work done by hand – Handarbeit
work environment – Arbeitsumfeld;
Arbeitsumgebung
work experience – Diensterfahrung
work extending beyond midnight –
Tätigkeit, mitternachtsübergreifende
work flow – Arbeitsablauf
work group – Arbeitsgruppe;
Arbeitskreis
work habits – Arbeitsverhalten;
Berufseinstellung
work history – Berufslaufbahn
work in small teams –
Kleingruppenarbeit
work level – Arbeitsstufe
work methods – Arbeitsmethodik

work order times – Auftragszeit
work overtime – Überstunden leisten;
Überstunden machen
work performance contract –
Werkvertrag
work performed on a public holiday –
Feiertagsarbeit
work period abroad – Arbeitsaufenthalt
im Ausland
work permit – Arbeitserlaubnis;
Arbeitsgenehmigung;
Arbeitsbewilligung
work placement – Vermittlung eines
Arbeitsplatzes
work planning – Arbeitsplanung;
Planung
work possibilities –
Arbeitsmöglichkeiten
work pressure – Arbeitsdruck
work process – Arbeitsprozess
work quality – Arbeitsqualität
work relationship – Arbeitsverhältnis
work results – Arbeitsergebnis
work satisfaction – Arbeitsfreude;
Arbeitszufriedenheit
work schedule – Arbeitszeitplan
work schedule and shift planning –
Arbeitszeit- und Dienstplanung
work schedule rule –
Arbeitszeitplanregel
work sheet – Arbeitsblatt
work shift – Arbeitsschicht
work short-time – kurzarbeiten
work situation – Arbeitssituation
work state tax – Arbeitsgebietssteuer
(USA)
work stoppage – Arbeitsniederlegung
work study – Arbeitsstudie
work success – Arbeitserfolg
work that spans two days – Arbeit,
mitternachtsübergreifende
work time event – Arbeitszeitereignis
work time event type group –
Arbeitszeitereignisartgruppe
work to rule – Dienst nach Vorschrift;
Streik durch passiven Widerstand
work together – mitarbeiten;
zusammenarbeiten
work tools – Arbeitsgegenstände
work's committee – Betriebsausschuss

work-related course –
arbeitsplatzbezogener Kurs
workaholism – Arbeitssucht
workday – Arbeitstag; Arbeitstag,
bezahlter
worker – Arbeiter; Handwerker
worker on temporary loan –
Leiharbeitskraft
worker participation – Mitwirkung der
Arbeitnehmer in der Geschäftsführung
worker participation in management –
Arbeiterselbstverwaltung
worker's compensation –
Arbeitsunfallversicherung
worker's control –
Arbeiterselbstverwaltung
worker's self-management –
Arbeiterselbstverwaltung
workforce – Arbeitskräfte, Belegschaft
workforce management – Management
menschlicher Ressourcen
workforce planning – Einsatzplanung
workforce requirements –
Personalbedarf
workforce requirements planning –
Personalbedarfsplanung
workforce structure –
Belegschaftsstruktur
working – berufstätig
working age – arbeitsfähiges Alter
working atmosphere – Betriebsklima
working career – Berufsleben
working class – Arbeiterklasse
working class family – Arbeiterfamilie
working climate – Arbeitsklima
working conditions –
Arbeitsbedingungen
working cycle – Arbeitstakt
working day – Arbeitstag; Werktag
working dinner – Arbeitsessen
working hour – Arbeitsstunde
working hours – Arbeitszeit
working life – Berufsleben;
Erwerbsleben; Lebensarbeitszeit
working lunch – Arbeitsessen
working man – Werktätige(r)
working materials – Arbeitsmittel
working speed – Arbeitstempo
working style – Arbeitsstil
working time – Arbeitszeit

working time lost (due to ...) –
Arbeitsausfall
working time model – Arbeitszeitmodell
working week – Arbeitswoche;
Wochenarbeitszeit
working woman – Werktätige
working world – Arbeitswelt
working-in period – Einarbeitung
workload – Arbeitsbelastung
workman – Arbeiter; Handwerker
works – Betrieb; Werk
works agreement –
Betriebsvereinbarung
works council – Betriebsrat
works handbook – Betriebshandbuch
works manager – Betriebsleiter;
Fabrikleiter
works meeting – Betriebsversammlung
works rules and regulations –
Betriebsvereinbarung
workshop – Werkstatt, Handwerksbetrieb
world of work – Arbeitswelt
worries about the future –
Zukunftssorgen
wrap-up session – Schlussbesprechung
writing ability – Schreibfähigkeit
writing impediment – schreibbehindert
writing sample – Schriftprobe
written application – schriftliche
Bewerbung
written examiniation – schriftliche
Prüfung
written form – Schriftform
written learning material – schriftliche
Lernunterlagen
written notice – Kündigungsschreiben
wrong – falsch; unrichtig

X

X-ray department –
Röntgenschirmbildstelle;
Röntgenstation

Y

year – Jahrgangsstufe
year book – Jahrbuch
year of apprenticeship – Lehrjahr
year of employment –
Beschäftigungsjahr
year-end bonus –
Jahresabschlussvergütung
year-end premium –
Jahresabschlussvergütung
years of service – Dienstalter;
Dienstjahre
young executives –
Managementnachwuchs
young managers –
Managementnachwuchs
young person – Jugendliche(r)
Young Persons Employment Act –
Jugendarbeitsschutzgesetz
young professional – Akademiker mit
erster Berufserfahrung
young programmer –
Anfangsprogrammierer
youngster – Jugendliche(r)
youth centre – Jugendzentrum
youth club – Jugendklub
youth representative – Jugendvertreter
Youth Training Scheme (YTS) –
Jugendausbildungsprogramm
youth unemployment –
Jugendarbeitslosigkeit

Z

zero-base budget – nicht
fortgeschriebenes Budget
zoology – Zoologie

The Educational System in England and Wales

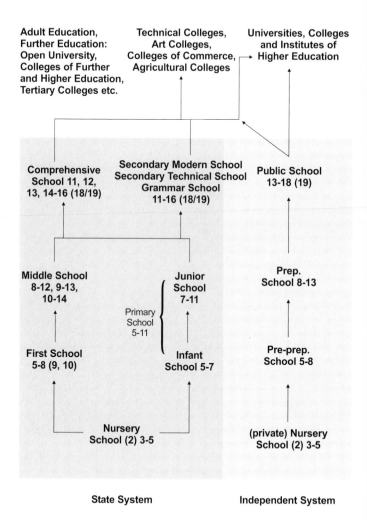

Adult Education, Further Education: Open University, Colleges of Further and Higher Education, Tertiary Colleges etc.

Technical Colleges, Art Colleges, Colleges of Commerce, Agricultural Colleges

Universities, Colleges and Institutes of Higher Education

Comprehensive School 11, 12, 13, 14-16 (18/19)

Secondary Modern School Secondary Technical School Grammar School 11-16 (18/19)

Public School 13-18 (19)

Middle School 8-12, 9-13, 10-14

Junior School 7-11

Primary School 5-11

Prep. School 8-13

First School 5-8 (9, 10)

Infant School 5-7

Pre-prep. School 5-8

Nursery School (2) 3-5

(private) Nursery School (2) 3-5

State System

Independent System

Nach: H. Händel/D.A. Gossel, Großbritannien; 3. Aufl., München 1994, S. 354